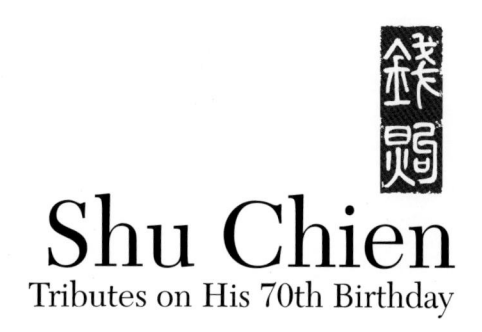

Shu Chien
Tributes on His 70th Birthday

Shu Chien
Tributes on His 70th Birthday

Lanping Amy Sung
Kuang-Chung Hu Chien
University of California, San Diego, USA

World Scientific

NEW JERSEY • LONDON • SINGAPORE • BEIJING • SHANGHAI • HONG KONG • TAIPEI • CHENNAI

Published by

World Scientific Publishing Co. Pte. Ltd.
5 Toh Tuck Link, Singapore 596224
USA office: 27 Warren Street, Suite 401–402, Hackensack, NJ 07601
UK office: 57 Shelton Street, Covent Garden, London WC2H 9HE

Library of Congress Cataloging-in-Publication Data
Shu Chien: tributes on his 70th birthday / [Edited by] Amy Lanping Sung & Hu
  Kung-Chung Chien.
    p. cm.
  Includes bibliographical references and index.
  ISBN 981-238-383-2 (alk. paper)
    1. Chien, Shu. 2. Educators--China--Biography. 3. Biomedical engineering. I. Sung,
  Amy Lanping. II. Chien Hu, Kung-Chung.

LA2383.C52C462 2003
610'.28'092--dc21
[B]
                                                                                          2003045087

British Library Cataloguing-in-Publication Data
A catalogue record for this book is available from the British Library.

Copyright © 2004 by World Scientific Publishing Co. Pte. Ltd.

All rights reserved. This book, or parts thereof, may not be reproduced in any form or by any means, electronic or mechanical, including photocopying, recording or any information storage and retrieval system now known or to be invented, without written permission from the Publisher.

For photocopying of material in this volume, please pay a copying fee through the Copyright Clearance Center, Inc., 222 Rosewood Drive, Danvers, MA 01923, USA. In this case permission to photocopy is not required from the publisher.

Printed in Singapore.

Shu Chien 錢煦

Three generations, 1937.

Three generations, 2001.

Contents

Foreword xxiii

Preface xxvii

Section I — Tributes Printed for the 23 June 2001 Celebration in La Jolla 1

A Few Words About Shu Chien and His Family
  Chi Hu 3

The Families
  Tsu Won Hu and Margaret Hu 6

On Shu's 70th Birthday
  Robert Chun Chien 10

Many Happy Returns to a Most Generous Brother
  Fredrick Foo Chien 13

Shu and I: Our Happy Marriage and Family
  Kuang-Chung Hu Chien 16

My Dad
  May Chien Busch 23

Professor Shu Chien, My Dad
  Ann Chien Guidera 27

My Brother-in-Law
  David Kuang-Jeou Hu 30

My Yih-Fu
  Mark Yong-Qiang Wang, Xiao-Chuan Wang and Tom Wang 33

Friendship for 54 Years
  Pei Pin Han 37

Shu Chien — A Truly Outstanding Alumnus
Peter Kuang-Hun Cheng 41

Lao-Chien is 70 Years Young Now!
Shaotsu Lee 44

Over a Half-Century of Friendship
Pung Show Liu 48

My Classmate Shu Chien
Chih Chao Chin 50

Talking about Shu Chien in Good Old Days
Hua Chin Chen 52

He is Always There!
H. Daniel Ou-Yang 54

Many Puzzlements of A Treasure
Mamie Kwoh Wang 56

The Seemingly Contradictory Characters of Shu Chien
Hsueh Hwa Wang 59

To Shu Chien, a Family Man, from ABMAC, One of His Families
Richard N. Pierson Jr. 62

Shu Chien at 70
Edward F. Leonard 66

Longevity!
Shunichi Usami 70

The Dimension of His Teaching
Kung-ming Jan 72

The World is Better Because of You
Lanping Amy Sung 76

To Professor Shu Chien for His 70th Birthday Celebration
Richard Y.C. Chen 81

Culture Clash
Herbert H. Lipowsky 83

Shu Chien, a Leader and Perfectionist in Physiology
and Bioengineering
    Geert W. Schmid-Schönbein 90

Tribute to Dr. Shu Chien
    Anne Chabanel 94

A Great Scientist Taught Me That…
    Steven D. House 96

From Columbia to UCSD
    K.-L. Paul Sung 98

A Mentor and a Friend
    Cheng Zhu 105

Happy Birthday, Most Esteemed One
    Anna L. Skalak 108

Sweet Memories
    Sheldon Weinbaum 111

I Shall Always Look Up to Him
    David H. Cheng 116

Shu — An Appreciation of His Achievements and Friendship
    Colin G. Caro 118

A Tribute
    Samuel C. Silverstein 121

A Tribute to Professor Shu Chien
    Van C. Mow 122

The Blooming of Taiwan Biomedical Sciences — A
Tribute to Dr. Shu Chien
    Cheng-Wen Wu 126

A Tribute to Professor Shu Chien
    Yuan T. Lee 129

"Li De; Li Yen; Li Gong"
    Paul O.P. Ts'o and Muriel Ts'o 131

Working Together with Shu
  Monto Ho 133

Some Small Role
  M. Lea Rudee 135

Counting Precious Pearls in the Family: What Makes Shu the Leader in So Many Fronts?
  Yuan-Cheng Fung 137

A Tribute
  Beatrice Zweifach 146

Shu Chien: A Diamond with Many Facets
  David Gough 149

Constraining Shu: Mission Impossible?
  Bernhard Ø. Palsson 152

Shu: You Can Really Cook!
  Robert L. Sah 155

A Tribute
  Shankar Subramaniam 158

A Tribute
  Gary Huber 159

The Making of a Bioengineering Dynasty: A Tribute to Dr. Shu Chien
  Andrew McCulloch and Deidre MacKenna 160

Reminiscences from a Physiological Perspective
  Paul C. Johnson 162

On the Occasion of Dr. Chien's 70th Birthday
  Peter C.Y. Chen 164

Leader of Bioengineering, Friend to Many
  Thomas C. Skalak 166

Dr. Chien's Impact on My Career
  Jeffrey H. Price 170

A Tribute to Dr. Shu Chien
  Richard Lieber 172

At the Top of My List
  Michael W. Berns 174

Bringing Life to Bioengineering
  John Watson 177

An Appreciation
  Larry V. McIntire 181

A Tribute
  Don P. Giddens 183

Redwoods are Tall, Mt. Everest is Big, and Sunsets are Beautiful
  Peter G. Katona 184

The Microcirculatory Worker's Microcirculationist!
  Aubrey E. Taylor 186

Life in Physiology
  Martin Frank 189

Establishing a Standard
  Michael J. Jackson 192

Shu Chien's Contributions to FASEB and Public Affairs
  Howard H. Garrison 194

What You May Not Know About Shu Chien
  Kevin W. O'Connor 197

Tribute to Shu Chien
  Wendy Baldwin 199

Working with Dr. Chien
  Song Li 201

Http://www.chienlab.com
  Julie Y.-S. Li 203

Happy 70th Birthday
  Gerard Norwich 205

A Tribute
  Ellen Tseng 206

A Tribute to Mark Dr. Chien's 70th Birthday
  Gang Jin 208

Professor Shu Chien — A Person with Great Wisdom and Kindness
  Yingli Hu 213

The Family of Dr. Chien
  Peter J. Butler 216

Happy Birthday, Dr. Chien!
  Yan-Ting Elizabeth Shiu 217

Happy Birthday to Dr. Chien
  Yingxiao Wang 218

A Tribute
  Roland Kaunas 221

Dr. Shu Chien, My Teacher and Collaborator
  Xiong Wang 223

A Salute to a Renaissance Man
  Koung-Ping Cheng and Lilly Cheng 226

The Golf Connection
  Ernest Chun-Ming Huang, Huei-Jen Su Huang and Natalie Huang 228

Section II — More Tributes for the 23 June 2001 Celebration **233**

A Fun-poem
  Chia Hui Shih 235

A Poem
  Suli Yuan 236

A Birthday Card
  Shlomoh Simchon 238

A Birthday Tribute to Shu
    Sidney S. Sobin 240

An Appreciation of Shu Chien's Work
    Kitty Fronek and Arnost Fronek 242

A Tribute
    Sangeeta Bhatia 244

Tribute to Shu Chien
    Ruth L. Kirschstein 245

Profiles in Physiology
    Ewald R. Weibel 247

Shu Chien and the International Union of Physiological Sciences (IUPS)
    Denis Noble 253

A Tribute to Professor Shu Chien on His 70th Birthday
    Rakesh K. Jain 259

A Sketch for the Lab People at Columbia
    Emily Schmalzer 262

My Role Model
    Gerhard M. Artmann 266

Calligraphic Tributes
    Kung Chen Loh and H.S. Fang 268

Section III — Additional Tributes Written by Speakers at the 23 June 2001 Celebration 273

A Tribute to Shu Chien
    Paul La Celle 275

Vascular Biology, Tissue Engineering, NIH and Shu Chien
    Robert M. Nerem 278

Tribute to Shu Chien
    Harry Goldsmith 281

Shu: You Are the Greatest
    Savio L.-Y. Woo 284

Professor Shu Chien: My Dearest Advisor
  Si-Shen Feng 287

The 70/30 Philosophy: A Tribute to a Teacher,
Friend and Colleague
  John Y.-J. Shyy 290

A News Report
  Daming Li 292

Tributes to Dr. Shu Chien on the Occasion of
His 70th Birthday Celebration
  Robert J. Dellenback 294

A Tribute to My Friend Shu Chien
  Chien Ho 297

Shu Chien's Influence on Our Lives
  Ann L. Baldwin and Timothy W. Secomb 299

Happy 70th Birthday to Our Dad — A Poem
  May Chien Busch and Ann Chien Guidera 302

To Shu and K.C. — A Drum Song
  Y.C. Fung and Luna Fung 305

Section IV — Tributes from the 8 August 2001 Celebration in Taipei 309

(Footnotes: English translations for Chinese characters) 311

To Shu Chien
  Sunney I. Chan 312

Academician Shu Chien — A Modest, Generous and
Decisive Intellect
  C.Y. Chai 314

Academician Chien's 70th Birthday Celebration —
Gratitude and Blessings
  Eminy H.Y. Lee 318

Gratitude and Blessing
  Lee Young Chau 322

Feelings and Thoughts on the Occasion of Academician
Shu Chien's 70th Birthday — Remembrance and Gratitude
    Tang Tang 325

My Teacher and Friend
    Danny Ling Wang 328

Forever a Boss in My Heart — Academician Shu Chien
    Annie Su-Chin Lin 331

Special Birthday Wishes to Chien Hsian-Sheng on His 70th
Birthday from IBMS Administrative Staff
    Nicki Shu-Chuan Chang 334
    Ivy Su-Ching Tu 335
    Shirley Tsae-Shene Hu 336
    Amy Hsiao Hsuan Chu 337

The Academician Shu Chien That I Know
    Winston C.Y. Yu 338

To Academician Shu Chien
    Jackson Chieh-Hsi Wu 340

On the 70th Birthday of Academician Shu Chien
    Jin-Jer Chen 343

A Tribute to Academician Shu Chien on Your 70th Birthday
    Jerry Hsyue-Jen Hsieh 345

Tribute to Academician Shu Chien on His 70th birthday
    Shing-Jong Lin 348

My Extraordinary Benevolent Teacher
    Lang-Ming Chi 351

Longing for Professor Chien
    Cheng-Den Kuo 353

On the Occasion of Academician Shu Chien's
70th Birthday: A Tribute
    Yu-Lian Chen 356

Thoughts on the Occasion of Academician Shu Chien's
70th Birthday
   Phoebe Yueh-Bih Tang 358

Twenty Seven Years Ago
   Yuan-Tsung Chen 361

A Poem for Shu Chien
   Sho-Tone Lee 363

Tribute to a Role Model
   Tommy Yung-Chi Cheng 364

Heartfelt Greeting and Blessing
   Yuan-Tsun Liu and Pauline Fu 366

My Outstanding Student
   H.S. Fang 369

A Tribute from the Mainland
   Zhu Chen 372

Section V — More Tributes from Colleagues Worldwide 375

Expressing Our Feeling at Your 70th Birthday
   Yun Peng Wu and R.F. Yang 377

A Tribute to Professor Shu Chien
   Fengyuan Zhuang 379

Sincere Personal Congratulations from Gothenburg

   P.I. Brånemark 382

Section VI — Shu Chien's Life 387

Appreciation, Gratitude, Reflections and Prospectives
   Shu Chien 389

Grandfather's Calligraphy, 1957
   Chao-ching Chang 490

Father's Calligraphy, 1981
   Shih-liang Chien 492

Medals	494
Curriculum Vitae of Shu Chien	497
Section VII — Programs of Celebrations	
A. La Jolla, 23 June 2001	549
B. Taipei, 8 August 2001	559
Postscript	567
Artwork by Mrs Kuang-Chung Hu Chien	569
Author Index	570

Foreword

Dear Shu,

On the happy occasion of your 70th birthday, your friends wanted to celebrate with you. Seventy-eight people who are more diligent with their pens and computers wrote tributes, which are collected here in this volume. Many friends who worked with you in the fields of physiology and bioengineering held a whole-day symposium on 23 June 2001 in your honor. They looked forward to the new bioengineering and physiology of the 21st century. They also looked backward to see how we arrived here. The program of the symposium is appended here in this book.

The book begins with a sketch of the history of the Chien and Hu families. Then your brothers, your wife, daughters, and a brother-in-law presented some reminiscence. They gave us invaluable sketches of your roots, the way you were cultivated as a sapling, and the way you as a majestic mature tree are regarded by other majestic trees in the same stand. K.C.'s article is especially precious. These articles are a big treat to your friends.

Then come the remarks of your college classmates, your students and your colleagues in New York. Your emergence into national prominence is described by many interesting stories. The mid-1980s was a special period. You went to Taiwan and caused the Taiwan biomedical sciences to bloom. The tributes by Drs. Chen-Wen Wu, Yuan-Tseh Lee, Paul Ts'o and Monto Ho are moving and inspiring. Then you came to La Jolla. Your career at UCSD is fabulous. The expansion of your influence to the whole country and the whole world followed. The written tributes are presented in approximately this chronological order. Many more people spoke at the banquet. We know that we were speaking also for your friends throughout the world, who have the same love and admiration for you as we do.

Our sentiment resonates with two of the oldest Chinese poems, which were collected by Confucius (circa 551–479 B.C.) in 詩經, the "Book of Poetry" or "Book of Songs." The first talks about gift-giving. The second talks about working together with mutual respect. Let me quote them below for your enjoyment.

Love and best wishes from all of your friends!

Yuan-Cheng Fung
馮元楨
(23 June 2001)

1. 國風衛風之十

    投我以木瓜,報之以瓊琚.
      匪報也,永以為好也.
    投我以木桃,報之以瓊瑤.
      匪報也,永以為好也.
    投我以木李,報之以瓊玖.
      匪報也,永以為好也.

2. 國風齊風之二

    子之還兮,遭我乎峱之間兮.
      並驅從兩肩兮,揖我謂我儇兮.
    子之茂兮,遭我乎峱之道兮.
      並驅從兩牡兮,揖我謂我好兮.
    子之昌兮,遭我乎峱之陽兮.
      並驅從兩狼兮,揖我謂我臧兮.

1.

She threw a quince to me;
In requital I gave a bright girdle-gem.
No, not just as requital;
But meaning I would love her forever.

She threw a tree-peach to me;
As requital I gave her a bright green stone.
No, not just as requital;
But meaning I would love her forever.

She threw a tree-plum to me;
In requital I gave her a bright jet stone.
No, not just as requital;
But meaning I would love her forever.

2.

How splendid he was!
Yes, he met me between the hills of Nao.
Our chariots side by side we chased two boars.
He bowed to me and said I was very nimble.

How strong he was!
Yes, he met me on the road at Nao.
Side by side we chased two stags.
He bowed to me and said, "Well done."

How magnificent he was!
Yes, he met me on the south slopes of Nao.
Side by side we chased two wolves.
He bowed to me and said, "That was good."

Both poems were translated by Arthur Waley in his "The Book of Songs." Grove Press, New York, 1937.

Preface

When the organizing committee for the celebration of Shu Chien's 70th birthday asked for a volunteer in late April to put together a tribute book for Shu, I raised my hand. At that time tributes had just begun to arrive in response to the committee's invitation. Little did I know that this book was going to be more than 500 pages containing 131 tributes and over 200 photographs. The organizing committee was chaired by Geert Schmid-Schönbein and consisted of Shunichi Usami, Paul Sung, Y.C. Fung, Julie Li, Jennifer Griffen and myself. We met regularly. Originally there was an idea of having a scrapbook for Shu. Later came a suggestion to make a formal book with complete information about authors. Then I looked at the tribute book that Shu put together in 1993 for Dick Skalak for his 70th birthday, and thought I probably could do something like that.

But 1993 was in a different era. There was no e-mail; fax was considered a fascinating telecommunication medium. Today, with a PC of 60-gigabytes in the lab, plus zip disks, scanners, digital cameras, and CDs everywhere, it is easy to do more. Tributes that came in varied from handwritten birthday cards to typed letters to e-mail attachments. Photos ranged from black-and-white pictures to scanned jpeg files to digital camera images. I adopted the principle of uniformity and completeness that Y.C. and Shu have always practiced in editing to put together this tribute book. I treated some scientific illustrations that I received originally as artworks and inserted them in the text; only later were short captions added with citations. I also selected pictures of Shu at different stages of his life to bring out the background, the contrast, the richness, the happiness and the fullness of his life across the book. When there was a Chinese calligraphy or poem accompanied by an English version, I made sure that they were arranged to the best I could.

About two weeks before his birthday, Shu came back from a trip and told me that he should be able to find pictures for most, if not all, of the contributors who did not send in their photos. Articles kept coming in, through Jen (mostly), Geert and Shu (from his classmates, old friends and family members). Some far away from different parts of the world and some nearby within the same EBU1 building. Photos followed, through Shu, Geert and Peter Chen (who shot several medals, calligraphies and paintings). Some dated many decades ago and some were taken only four days before the celebration. The intensity of the work escalated exponentially toward the end. Fortunately, people in my lab were very helpful. Lynn Truong and Carlos Vera helped me overcome some technical challenges from time to time; and Thomas Kim worked with me side by side with great computer skills day and night for about two weeks. A few days before the June 23rd celebration my son Eric and daughter Wendy also chipped in. The Papa John's Pizza knew exactly where to make the delivery when I called.

A tribute book was not meant to have the person honored helping, but Shu did. He helped find contributors' photos, he identified events, individuals and the years of these photos, he provided citations for scientific illustrations, and he clarified certain facts for me in the text. Some of the Chinese poems or calligraphies he received did not come with an English version. So he provided the English translation. He even provided a typed version of Kung Chen Loh's marvelous calligraphy using the Chinese software in his computer at home. He did all these because he wanted all his friends to be able to enjoy them without the cultural limitation. As the editor, I appreciated greatly his help and (in spite of the authority he has over his own life) his respect of decisions I made in this tribute book.

A leather-bound, glittering tribute book was to be a gift for Shu at the birthday celebration. For that Paul and I visited several bindery shops. Dates for delivering printed pages for this special binding were set, but one by one I let the deadlines go by in order to accept newly arrived materials. Each time the decision was made based on Shu's saying that he appreciated more the content of the book than the cover itself. At the end I finally realized that Shu would really like to see his close friends, colleagues, students, family members, relatives and old classmates together with their tributes in this book on his 70th birthday, and thereafter for a long, long time.

The committee made 120 copies of the tribute book the afternoon before the day of celebration. All guests who attended had the 255-page red-cover paperback, black-and-white tribute books on 23 June 2001, and that made Shu very happy. His leather-bound, hard-covered, full-colored tribute book was still a gift from all of us, around and away, but it would have to be delivered to him after his birthday.

Shu told me at the party and afterward that he had heard many praises about the tribute book. I replied, "When the person we celebrated for is so great, stories and testimonies so marvelous, scientific illustrations so truthful, calligraphies and painting so beautiful, and photographs so telling and full of history, the tribute book is bound to receive such praises." It was indeed a tremendous privilege for me to work on this tribute book.

After celebrating Shu's 70th birthday in La Jolla and collecting additional tributes and precious materials, we printed a more complete, leather-bound personal copy with full-colored pictures as a gift to Shu. A number of his family members and friends also received this second version, desktop paperback publication of 381 pages in black-and-white. Across the Pacific Ocean, another celebration for Shu took place on 8 August 2001, in Taipei. At that time, a tribute book written mostly in Chinese by his colleagues at Academia Sinica, especially the Institute of Biomedical Sciences (IMBS), and his friends and students in Taiwan was put together by Eminy H.Y. Lee and presented to Shu. Thereafter, the idea of formally publishing the tribute book that would also include the English translation of these tributes in Chinese gradually solidified. Now, with the professional touches of Joy Quek and her team at the World Scientific Publishing Company, this tribute book has transformed into a true publication, including an artistic design for the cover and elegant presentations of the contents. We also cannot thank them enough for their patience in dealing with the changes made in the first, second and third readings of the book.

Y.C. wrote a wonderful Foreword for the tribute book before Shu's 70th birthday. Now this 604-page book includes tributes printed for the 23 June 2001 celebration (Section I), more tributes before and after the celebration (Section II), additional tributes from speakers at the symposium and the banquet (Section III), tributes from the 8 August 2001 celebration in Taipei (Section IV), more tributes from colleagues worldwide (Section V), and Shu's closing article (in Section VI on Shu Chien's life).

Shu's closing article grew from the "Appreciation and Gratitude" that he delivered in his closing remarks at the celebration, to include "Personal Reflections" as he read and resonated with more than 100 marvelous tributes, and the "Retrospects and Prospectives" as he reviewed his own life at the milestone of age 70. This invaluable, first-person article will be loved by his friends and family, as well as historians. In addition, the calligraphy done by Shu's father, Shih-liang Chien, in 1981, and that by Shu's maternal grandfather, Chao-ching Chang, in 1957, are also included. What treasures to have! At the end of the book, K.C., who had provided important inputs to the tribute book, wrote a deeply touching postscript and concluded this book with a beautiful painting she made in 2000.

Dear Shu, This is a birthday gift to you from all of us! Each piece of the tribute tells a story about you. Each piece of the tribute also represents a heartfelt wish to thank you for sharing your wonderful life with us.

Lanping Amy Sung
曾藍萍
(5 October 2003)

Section I

Tributes Printed for the 23 June 2001 Celebration in La Jolla*

---

*Pronounced "LA HO-YA." In Spanish it means "The Jewel." It is a town in the City of San Diego and where the University of California, San Diego is located.

A Few Words About Shu Chien and His Family

*Chi Hu**

Dr. Shu Chien is a scholar well known nationally and internationally. Shu and Kuang-Chung are sincere, honest, considerate, amiable and loving. Shu's success is of course attributed to his innate character and ability and his wonderful wife, but his family background is also an important factor.

When Shu's father Shih-liang completed his Ph.D. (Chemistry) studies at the University of Illinois and returned to China in 1934 to teach at the National Peking University, he and his wife Wan-tu rented a part of the house that my father owned in Beijing. Wan-tu's elder brother T.K. Chang also rented part of the same house. I knew Shu when he was only a little over a year old, and I became his godfather. Like his father, Shu had a relatively large head, which tended to hit the ground whenever he fell.

Shu's grandfather Hong-Yeh Chien was an honest and decisive judge. During the early phase of World War II, although the Chinese government had moved to Chungqing, it still had jurisdiction in the international settlements in Shanghai. Judge Chien was the Chief Criminal Judge in the Special District Court and was quite strict with cases of Chinese turncoats who were prosecuted. As a result, he was assassinated in 1939.

---

*Shu Chien's godfather.

Shu with Mr. and Mrs. Chi Hu, 1952.

As a result of Japanese occupation in eastern China, National Peking University, together with Tsinghua and Nankai Universities, moved to Kung-ming (near Chungqing) to form the Southwestern Union University. Shih-liang moved with the University to Kung-ming; he travelled back and forth to be with his family in Shanghai during summer and winter vacations. After the passing of Judge Chien, Shih-liang stayed in Shanghai and work in a pharmaceutical company until the end of World War II. National Peking University moved back to Beijing after the war, with Dr. Hu Shih as the President. Dr. Hu invited Shih-liang to serve as Professor and Chairman of Chemistry. In 1949, when Beijing was under siege due to the civil war, the Nationalist Government sent a couple of airplanes to evacuate university professors. Shih-liang and his family took one of the last airplanes out of Beijing to go to Nanjing, Shanghai and then Taipei, where he accepted the position of Dean of Academic Affairs (Provost) at the National Taiwan University (Taita).

Following the sudden passing of Taita President Sze-nien Fu, there were a large number of senior people interested in this top education position in Taiwan. Dr. Hu Shih wrote a strong letter of recommendation to the then Premier Cheng Chen. Premier Chen, after consultation with President Chiang Kai-shek, appointed Shih-liang as President of National Taiwan University at the age of 41. Despite his extremely important position, Shih-liang he treated everyone with the same level of respect, regardless of his/her social or economic status.

He handled all matters with a high sense of fairness and always sought solutions with equity and principle. Shu interacts with people and handles affairs in the same way as his father.

Wan-tu's father Mr. Chao-ching Chang served as County Executive of Tuh-Yang County in the Sze-chuan Province about a century ago. On one occasion, the city was in serious danger of being overrun by renegade rascals. Wan-tu's mother went up to the city wall to encourage the soldiers. She cut her ring finger, removed her ring and gave it to the soldiers as a symbol of encouragement. This greatly enhanced the morale, and the city was saved. This story is just one example of the excellent genetic and cultural tradition received by Wan-tu from her family.

I would like to share these thoughts with everyone because I believe, while Shu's remarkable accomplishments and marvelous personality are attributed to his outstanding talents and tremendous efforts, the roots of these can also be traced to his parents and grandparents.

The Families

Tsu Won Hu and Margaret Hu[*]

The Hu family and the Chien family have been close for many years. It started when T.W.'s father[*] and Shu's father taught at the National Peking University in the thirties. When the Sino-Japanese war started in 1937, father and T.W. left Peking and eventually end up in the U.S., whereby mother and younger brother and the Chiens ended up in Shanghai. After the war the two families met again in Peking, and most of us settled down in Taiwan one way or another eventually.

The friendship between the two families exceeds half a century. We witnessed the passing of Shih-liang and Wan-tu and the growth of the three Chien brothers from childhood to become seniors. Being located far away, we do not meet often, but fortunately we still keep in touch by mail and phone. As Margaret recalls the time she lived in Chien's house in Taipei, she can still feel the warmth in her heart. At that time Shih-liang was President of National Taiwan University, and his three sons were students at the University. Although the economic situation was not that good, the family of five lived happily and enjoyed themselves. Shih-liang loved cats and he liked to give nicknames to members of his family and to his pet cats. His three sons were called Tah-Hsing, Er-Hsing and San-Hsing (first, second and third Hsing, which could be translated as prosperity). Shu was also called

---

[*]T.W. Hu's father is Dr. Hu Shih, a great Chinese scholar and philosopher in the 20th century. He served as the Chinese Ambassador to the United States during World War II and President of Academia Sinica in Taiwan, 1958–1962 (a position also held by Shu's father later).

Mrs. Hu Shih (T.W.'s mother), Foo Hu (T.W. and Margaret's son) and Margaret, 1961.

K.C. and Shu with T.W. Hu, 1997.

On Shu's 70th Birthday

Robert Chun Chien[*]

Shu is my younger brother. During our formative years, we were very close as we shared the same bedroom from kindergarten to college. I know him so well since our childhood that I dare to say no one knows him better than I do until he got married. He is clever, generous, kind and smart. He learns fast and works hard. Shu is a renowned scientist, one of the top in his chosen field, a dedicated educator, and last but not least an excellent administrator.

When Shu was 16, a junior in high school, he was already confident that he was ready for college. Physics is one of the subjects that were tested in the college entrance examination. However, at that time Shu had never taken any lessons in physics because it was a course for high school seniors. Knowing that there were no alternatives, I volunteered to serve as his tutor even though I myself had only taken a year of physics during my high school days. Time was running short, there was only one month left before the entrance examination would take place. His cramming of the subject started with the help of an amateur tutor who himself knew very little about the subject. However, Shu was so brilliant and hardworking that he scored a B+ on physics and was admitted in the premedical department of the National Peking University, one of the best

---

[*]Minister of Finance (1985–1988) and Secretary General of the Executive Yuan (1988–1989), Republic of China. Shu Chien's elder brother.

universities in the Republic of China at that time. That is what I call smart and clever.

After he completed his medical training in Taipei, Shu was awarded with a Li Foundation scholarship to pursue further studies in Columbia University. A year or so later I went to Minneapolis for my graduate studies at the University of Minnesota. He knew that I was short of funds and had to work my way through college. Without my knowledge he sent me a check, in the amount equivalent to almost half of his scholarship to support me. How generous and nice he is. In 1958, I made my first trip to New York City where Shu lived. He asked me to stay with him and I did. However, we could not share the same room again because he was married. During the time I stayed with him, he showed me around the largest city in the world. Even though Shu had an extremely tight schedule, he took me to see the Radio City Show, the rehearsal of the Ed Sullivan Show, and the circus at Madison Square Garden. Since then I had visited

Robert and Ruth with Shu and K.C., 1996.

New York City numerous times, but I never had the opportunity to revisit those places again.

Being a student of economics, I am a layman as far as physiology and bioengineering are concerned. I learned from many mutual friends who are experts of medical sciences that Shu's achievements in his field are phenomenal. The fact that he was elected to be the president of the American Physiological Society and the Federation of American Societies for Experimental Biology in the early 1990s illustrated that he had not only achieved academic excellence but was also well liked by his peers. There is an old Chinese saying that life starts at 70. Shu will turn 70 on June 23rd of this year. May we all wish him the very best in his marvelous research and teaching work.

Families of the three Chien brothers, 1995.

Many Happy Returns to a Most Generous Brother

Fredrick Foo Chien[*]

My brother Shu is going to celebrate his 70th birthday soon. His friends and students all over the world would be writing to express their best wishes to a real jolly good fellow.

Since his birth, Shu became the darling baby of our mother. When he attended schools and universities, all his teachers enjoyed having such a brilliant student. He has been teaching for the last 40 odd years and for many times he was selected by his students at Columbia Medical School as "Professor of the Year." All his colleagues love and respect him.

From the day he married my sister-in-law, Kuang-Chung (K.C.), he has always been a devoted and faithful husband. To May and Ann as well as their families, Shu is a most thoughtful father and grandpa.

I guess that, probably, I am the only person who can claim him to be a most generous brother, because I am fortunate to be four years younger than Shu. When I completed my college work and ROTC in 1958, a very kind teacher of mine, the late Professor David Nelson Rowe, helped me to obtain a full scholarship at Yale Graduate School's International Relations program. A travel grant was also provided by the Asia Foundation with a round trip air ticket between Taipei and New York. But I still needed to get some funds to support my daily

---

[*]President, The Control Yuan, Republic of China. Shu Chien's younger brother.

expenses. My parents were not in a position to do that. There was no such loan available at that time. When Shu learned about it, he immediately wrote: "Don't worry. Do come. I can take care of that." So with that assurance, I proceeded to New Haven. Each month a check of $100 was mailed to me by Shu for the next nine months allowing me to complete my first academic year. I knew that was not easy for him, as he was just starting his teaching career and May and Ann were still babies. But Shu did it willingly and without any hesitation. Such a generous act enabled me to dedicate my entire time for studying and after nine months received my M.A. degree.

This little story is, most likely, not too different from what other contributors might write in this tome. But as far as I am concerned, I owe a lot to Shu for what I have done during the last forty years.

The Chinese sage, Confucius, once remarked: "At seventy I could follow whatever my heart desired without transgressing the law." Shu lived the last sixty-nine years mainly to help and serve others. Now he has reached seventy. I sincerely hope that he can follow whatever his heart desires.

Julie and Fred with K.C. and Shu, 1994.

About fifteen years ago, I convinced Shu to move West for better climate. He lives only couple minutes away from Torrey Pines Golf Club. My fervent hope is that he would go out to play more golf and before too long he could shot his age.

Many happy returns, my generous brother Shu!

Three brothers getting ready for golfing, Taipei, Taiwan, 1970.

Three brothers singing, La Jolla, CA, 1995.

Shu and I: Our Happy Marriage and Family

Kuang-Chung Hu Chien (K.C. Chien)

My marriage with Shu is truly wonderful. Our marriage has made me a better person and the happiest person. My mother (MaMa) said I have the most marvelous marriage. Time flew past; Shu and I have now been married for 44 years.

I first heard about Shu in 1948 when I was in my last year of high school. My great aunt Mrs. Hu Shih, who often came to play mahjong with MaMa, praised Shu on many occasions, saying how wonderful a student he was at the medical school of National Taiwan University (Taita). She referred to Shu as the second son of the Chien family (Chien Chia Lau Er).

I saw Shu for the first time later that year. One day when I came home from school, I saw a neat and handsome young fellow sitting comfortably and contented at the entrance alcove of our home (later I found out that he was asked by Mrs. Hu to deliver a set of mahjong to MaMa). He looked up and saw me coming in from the door with my bike. He stood up to greet me and let me put my bike in the doorway. He sure was handsome and attractive. I briefly said hello and went in. I don't remember how he left, and nor can he.

In 1949, I passed the exam to enter Taita medical school, and I wanted to find out how good Shu really was. On the school bulletin board, the names of students in the top 4% in each semester were listed along with their academic scores. Shu's name was listed consistently there as number one in his class.

Wedding, 7 April 1957.

The university students in Taiwan at that time were quite conservative in dating. Shu and I participated in many group activities together such as ping-pong, running, outings, etc. We did not go out by ourselves, however, until he graduated from medical school and received the mandatory one-year military training (ROTC) in 1953–1954. We had many wonderful times going to movies, watching sports events, rowing, etc. and we talked seriously about our future.

In 1954, Shu received the Li Foundation Scholarship to pursue his Ph.D. study in physiology at Columbia University College of Physicians and Surgeons (P&S) in New York City. Because I would not graduate until 1956, I had to stay in Taiwan for two more years while Shu studied in the U.S. I missed him very much during that period, but this allowed me to focus on my study and training. We wrote to each other frequently. (In those days, it was very difficult and too expensive to make telephone calls. Of course, there was no fax or e-mail.) During that period, I was often invited by Shu's parents to their home. Shu was his mother's (Shu called her Nian) favorite, and I too received special love from her, even before we got married. Nian and I had many conversations, and the subject was often about Shu. Nian proudly told me that Shu could perform well under all circumstances; he could achieve better than 100% in tests even on subjects he did not have full knowledge of. She cared very much about us, including our future financial well-being. She suggested that I encourage Shu to pursue clinical medicine rather than become a professor, because that would make our life easier fiscally. I said that I would help Shu do whatever he likes, and she agreed with that.

In 1955, I applied for internship in several hospitals in New York City. I decided to go to the Mother Cabrini Memorial Hospital because it was only a few blocks away from P&S. In the summer of 1956, I started to get ready for my trip to the U.S. Shu sent me U.S.$400 to buy the ticket to travel by sea. I took a cargo ship, which routed through many countries in Southeast Asia before heading for the U.S., and the journey took 60 days. Shu had written me saying that "time moves fast once you have a date." That made me feel much better. On 2 November 1956, our ship finally landed in Portland, Oregon, and I took the train to New York City. Shu met me at the Grand Central Station wearing his beige raincoat. I can still vividly recall how happy and excited I was to see him!

I arrived in New York on Election Day, and I started working at the hospital the very next day. The work as an intern was very busy, but it went smoothly. When I was not on call, Shu would bring me to visit Mrs. Hu Shih or out sightseeing. Two months after my arrival in New York, Mrs. Hu gave an engagement party for us in her apartment. It was a wonderful event, with many friends and relatives coming to celebrate with us.

Our wedding was on Sunday, 7 April 1957 in the Madison Avenue Presbyterian Church in New York City. I still remember the wonderful feeling I had. If I can be married to Shu, what else matters? None! Being married to Shu is my happiness, and of course it is his too (though I never had to ask him).

Mrs. Hu arranged the church and the reception at her friend's home, and we brought the food and drinks to the reception early that day. In addition to our relatives and friends, Shu and I invited our superiors (professors and attending physicians) and colleagues. It was a marvelous wedding. Our schoolmates Chi-chien Chu and Shao-nan Huang, who almost had a double wedding with us, had their wedding on the day before. So I was able to borrow the entire set of wedding

Honeymoon, 1957.

May and Ann, 1962.

dresses from Chi-chien, including the wedding gown, bridemaid's dress, and many other accessories.

Back in 1957, interns in some hospitals were not supposed to get married and therefore had no wedding holidays. I had to go back to work the day after the wedding. At that time Shu was also very busy preparing for his Ph.D. defense examination. So we went on our honeymoon trip after the exam. Marriage, graduation, work and homemaking seemed to move along smoothly and naturally. A little over a month after our wedding, my classmates Sin-Ping Chueh and Genesia Liu came to New York for their internship training. When they came to our apartment, they said with surprise that we appeared to have been married for a long time.

When I was an intern and later a pediatrics resident at the Metropolitan Hospital, training was my priority, i.e. work came first. Shu and I were extremely happy when May was born in 1959. Because I was working, we had to arrange nursing care for May. We went through several different arrangements, and I began to reprioritize my life. I stopped working when I became pregnant with Ann. The birth of Ann in 1961 was another extremely happy event. Our two lovely children are the crystals of our love, and they are our pearls. I stayed home over the next three years to bring them up myself. That was a very happy period. Besides taking care of May and Ann, I enjoyed entertaining. I often cooked for some 40 people in our New York apartment, including relatives, friends, co-workers and visitors from out of town, such as our teachers from Taiwan. In 1964, I went back to work as a part-time researcher at P&S, and at the same time I studied for the medical licensure examination in New York. After passing the exam, I worked in the child health clinic of the Department of Health in New York City and became a Physician-in-Chief. It was very rewarding to help the children of indigent families.

In January 1987, Shu went to Taiwan for 18 months to establish the Institute of Biomedical Sciences. I resigned from my position in New York City to go with him. The staff members in the Institute asked me how could I leave such an important job. My answer was that nothing else matters when you love your man.

When UCSD was recruiting Shu, we had many long discussions. He was troubled by the fact that the State of California would not reciprocate my medical licenses of New York and New Jersey. I was also concerned about being far away from our daughters. Finally we decided to give it a try. With Shu's encouragement, I took and passed the California medical licensure examination more than 30 years after

Shu, K.C., May and Ann with Shu's parents, New York City, 1962.

my graduation from medical school. This not only allowed me to continue working in pediatric clinics helping the sick children, but it has also given me a great deal of confidence and courage. I am very happy that we made that decision to come to San Diego.

Shu is a most understanding, supportive, loving and caring husband. Just like what was said in the marriage vows, he and I love and respect each other. He never puts me down; instead he always encourages me to achieve my goals and to follow through on what I do. I have learned from Shu to work hard and play hard. Shu makes life simple and enjoyable. He can change complicated matters to simple matters. He can work on a multitude of things as if they were nothing.

Shu and I have a wonderful marriage and family. Together we have an extremely rich life. Nian, I am happy to say that we are very rich in our way. I am very thankful and grateful for everything. MaMa is right: My marriage with Shu is indeed marvelous.

My Dad

May Chien Busch, M.B.A.[*]

I feel truly fortunate to have Shu Chien as my father. When asked whom I most admired during a high school assignment, I chose my Dad. I have 42 years of particularly rich data that supports my thesis and I am pleased to have the opportunity to share some of the highlights with you.

First, my father has always been there for me, providing wholehearted support without ever interfering. He has always had time for me.

- I liked to barge into Dad's study at home where he was usually at his desk working on a paper or grant proposal, or arranging slides for a lecture (and I am in fact writing on that same desk right now!). I would ask him all sorts of questions: "are you busy?" (he would say no, but of course he was), "what are you working on?" (he would patiently explain in a way that made me feel that I actually understood what was going on with those RBCs), and then I would move onto whatever was worrying me that day.
- He and my mother went to every one of my ballet, violin and piano recitals, and they drove me and my sister to my piano competitions with plenty of encouragement to stave off that really nervous feeling as we traveled the stretch of road leading to the competition hall.

---

[*]Managing Director, Morgan Stanley. Shu and K.C. Chien's elder of two daughters.

- When I was sick, he would come upstairs to sit with me when he got home and tell me Chinese classics and "wu-xia" (martial arts) stories such as those about Zhu Ge-Liang and the monkey king, which I loved (thanks to Mom for sending him up!).
- When it was time to go clothes shopping, Dad drove us to the mall and sat outside every relevant dressing room in Bergen County, waiting for three females to finish shopping — it took hours. Of course, he would use the time wisely, either working on his papers or, on the odd occasion, sleeping (I think it's now called taking a "power nap").

Front: Renee, Natalie and Kristen Busch.
Back: May Chien and Leonard Busch,
in Chiens' backyard, 2001.

- I had hurt my knee just before arriving in La Jolla this past April and Dad arranged an appointment with the UCSD sports medicine team and accompanied me to the visit. It was just like old times, sitting between Mom and Dad in the waiting room. Except for my knees, I felt like I was a teenager again!
- And, of course, there was my senior honors thesis where I had misjudged just how long it would take to write, edit and type it. I was camped out in Dad's lab all weekend with my sister, Dad and I alternating typing, me and my Dad alternating editing, and Mom bringing food. Dad never expressed any criticism, but took one look at the state of my thesis, identified a realistic plan, and kicked us all into high gear to execute.

Second, Dad made growing up fun for us. Thanks to him, my sister and I certainly had as much exposure to sports as any family of boys. We cheered the Knicks in the days of Clyde, Dave Debusschere, Bill Bradley, Earl the Pearl Monroe, Willis Reed and Phil Jackson (before he became a coach); we ate Almond Roca and entire bags of Fritos corn chips as we watched. We played basketball in the driveway, becoming quite proficient at moves such as the "give and go" as well as the reverse lay up. Dad taught us ping-pong Chinese style with forehand and backhand topspin smashes. He was such a great player that he could feed the ball consistently to the right spot to make it easy for us to hit smash after smash. It was only when I had a different opponent that I realized that Dad actually made me appear to be a better player than I really was. We were included in Dad's lab parties, and he and Mom also held tennis parties where we all had a really great time. In fact, I remember one time when Dad didn't want to stop playing until he finally got a cramp in his leg and had to be helped off the court!

Most importantly, Dad taught us about the important things in life and for living. We learned so much from living with our parents, and their complementary wisdom has been the foundation for our lives. Looking back, it is very clear that those seemingly small lessons that we learned through daily living in the Chien household, sometimes by simply having observed our parents in some interaction or situation, have helped us later in life in a meaningful way.

- **Only positive words:** He rarely criticizes, but instead finds a positive way to convey the message. I remember him helping me with a presentation, and advising me to "use only positive words," which I tell all of our people at the office as well. My father is still my role model as I often handle business situations, as I would envision my father handling them.
- **Perception is reality:** I remember having an argument with Dad when I was a teenager, and insisting, "that isn't what I said!" to which Dad calmly responded that what is relevant is how the recipient perceived it, not how the speaker thinks she said it. This, of course, is so true because you have to deal with each other's perceptions, and this becomes the reality.
- **Chance favors the prepared mind:** Very useful in making the most of those unexpected meetings with top management and other benefactors. Because he is always thinking ahead, my Dad has an uncanny ability to recognize and capitalize on opportunities.
- **Don't worry about what others think:** I can still hear Dad comforting me after some perceived "embarrassing crisis situation", saying that "a year from now, you won't even remember this incident; take away the lesson to be learned, and move on. Always look forward — that's why the eyes are in the front of your head!"
- **Be realistic and be calm under pressure:** Dad always knows how much time he has to do a project and is realistic about what can be done in that time. He uses his time incredibly efficiently and uses every minute. In fact, the more he has to do, the more he seems to accomplish.
- **Do things sincerely:** If something is worth doing, give it all you've got, do it 100% and think through every aspect.
- **Swing out:** Always give it your best shot and don't be afraid to fail. Win or lose, it is much better to swing out than to be tentative and wonder what might have been.

My father is rich in all the ways that matter and I am grateful that he has shared so much of his experience, wisdom and love with me. All the best wishes for his 70th birthday.

Professor Shu Chien, My Dad

Ann Chien Guidera, M.D.[*]

My dad is a man with tremendous talents and great achievements of whom we are all extremely proud. It is a wonderful thing that my interest in science and medicine allowed me a firsthand look into his world of research and teaching.

I worked in his lab as a high school student and saw that he was not only a brilliant scientist, a great communicator and a natural leader, but also the creator of a true lab family. His office door was always open. He was known to be the source of valuable advice and guidance. The atmosphere was both dynamic and productive. People at every level were treated with respect and encouraged to do their best. There was a true sense of comradry; whether it was during morning meetings or the softball game at the lab picnic (the volleyball games were a bit more competitive).

My dad was one of my physiology professors during medical school. He taught with energy and enthusiasm that were contagious. His teaching style was at once engaging and enlightening. I still remember his analogies of the flow of cells through capillaries to the traffic of cars across the George Washington Bridge. We all enjoyed his physiology demonstrations with class participation during which one of my classmates (who was to become my husband) hyperventilated oxygen and then was able to win the breath holding contest. (That's

---

[*]Ophthalmologist. Shu and K.C. Chien's younger of two daughters.

*Front: Jenny, Laura and Katie Guidera.
Back: Ann Chien and Steven Guidera,
in Chiens' backyard, 2001.*

not what made me marry him, but I do wonder now if there was any adverse effect.) My classmates clearly shared my feelings when we elected him Teacher of the Year in 1984. He was also given this honor by multiple other classes of medical and dental students.

But through the years and his many impressive achievements, too numerous to list here, to me, he is after all, just my dad.

My dad is a man with tremendous wisdom. His wisdom has gently guided me. Among the many lessons that have enriched my life, his words of wisdom have taught me that one is responsible for ones own happiness. He taught me to make the most of the present and what is presented to you, to be flexible.

My dad is progressive. I was impressed with his open-mindedness while raising two girls in a culture so different from the one in which he grew up. I saw this progressiveness in his professional life when he brought molecular biology into his lab and rapidly became more adept in this new field than many of us who trained later.

My dad is inspiring. His continued personal growth and professional productivity will no doubt continue through his 70s, into his 80s, and beyond. He shows the same enthusiasm now for more recent hobbies such as golf and karaoke as he did in years past for tennis and ballroom dancing.

My dad is generous with his time and talent. Through the years when I was growing up, even when finances were tight on a young professor's salary, he and my mom helped support and integrate many friends and relatives who came from China to stay with us. They helped provide the opportunity to pursue the American dream.

My dad is loving. He is a fantastic father and husband. Those who know my mom understand when I say, "to know Kuang-Chung Hu Chien is to love her," and my dad loves her best. Together they are a fabulous team in family, society and life. I thank God for having blessed me with these two people to be role models for marriage, parenthood and life.

Thank you, dad for sharing all your wisdom, love and experience with me. These will stay with me throughout my life. Happy 70th Birthday! We love you, dad.

My Brother-in-Law

David Kuang-Jeou Hu[*]

To others, Dr. Shu Chien is a world-renowned scholar, but to me he is also a wonderful person and a special brother-in-law.

He is a very humble person who has not, in my recollection, ever mentioned in front of others about his achievement in the academic world. I often read about it from the newspapers and felt very proud of him. Occasionally I may overhear it from my sister in a casual conversation.

He is a very sincere and polite person who is very easy to talk to, makes you feel at ease, genuinely listens to you, and gives you good advice.

He is a person who has the ability to focus his attention in reading, writing, and learning at any given time intervals and in any physical environment. Yet, he is also able to relax, rest and sometimes even doze off in a similar situation.

He loves my sister. Their mutual respect of each other exemplified in their daily life is something for us to learn and admire.

With all these good qualities as an individual, Chien Shu is also a very generous brother-in-law. He and my sister had helped me through my college education. They extended their assistance without any thought of what they would receive, and by doing so they changed my life.

A few months after my arrival in Los Angeles in 1956, I began to feel very ill. The doctors' diagnosis was tuberculosis — at that time a

---

[*]Vice President European Operation, Kingston Technology Europe Ltd.

death warrant. It was a fight for my life! After seven months in the hospital, I was well enough to be discharged. After a hard fought battle, I had managed to overcome the disease. It was at this time of great need that Chien Shu and my sister started what I would call, a mission of mercy. Each month, they sent me $100. Back then, that was a substantial amount, especially given that they themselves were just starting and struggling to make a life and to start a family together. Without their help, attending college would have been prohibitively expensive.

This generosity continued throughout my college years, which is longer than I would like to admit. Never once was there talk of how the money would be spent. It was definitely the best loan I have ever received — no strings attached and even more importantly, there was never any discussion of repayment. In the years since, I have tried to repay their kindness, but each time to no avail. A life well lived was thanks enough for them, they said. It is this kind of big-heartedness and sacrifice that I will always remember and cherish.

K.C.'s siblings: Elder sister Katherine Kuang-Min, elder brothers Charles Kuang-Zuai and Robert Kuang-Qi, younger brothers George Kuang-Chi and David Kuang-Jeou and sisters-in-law Eileen, Qi-Feng, Theresa and Winnie, respectively. From left to right: Charles, Eileen, George, Katherine, Theresa, Qi-Feng, Robert, K.C., Shu, Winnie and David, 1999.

Dr. Shu Chien is indeed a wonderful person and a special brother-in-law. There are always some things for me to learn from him. I am very proud of him and I am very grateful to have him as my brother-in-law.

My Yih-Fu

Mark Yong-Qiang Wang,[*] Xiao-Chuan Wang and Tom Wang

There is a portrait on the wall of our living room. It was taken in the late 1980s during the anniversary party for my "Yih-Fu" (husband of my mother's sister) and "Yih-Ma' (my mother's sister). It has them in the middle, with my wife Xiao-Chuan and I on their sides, and our son Tom standing in front of them. This is a lovely picture, which Xiao-Chuan had framed very nicely.

Whenever our friends come to our house, they would ask about the picture: "Is this you uncle and aunt?" I would correct them in Chinese, saying that "No, they are my Yih-Fu and Yih-Ma." They would ask, "What is the difference?" It is true that uncle and aunt in English are equivalent to Yih-Fu and Yih-Ma, respectively. But I always like to say Yih-Fu and Yih-Ma, rather than uncle and aunt. Why? I don't know. I cannot be certain why I feel so strongly about the subtle differences. Often, we would move onto a discussion of a comparison between the Chinese and English languages. Then, I would tell my friends the following story.

In the mid-1980s, Xiao-Chuan and I visited Yih-Fu and Yih-Ma one night in their house in Englewood, New Jersey. May, Ann and Daniel and Evelyn Ou-Yang were also there. During the conversation,

---

[*]Mark Yong-Qiang Wang is the second son of K.C.'s elder sister Katherine Kuang-min Hu. Mark is a Senior Network Engineer for Parsons Brinckerhoff, Inc. at Newark, NJ.

Shu and K.C. with Yong-Qiang, Tom and Xiao-Chuan Wang, 1987 (portrait mentioned in the article).

the topic of the differences between Chinese and English languages was brought up. I remember Yih-Fu said, "One of the differences between the two languages is in their poems. While both Chinese and English poems have their rhythms that make them sound beautiful when read out loud, Chinese poems have a feature in its verse structure. Usually the verses in Chinese are fixed in their length, such as five or seven characters each. Furthermore, in some cases it is possible to retain the main meaning of the poem after removing equal numbers of characters from each verse, although the complete poem gives it a greater richness." He used the example of the Chinese poem on "Tomb-sweeping day" as an example. This poem is composed of four verses of seven characters each, but the basic meaning can be retained when the length is reduced to five, three and even one character per verse. Although this conversation occurred nearly 20 years ago, it left a deep impression on me.

I recently talked to Yih-Fu about this and interpreted his remarks as saying that such kind of treatment is not possible in English poems. He told me that one can also do that in English and gave me the following translation:

First, he did it in seven words per verse and was able to keep a rhythm:

> Rain drizzling on the tomb-sweeping festival day.
> A person is entranced on his way.
> He asked where is a tavern bar.
> Shepherd-boy pointed to the Apricot Village afar.

When reduced to five words each, it becomes:

> Rain drizzling on tomb-sweeping day.
> Person entranced on his way.
> He asked for a tavern-bar.
> Shepherd-boy said Apricot Village afar.

When reduced to three words each, it becomes:

> Rainy tomb-sweeping day.
> Person on way.
> Asking for tavern-bar.
> Apricot Village afar.

When reduced to one word each, it becomes:

> Rain.
> Person.
> Tavern-bar?
> Village!

Even though something is lost during the successive shortening, it is remarkable how the fundamental message can be retained.

I learned from this that poem is an art of language. The language of poem is the most refined language. Yih-Fu is a scientist, but he has a deep insight into the two languages (Chinese and English), as illustrated by the example of the poems. He is a true scholar with a great breadth in knowledge that transcends East and West. On this marvelous occasion of his 70th birthday, we would like to use this article to express our families' sincere admiration, love and blessing.

Our best wishes for a Happy Birthday and Excellent Health to Yih-Fu and Yih-Ma.

比如这样一首七言中文律诗：
　　清明时节雨纷纷，
　　路上行人欲断魂。
　　借问酒家何处有，
　　牧童遥指杏花村。
你可以把它断句成一首五言中文律诗：
　　清明雨纷纷，
　　行人欲断魂。
　　酒家何处有，
　　遥指杏花村。
你还可以把它断句成一首三言中文律诗：
　　清明雨，
　　人断魂。
　　问酒家，
　　杏花村。
甚至可以把它断句成一首一言中文律诗：
　　雨，
　　人。
　　酒？
　　村！

Friendship for 54 Years

*Pei Pin Han**

In a recent telephone conversation with Shu, he mentioned that I am the one that he has known the longest among all his friends. We have known each other for 54 years since we first met in 1947. Young classmates then, grandfathers today. The rapid passage of time is amazing.

In 1947, two years after the end of World War II, I graduated from the Northwestern Normal College High School in Lanchow and took the entrance examination for National Peking University (Beida) Medical School. Right after the exam I traveled by boat from Shanghai to Tientsin, and then took the train to Peking to see my grandmother. When I got to Peking, I was extremely happy to learn that I had passed the exam for entrance into Beida Medical School.

Beida Medical School had a seven-year program. The courses for the premedical program in the first two years were taught in the College of Sciences. As we started the classes, I heard that we had an outstanding classmate Shu Chien, who was only 16 years old and had two scholarships. Our class had over 70 students and most classes were taught in small sections. Shu and I were in the same class only for chemistry and biology, which were taught in a large amphitheater. I first met Shu in a chemistry class. I still remember that there were wood-carved large ancient Chinese characters from *The Book of Moderation* — "Learn with Breadth, Question with Analysis, Think

---

*Shu Chien's classmate in Beida Medical School.

with Care, Decide with Clarity, and Act with Dedication." I was able to recognize all of the 15 ancient Chinese characters only after Shu's help.

Shu was kind, gentle and modest. He wore a long Chinese gown and a pair of glasses, and had a scholarly appearance. At that time, it was customary for male college students in Peking to wear long gowns. With a pair of slacks and leather shoes, plus a scarf in the winter, Shu looked very handsome. I tried to wear the gown for two days, and gave it up because I found it difficult to reach the slack pockets under it. As far as I can remember, however, Shu always wore that.

Shu's father, Professor Shih-liang Chien, was the teacher of our sophomore class on qualitative analytical chemistry (Shu was in the same class). Professor Chien was kind, gentle and amiable. He was an excellent teacher who was greatly appreciated by his students. Shu obviously got most of his character from his father.

In the years 1947–1948, student anti-government movements were rampant in Peking. Beida students frequently had strikes and parade demonstrations. Shu was living at home, and he always stayed home studying when the strike was going on. Hence he was criticized by some that he did not pay attention to national affairs. Since I was living in the dorm, I had to follow the parade, waving flags and yelling, and as a result wasted a lot of precious time.

Near the end of 1948, I went to Taiwan with my family. Shu went to Taiwan with his family in 1949. I don't remember how I knew the date and time of his arrival in the Keelung port from Shanghai. I got there in time to wave a placard saying "Welcome Shu Chien" from the platform on the second floor of the pier. We were extremely happy to meet again so far away in a new land.

Shu transferred to Taiwan University (Taita) Medical School. I was also admitted for transfer, but I had to give that up and make new plans for my life because of family responsibilities. When I went to Taiwan from Peking, I thought the civil war would end soon and I would be able to return to Beida to continue my studies there. Therefore, I brought some books, syllabi, etc. with me for review, including the test papers from the quizzes and exams from the analytical geometry course. In Taipei, Shu told me that his father would like to have these materials for his reference, and I happily provided them. Thereafter, Shu mentioned many times how his father appreciated that. I gather that these teaching materials were useful to Professor Chien.

Shu's wife Kuang-Chung Hu was a classmate of my wife Chi Kuang Kuo at the Taipei First Girls' High School. They both attended Taita and were close friends. Therefore, our two families have multiple relations. When our sons Allen and Peter studied on the East Coast, they frequently spent their holiday times at the Chiens' home in New Jersey. They were very fortunate to have the opportunity to experience first-hand Shu's modest, broad-minded character of a true scholar.

In June 2000, the Taita Early Graduates Alumni Association held its annual meeting in Los Angeles. When Chi Kuang and I heard that both Kuang-Chung and Shu would be there and that Shu would be the keynote speaker, we decided to go. Although we had not seen each other for many years, Kuang-Chung and Shu still looked so young. We spent a wonderful time together during the days of the meeting, and

With Chi Kuang and Pei Pin Han, 2000.

we had a marvelous dinner hosted by Kuang-Hun and Ming Yue Cheng. The trip to Los Angeles was indeed most enjoyable.

On this joyous occasion of Shu's 70th Birthday, I would like to write down some precious memories of our schooldays at Beida more than half a century ago and our continued friendship over all these years. I would like to present this to him and send my best wishes for his health and happiness.

Shu Chien — A Truly Outstanding Alumnus

*Peter Kuang-Hun Cheng**

National Taiwan University (Taita) has many outstanding alumni who are well known in Taiwan and/or internationally. Being widely known, however, is not identical to being outstanding. In my view, Shu is an outstanding alumnus who transcends time and space.

Shu transferred to Taita from National Peking University in 1949. During his four years of study at Taita Medical School, he was always at the top of his class, but he never behaved like most top students would. He was sincere, humble, impartial and easy to get along with. After his graduation in 1953 and one year of ROTC training, Shu received the coveted Li Foundation Fellowship to pursue graduate studies at Columbia University. He received his Ph.D. in Physiology in 1957 and joined the Department as a faculty member.

The word "Doctor" in Ph.D. is translated in Chinese as "Scholar with Breadth." I had been puzzled by this translation because I thought it should have been "Scholar with Specialization." Several years after Shu received the degree, I finally understood the meaning of this translation from his range of knowledge and scholarly pursuits. It is my sense that the goal of Ph.D. education is to have every scholar with specialized knowledge contribute positively to the enhancement of human welfare. Shu is a medical doctor, and in my view he is a truly representative Doctor of Philosophy. His subject of study in

---

*Chairman of the Board, Chung Kuo Insurance Co. (ret.).

medical sciences is physiology, which provides the foundation for the understanding of the mechanisms of all diseases. Through his research on blood circulation and biomedical engineering, which has been published in many first-rate biomedical journals, he has contributed to the understanding of pathogenesis of diseases and the improvement of medical technology for the treatment and diagnosis of illnesses. Shu has provided the foundation for physicians to better treat their patients, thus realizing the functions of "helping the world." Although Shu did not study philosophy as a subject, he possesses a special gift in philosophy, which Buddhism refers to as "intellectual root." This allows Shu to attain his outstanding accomplishments that transcend time and space.

Everyone who knows Shu is impressed by his sincerity, humility and easy accessibility, as mentioned above. These characteristics stem from his respect and consideration for others, and his selflessness from within. As a result, he has the outstanding ability to generate affinity and cooperation. These are relevant to his "intellectual root" that encompasses a superior intelligence.

The "intellectual root" Shu exhibited in Taita was further developed after he went to the United States. Shu is able to combine the strengths of the Eastern and Western cultures and fully express his "democratic personality." When discussing any problems, Shu always respects the opinions of others, while expressing his own. If the discussions result in the adoption of Shu's idea, he would show his natural appreciation and consideration. If his views were not accepted, he would gracefully respect the opposite point of view and endeavor to find points of commonality. The way Shu treats people and matters, together with his accomplishments in biomedical research, has led to his election to many leadership positions in professional organizations, e.g. the American Physiological Society, Federation of American Societies of Experimental Biology and the American Institute for Medical and Biological Sciences. Shu used his sabbatical time to establish the Institute of Biomedical Sciences in Academia Sinica in Taiwan, and he is the Chairman of the Advisory Board of several scientific organizations in Taiwan. He has also been active in assisting biomedical research in mainland China, Japan, Europe and the United States. Therefore, Shu's contribution to the society has no geographical boundaries. Shu's interactions with people are unaffected by their socioeconomic status. What he does is not limited by the element of time. His ability to transcend time and space is way beyond what a general Doctor of Philosophy can do.

I often thought why Shu could have such outstanding accomplishments. Of course, I would first think of his father, the late Taita President Shih-liang Chien. In Shu's obituary, he called his father "the perfect person"; this reflects the role model effect his father had on him. Later, Shu was also influenced by Dr. Hu Shih. The influences of these great people help to make Shu who he is and let others feel like being bathed by the spring wind when meeting him. In my mind, Shu is an outstanding alumnus, and like his father, very close to being a perfect person.

With MingYue and Kuang-Hun Cheng, 2000.

Lao-Chien is 70 Years Young Now!

Shaotsu Lee, M.D.[*]

Time flies! It was 1949. It seems just yesterday that Shu and I trekked back and forth between amphitheaters and classrooms at the medical school and were sharing a bunk bed living in the intern's dormitory at the Taiwan University (Taita) Hospital. We played basketball and ping-pong together. In class, we were trying to understand most of our classmates who spoke either Taiwanese or Japanese and also had to decipher the meaning of the Taiwanese lectures using Japanese, German or English terminology. Amidst all the confusion we, the mandarin speaking few, were not too concerned, as we could always depend on the lecture notes recorded faithfully by Lao-Chien (Good Old Chien) in the classes of different professors. Shu was often the intermediary between the professors and us. It was an integral part of our education. If situations required for us to skip classes, help in the form of copying one's signature for attendance requirements was always available. With practice, we became quite proficient at this.

Thanks to the generous hospitality and the ever kind presence of Chien-bai-mu (Aunt Chien), we have had many wonderful times at Chien's home. Canasta anyone?

---

[*]Anesthesiologist (ret.). Shu Chien's classmate at National Taiwan University College of Medicine from 1949 to 1953.

With Shaotsu Lee, 1993.

Once Shu and Kuang-Chung moved to New York City, their home became the center of gatherings for friends all around. In 1957, when I was traveling from Europe to Massachusetts, my first ever stop in this new world was at Chiens' apartment near the Columbia Presbyterian Medical Center in Manhattan. Vividly I remember the welcome, cozy and friendly air, delicious Chinese food and a flickering TV set.

In the early 1960s, more alumni from Taita inhabited the New York area. Again, there were happy gatherings of friends, old and new, at the Chiens'. There are fond memories of the birth of May, Fred's encounter with his newborn niece, his nervousness when holding her, etc. Even though we lived just across the river in another borough, I hated the drive to crowded Manhattan except to visit the Chiens. Occasionally I visited Shu's circulatory lab at Columbia. I had followed Shu's work in the area of blood rheology, microcirculation, blood volume, etc. with special interest and pride in all his accomplishments.

Even during our school years, we all appreciated Shu's organizational and administrative abilities. After completion of the founding tenure at the Institute of Biomedical Sciences in Taiwan, Shu took over the leading task of further developing bioengineering at UCSD, where he steered his work towards the modern trends and advances in molecular biology. I envy Shu's ability and determination to pursue new endeavors. In time, an Institute of Biomedical Engineering was established, and he was appointed as its director. Through the years, every one of our old friends has been watching Shu's advancement and accomplishments. All of us are very proud of Shu. We are often identified as one of "Chien Shu's classmates" during alumni gatherings in recent years. How else would the younger alumni know the rest of us after 48 years?

In recent years, it has been wonderful for me to be physically close to Shu again, since my relocation and retirement to the West. It has always been a pleasure to visit the Chiens in San Diego. Their hospitality has been gracious as always.

For more than 50 years, I have had the privilege and fortune of knowing Shu, Kuang-Chung and all of Chien's family. It has certainly been an integral part of my life, to say the least. From the deepest depths of my heart, my dear friend, I congratulate you, Shu! You have done and accomplished very much in your life, Shu! There are few, if any, in our generation who can even come close in comparison.

On your 70th birthday, Peggy and I wish you the very best for the years to come!

A Very Old Friend,

Shaotsu

Outing with Classmates in Taipei, 1951. Kneeling in right front is Pung Show Liu (p. 48). Standing second from left is Chih Chao Chin (p. 50). Alone in third row is Shaotsu Lee (p. 44). Shu is immediately behind Shaotsu.

Over a Half-Century of Friendship

Pung Show Liu, M.D.[*]

It was 1991 in Williamsburg, VA when I last saw Shu "San-Mao" Chien and my emotions ran high as I recalled the past 50 years that I have known him:

I first met Shu in school and I am sure that it comes as no surprise that Shu was one of the best students not only at the university, but also nationally in Taiwan. However, it may surprise some that he was also an avid sports fan and excellent athlete in basketball, table tennis and soccer. Although he was slight of stature, he was wiry and speedy, receiving his nickname, "San-Mao" (Three Hairs), as his hair stood on end as he darted around the soccer field. Back then, he was prematurely balding, with virtually three hairs (even his mother teasingly called him by such on occasion).

Shu, his brothers and his parents were a wonderful loving family who opened their hearts and home to everyone, including myself, a young man from China, alone in Taiwan. His family instilled a sense of generosity, curiosity and honor that Shu and his wife have retained in America. Not many know how Shu met his wife — the two were children of long-acquainted families who wanted an arranged marriage. However, Shu was wary, being of the "new generation." Then one day as the families were playing mahjong, Shu's mother asked him to

---

[*]Psychiatrist (ret.). Shu Chien's classmate at National Taiwan University College of Medicine from 1949 to 1953.

serve tea. Unsuspecting, he served tea, saw a beautiful girl, and the rest, as they say, is history. Shu and his wife have raised a lovely family, passing on the principles that his family instilled in him.

After graduation, as Shu and I took different paths to the United States, we have remained friends. I wish him a Happy 70th Birthday and many more to come, and I congratulate him on his many contributions and accomplishments in his career.

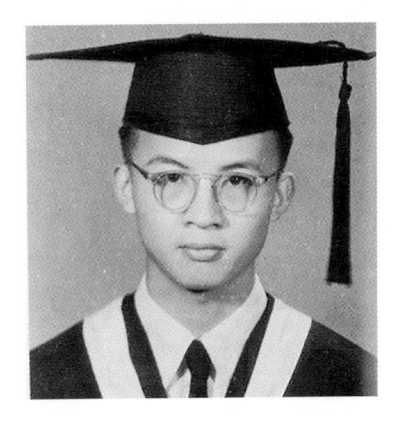

Shu: Graduation from National Taiwan University, College of Medicine, 1953.

My Classmate Shu Chien

Chih Chao Chin, M.D.[*]

In anticipation of their 44th wedding anniversary, as well as Shu Chien's 70th birthday celebration later this year, Kuang-Chung asked me recently to write a few words recounting the years of friendship that I have had with them dating back to the days when we were studying together at Taiwan University Medical School. I am obviously honored to do so for the benefit of such an important occasion, despite the fact that I am not such a great writer of prose.

I first met Shu Chien during my second year in medical school. We had heard that the son of the dean of our university was transferring his studies from Peking University. When I first saw Shu, I remembered that he struck me as having a relatively small build and a very youthful face — looking very much like a high school student himself. As we got to know him better, it quickly became clear to all of us that his appearance belied an individual with profound maturity, kindness and sensitivity. Additionally, we soon came to respect his intellect and more importantly the humility that he demonstrated to his fellow classmates. When Shu Chien began dating Kuang-Chung, who at the time was a few years behind our class in medical school, we all marveled at how they made such a perfect couple.

Over the years since graduation from medical school, most of us went our separate ways. I lived in Hong Kong for a number of years after my postgraduate training in Japan and my family eventually

---

[*]Ophthalmologist (ret.). Shu Chien's classmate in National Taiwan University College of Medicine from 1949 to 1953.

K.C. with Chih Chao Chin, Rebecca Li, Luna Fung, Ning Li, Jane Hu and Lily Meng, 2000.

immigrated to San Diego in 1977. While we had intermittently stayed in touch by mail over the years, it was not until when Shu Chien moved with his family to San Diego 13 years ago to accept a professorship at UCSD that we actually had the opportunity to deepen our friendship from years past. What I have come to realize is that things really have not changed much — Shu Chien is still the understated yet thoughtful and caring individual that everyone looks up to. He continues to dedicate himself to his work and excel in his profession, and he and his wife are still an enviable loving couple. I am proud to be one of their lifelong friends and wish them the best at this important milestone of their personal and professional life.

Talking about Shu Chien in Good Old Days

Hua Chin Chen, M.D.[*]

In 1948, when I was about to graduate from Xi-An high school in China, I passed the examination to enter the Medical School of National Taiwan University. So, I became a student in Taiwan without my own family, but the parents of many of my classmates treated me as a member of their families, and especially Shu's mother Chien Shih Mu (Shih Mu means the wife of our teacher, President S.L. Chien). I learned a great deal about Shu from his mother.

Chien Shih Mu told me that after having Robert as their first child, she was hoping for a girl next, but Shu came as another boy, and then Fred came later. Her wish was fulfilled, however, in the next generations. Shu and K.C. have two daughters, who then have three daughters each of their own.

I was one class after Shu in medical school, and we were both on the editorial staff of the school's first journal "Camel's Bell." Shu was the youngest member in his class, and he was a key member of the school's basketball team despite his size. Because Shu did not have much hair, I gave him a nickname "San Mao," which was used by a few of us.

The year before Shu left Taiwan to study in the U.S., he was often absent in various group gatherings. It took us a while to figure out

---

[*]Pediatrician. Shu Chien and K.C.'s schoolmate at National Taiwan University College of Medicine.

K.C. and May with Hua Chin Chen (May's godmother), 1960.

that he was dating Kuang-Chung. I later found out that Shu often delivered various kinds of gifts and things to K.C.'s mother for Chien Shih Mu or for Madam Hu Shih, who was the wife of the famous Chinese philosopher. When K.C.'s mother finally understood the purpose of the visits, she told Shu that he could get together with K.C. directly without any delivery.

After Shu got to New York, he sent his friends in Taipei a large box of chocolates in response to our requests. I was elected to pick it up from the post office. Because of the hot wheather in Taipei, by the time I cycled back to the University Hospital, all the pieces melted into a large lump, which became almost inseparable. This symbolized the love and togetherness between Shu and K.C.

Shu was the best student in our school. Although he is now a famous professor, the seriousness with which he pursues his work seems to be exactly the same as when he was a student. He is very sincere and courteous to people; he would nod his head frequently during telephone conversations, even though the person on the other end of the line could not see it. Madam Hu Shih told me that Shu would get properly dressed before answering the phone; this reflects his respect for whoever he speaks to.

Half a century passed by quickly, but we still remember vividly all the good old times we had when we were students. Our feelings and our hearts have not changed with time.

He is Always There!

𝓗. Daniel Ou-Yang, Ph.D.[*]

So many things can be said about Shu, his stellar professional achievement, his leadership, his dynamic and yet affable personality, his enormous generosity, his restless pursuit of perfection, his endless energy, and his being a great teacher and mentor, loving husband and caring father. The list goes on and on, you don't know where to start saying it, and you wonder how he can accomplish all that.

With all his professional accomplishments, he does not hesitate to help others. With all the leadership positions he has had, he does not have an entourage around him. With all the visionary actions he has

With Daniel Ou-Yang, 1998.

---

[*]Associate Professor, Department of Physics, Lehigh University, Bethlehem, Pennsylvania.

taken, he does not have an unusual eyesight. And, although he is not taller than most people around him, he does stand taller than most of us. You wonder what makes him unique.

Although I had heard about Shu Chien, almost since I can remember things, I did not have a chance to meet him until I completed my formal education. When I first moved from California to the East Coast, I had the good fortune to stay with K.C. and Shu in their New Jersey home before I found an apartment near my job. Since then, I have been in contact with the Chien family and met them on many occasions. Whenever I can, I'll seek advice from Shu both for the development of my professional career and as a person. Shu is always there to help me, and his advice has guided me well.

So many things can be said about Shu, you wonder where you begin. I will just say, "When you have a need, he is always there, no matter whether you are next to him, or you are three thousand miles away."

Many Puzzlements of A Treasure

Mamie Kwoh Wang[*]

I observed your reverential attitude,
your demeanor toward your parents and elders,

So I sensed your parents' legacy.

I heard you excelled in all courses in schools and universities,

So I witnessed the power of youthful determination.

I learned of your championship in table tennis,
your mastery of the various dialects, your perfection in
modern ballroom dancing, in golf sportsmanship and
in the challenging games for the young and old,

So I learned what wholesome development is.

I learned of the Columbia Medical students' throwing roses at
your feet after your lectures,

---

[*]Faculty in nurse education at the Cornell Medical School (ret.). Her late husband Dr. S.C. Wang was Pfizer Professor of Pharmacology at Columbia University College of Physicians and Surgeons and Shu's teacher.

Mamie Kwoh Wang 57

So I recognize the respect and affection from the
young to a rare treasured teacher.

I remembered your choice of a mate and your wedding with
a lady of inner beauty and unusual strength,

So I learned what wisdom is.

I watched with amazement the love and harmony of your marriage,
the inspiration, you the couple, spread to your friends,
families and colleagues,

So I am humbled by you, the inspiring young.

I listened to the chattering of your little grandchildren,
their company with you in work and play,

So I admired your boundless tenderness and nurturing even
to the tiniest budding souls.

I heard from the young and old of your clarity in
scientific investigation, your integrity, your compassion
for your students and colleagues, your helping, guiding hand in
every endeavor of their scientific growth,

So I recognized what a national TREASURE you are.

I stand in awe by the dignity of you,
marching abreast with the leading academicians,
laureates and royalty, always humble, always smiling,

So I know you are an international TREASURE.

Rarely the world enjoys the luster of human,
his guiding voice, his weighty, constructive words,
his positive influence, his sincerity, his love of life and others,

It remains a puzzlement to me — what makes a young human being like you becomes such a TREASURE,

Chien Shu

You are endowed with the gene, you found the formula in your soul. I am deeply honored to cross your path.

Respectfully, and fondly,

Mamie Kwoh Wang
(Mrs. S.C. Wang)

Dr. S.C. Wang (center) receiving Research Achievement Award from the American Chinese Medical Society, with Mrs. Mamie Kwoh Wang at his left, 1976.

The Seemingly Contradictory Characters of Shu Chien

Hsueh Hwa Wang, M.D.[*]

I have known Shu for over 40 years. He is both a colleague and a very dear friend. We are very close and I rather like to regard him, affectionately, as a kid brother.

What shall I write about Shu? Shu is not an ordinary person. I can think of at least three aspects about him that are seemingly contradictory.

First, he is a very methodical and rational person. He is very well-organized in whatever he does, be it paper publication, lecture presentation, or meeting chairing. But I remembered that when I ran to the 11th floor of his office in the Columbia P&S building (for whatever work we had to do together), I could see that his desk was always piled sky high with papers. Often there was no room to work and a separate table had to be brought in. I smiled secretly, to see someone's desk messier than mine. This also reinforced my belief that appearance cannot be equated with substance.

Second, he is slightly built and gentle, with the appearance of a desk-bound scholar. But he is in fact an excellent athlete. During one of our trip, to Taipei I watched with disbelief his deft and aggressive game at table tennis. We had also played tennis together. He was

---

[*]Professor Emeritus of Pharmacology, Columbia University College of Physicians and Surgeons, New York City.

better but would accommodate me when I could not find a partner. My tennis had improved since, but he is now turning to golf.

Third, he is a courteous man, obliging, easy-going and never hesitates to give a helping hand. He looks like a pushover (and could indeed be one to his women folks). Underneath the mild façade, however, is an unquestionable decisiveness and a will of steel. He would spring to action at a moment's notice. I remember once serving as an examiner in the oral defense of one of the graduate students in the Department of Physiology. During the questioning, Shu sat there quietly. Then one of the senior professors started to get picky. At the executive session, he questioned whether the student should pass. Without raising his voice, Shu first summarized the merit of the thesis work, and then he compared it to that done by one of that professor's students who was considerably more inferior but did pass the exam recently. End of discussion, the committee unanimously passed the student being examined. Another instance is related to the memorial service for my late husband, Shih-hsun Ngai, which was planned by

With Dr. H.H. Wang and her late husband Dr. S.H. Ngai, Professor of Anesthesiology at Columbia Medical Center, 1989.

friends from the anesthesia community and from Academia Sinica in Taipei in December 1999. I received the program about ten days prior to the event and noted that Shu was to give the eulogy. I asked whether Shu knew about this and found out that he was not yet informed (so much was being taken for granted)! The Program, which included some scientific presentations of Shih-hsun's disciples, was to run from 9 a.m. to 1 p.m. without a break. I had argued about the length of the ceremony with no avail. I phoned Shu (he had not left San Diego for Taipei yet) and told him about the eulogy and apologized for not informing him. "That's all right" "Will you do it?" "Of course I will" (he was so obliging). I told him the service appeared very long and I will fax the program to him. He called back as soon as the fax went through. "It is too long. There must be a break," said he "I will take care of this." He immediately phoned Taiwan and must have woken up people because of the 15-hour time delay. With several calls the service was nicely rearranged in two parts, with the funeral service first. The scientific presentations were moved to part two, and there was an intermission with reception between the two parts. Thus, people would have time to get comfortable and to visit with each other during the intermission, and to skip the second part if they so wish. It was a wonderful ceremony, thanks to Shu's intervention.

Happy Birthday, Shu! Life begins at 70 and what do you plan to do now? I think I know. He will do what he always does! With a keen mind, good health and many, many scientific prospects, why should he be thinking about retirement?

To Shu Chien, a Family Man, from ABMAC, One of His Families

Richard N. Pierson Jr., M.D.*

Shu Chien is a family man, with the word "family" written in large letters; may I borrow the flavor of the Chinese word "jia-ting" to suggest the bonding, connection and service which family members give to one other, and to suggest a model by which to describe his life of many roles, to the several families which he has served, all of them with distinction, honor, depth, perception and kindness. Perhaps in the distinction of the larger Chien one could expect no less. Others who have been closer than I to the intricacies of the Physiology Department at Columbia, the Bioengineering Department at UCSD, and the IBMS in Taipei will trace the influences of this Family Leader in these enterprises, in the latter two of which the adjective "Founder" belongs. Others will speak for the large worlds of rheology, the American Physiological Society, and the FASEB. Still others will tell of his genetic family, a source both of pride and of pleasure, and surely to extrapolate a little, of honor.

My personal invitation is to give recognition to one of the very peripheral "families" which Shu has served, a family in one of the outermost orbital rings — perhaps the "M" shell — for this many-dimensioned leader, whose high atomic number deserves many orbitals.

---

*Professor of Clinical Medicine, Columbia University College of Physicians and Surgeons, NYC.

The American Bureau for Medical Advancement in China was founded in 1937 (Shu was six years old), at a time when China was ill-prepared for a Japanese invasion, and an outpouring of Aid of many dimensions was organized in New York and Los Angeles by a small group of Chinese-American and American-Chinese leaders who quickly concentrated their efforts on Medicine and Health. Donald van Slyke, Frank Meleney and Magnus Gregersen, were among the American leaders whose peripheral visions and energies had propelled to serve in the Rockefeller Foundation and the Peking Union Medical College. These high sponsorships gave ABMAC gravitas among Relief agencies. The great Chinese physician leaders Robert K.S. Lim, Dick (C.T.) Loo, J. Heng Liu, and later C.Y. Chai, Huo Yao Wei and Shu Yeh were exquisitely impedance-matched to provide the linkages between these American leaders and their Chinese institutions, which made ABMAC a tiny, influential organization, capable of surviving the political complexities of the transfer to Taiwan in 1949, the re-building of a Chinese medical service and education system in Taiwan, the political transitions of de-recognition and later a "new stability" between the small, fragile, Republic of China of Taiwan, and the U.S. government. Later "giants" in the ABMAC family, George Humphreys and Gerard Turino, brought the skein of cooperation and service into and through the second half of the 20th century. Everyone of these leaders in ABMAC, and the many in between, would turn to Shu for advice, a contact, for another loyalty.

The Chien family leader, Shih-liang, already a senior and respected educator in Peking and Kun-ming, brought his family to Taiwan, and proceeded to serve his country first as President of Taita, and then as President of Academia Sinica, a giant family man, short in height but very great in stature. His three sons over the next 30 years became leaders, the eldest Robert, a leading banker, the third Chien Fu, a leader in the Diplomatic Corps. Our colleague Shu graduated from Taita Medicine in 1957, en route to a career of academic medical leadership, first under S.C. Wang at Columbia, with his nurse-educator wife Mamie, strong names in the history of ABMAC. I was in Taiwan from 1956 to 1958, then unaware of the existence of Shu or his family, or indeed of ABMAC. Among his first papers were several in

Body Composition, a topic which later became my magnet into a career in research. Shu presumably first became aware of ABMAC when he came to Columbia to take his Ph.D., where Magnus Gregersen the Chair of Physiology was also an early President of ABMAC.

The four decades of growth and development of our honoree have been described elsewhere in these pages; his personal vector has been constant, in pitch and in direction. The vector for ABMAC, always of necessity a "spare time" issue for its American leaders, was often troubled, by the great political turbulence which describes the triangle between China, Taiwan, and the U.S., and their changing relations. The role played by ABMAC could only be effective if its leaders in New York and the U.S. were very well linked with those in Taiwan; if both the American and the Chinese leaders maintained "convening power" within their own constituencies. This principle has been primal in the history of every three-cornered political battle; make sure that your proponents are leaders, in everyone's eyes, whether in Medicine, in American politics, in Chinese politics, or in Academia. This is where Shu came in: a loyalist to ABMAC, because he worked in a rich network of training for young Chinese physiologists, he could speak with his father and with his brothers "outside of the box," being in touch with diplomacy, banking, and academia in Taiwan. By the quality of his own contributions to Columbia University, and later to UCSD, he maintained a convening power, which underlay many proposals and fellowships.

The founding of the IBMS must be, on the scientific side, his pinnacle of achievement, taking a sabbatical year from Columbia to launch a "Chinese NIH," which at once attracted large fiscal support from the government in Taiwan, and a swelling trickle of top-quality Chinese-American scientists to transfer their laboratories and their careers back to an increasingly prosperous and promising Taiwan. That full story will receive a larger chapter, but it is just one example in which the leadership provided by this "family man" expressed with dignity, sincerity and energy "just the facts," in such a compelling manner that the IBMS is well launched. I suggest that Shu's role as "trigger" to the founding of the IBMS will be honored for many decades.

My colleagues in ABMAC join with me in celebrating this giant "family man," who has helped with innumerable projects and fellowships, lending the luster and the wisdom of his wide connections to the fabric of connection between China and the U.S., in the splendid settings of biomedical sciences, where "progress" goes in just one direction.

College of Physicians and Surgeons of Columbia University in the City of New York at 630 West 168th Street (between Broadway and Fort Washington Avenue, east of the Hudson River).

Shu Chien at 70

Edward F. Leonard, Ph.D.[*]

My memories of Shu Chien are unremarkable in the large world, devoid of moments of high drama, but quiet, eminently trustworthy, and full of his insight and immense capacity for empathy. In this way, I have known Shu for 40 years since our contemporaneous arrival at Columbia. Together, we were influenced by Magnus Gregersen, who always seemed to be living up to his given name. I could admire, up close, Shu's unraveling and rationalizing of blood rheology in Columbia's physiology laboratory. I suppose it was in this context that I first came to see the unique combination of grace and strength of will, of openness to others but with a firm set of internal standards, that made Shu the real and deep contributor that he has always been. In these adventures and in many more that came later, I came to see Shu's greatest strength: his ability to take something constructive from every collaborator and every encounter, and to spend time with fault and shortcoming only for so long as he could do something to fix it. Shu spends his energies exclusively, unerringly on expanding truth and goodness.

Shu taught me physiology. He was part of a superb team then at Columbia and — true to his nature — he played the course as a team member. He strove for the greater goal: to make not just his lectures but the whole course a success.

---

[*]Professor of Chemical and Biochemical Engineering, Columbia University, NYC.

Shu has always helped me. He has sat by my hospital bed, encouraged me when I was tired or pessimistic, seen and empathized with difficulties that I could not even articulate, counseled me on innumerable matters, scientific and collegial. Shu's unfailing perspective was to deal with wrong and fault only long enough to do what could be done and then to work to bring triumph and happiness to those around him. And I am a happy beneficiary.

With Edward Leonard, 2001.

Shu and I have long shared a love of Columbia University. When it was left for me to remain there and for Shu to begin and sustain his new wave of contributions at La Jolla, he never once flagged in his concern for Columbia. His queries about our progress were urgent and intense, and his generous concern remains to this day.

There is no neat finish here, nor do I want one. I look for more of the same for many years to come.

68 Tributes to Shu Chien

Magnus I. Gregersen, Ph.D. (Shu's Ph.D. Advisor; Chairman, Department of Physiology, Columbia University College of Physicians and Surgeons, 1941–1963).

Edward F. Leonard 69

Physiology faculty at Columbia, 1958. Front (from left): William Nastuk, Louis Cizek, William Walcott, Magnus Gregersen, Walter Root, Nicholas DiSalvo. Back: Shu Chien, Ruth Rawson, Robert Dellenback, Leonard Levine, Mero Nocenti.

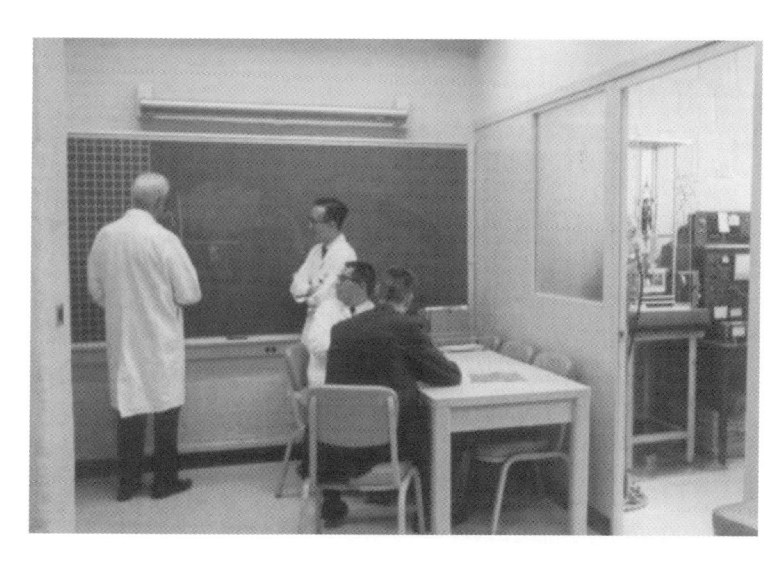

With Magnus Gregersen, Shunichi Usami and visitor in the Laboratory of Hemorheology, P&S, 1965.

Longevity!

Shunichi Usami, M.D., Ph.D.*

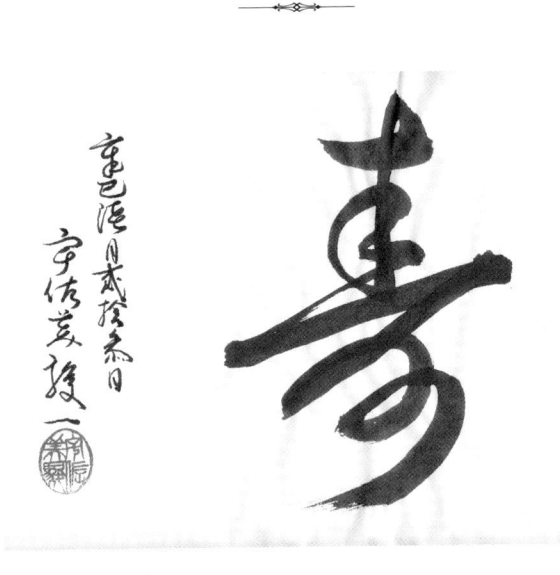

With Magnus Gregersen and Shunichi Usami, 1965.

---

*Research Bioengineer, Department of Bioengineering, UC San Diego.

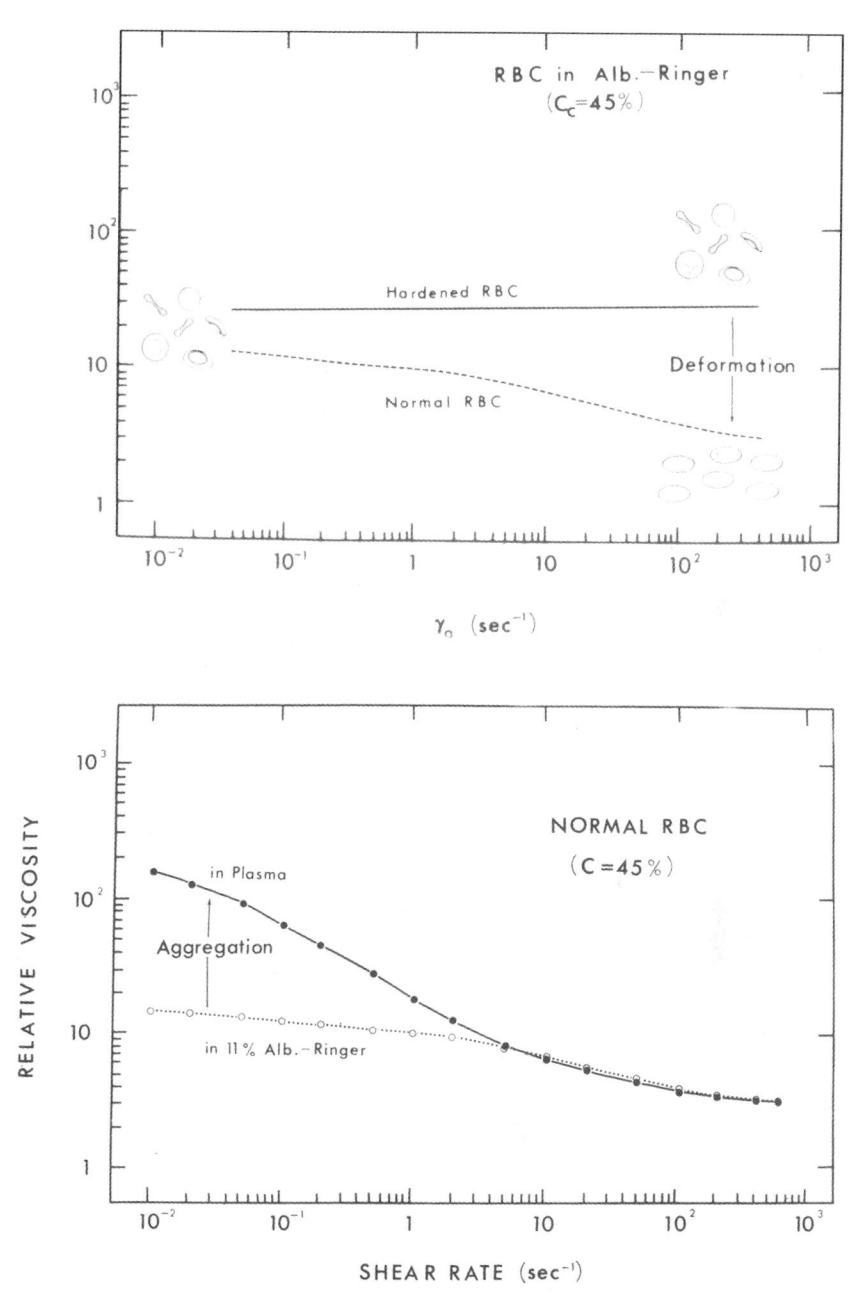

The roles of red blood cell deformation and aggregation as determinants of blood viscosity. From S. Chien, 1975 in Publication C20.

The Dimension of His Teaching

Kung-ming Jan, M.D., Ph.D.[*]

"There is one thread that runs through my way."
— *Confucius 551–478 B.C.*

I was a medical student at National Taiwan University when I first met and was taught by Dr. Chien in 1965. I had the privilege to become his first Ph.D. graduate student at Columbia University between 1968 and 1971. Dr. Chien's guidance began with cardiovascular physiology followed by research on blood rheology. When I finished my thesis entitled "Role of the electric charge in red blood cell interaction," I was educated in not only performing experiments but also theoretical modeling. The scope of research was not only in macro-rheology but also in micro-rheology. My graduate learning that proved to have wide applications included electron microscopy, physical chemistry on thermodynamics and quantum mechanics, mathematics on differential equation and vector analysis, engineering theories on fluid mechanics and colloidal chemistry, and most importantly molecular biology. Research in Dr. Chien's laboratory was conducted by integration of experiments and theories and consisted of macroscopic

---

[*]Associate Professor, Department of Medicine, College of Physicians and Surgeons, Columbia University.

and microscopic views. Over the years, the progression and expansion of my theories of the universe transcended physiology into other medical sciences, continuously overlapping into non-medical sciences and humanities.

There is a myriad of principles in physiology. However, there are also numerous principles in other sciences. Remarkable similarities exist between them, but they are expressed in different languages. Ultimately they all share the same principles. For example, in Dr. Chien's laboratory I learned about blood rheology of the human body and its principles can be readily applicable to city traffic. I also learned about the equilibrium equations of physiology in the laboratory and they are almost identical to the equations used in economics. The balanced human body often sets good examples for other systems in our daily life to follow.

The chronological list of Dr. Chien's earlier graduate students at Columbia University Department of Physiology includes Ronald Carlin,

With Professor Chia Liu Yuan and Kung-ming Jan (standing), Rex Jan, Steve Jan, Professor Chien-Shiung Wu (Mrs. Yuan) and Connie Jan, 1983.

Syngcuk Kim and Shlomoh Simchon. We learned to unify different levels of experiments and theories to discover that a higher level of dimension exits. When we asked Dr. Chien questions, we always obtained satisfactory answers. He not only solved our questions, but also allowed us to seek the master principle that lied beyond our questions. When we had that ability, we then were able to answer our own questions. Seeing things from a higher dimension has given us a more thorough understanding of the universe.

Yen Yuan described Confucius, *"The more I look up at it, the higher it appears. The deeper I drill down into it, the harder it becomes. I see it in front; but suddenly it is behind. The master is good at leading one on step by step. He broadens me with culture and brings me back to essentials by means of the rites. I cannot give up even if I wanted to, but, having done all I can, it seems to rise sheer above me and I have no way of going after it, however much I may want to."* — *The Confucian Analects Chapter 9.*

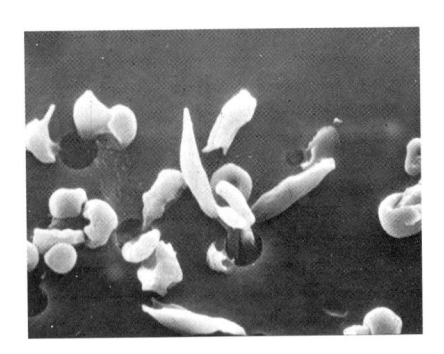

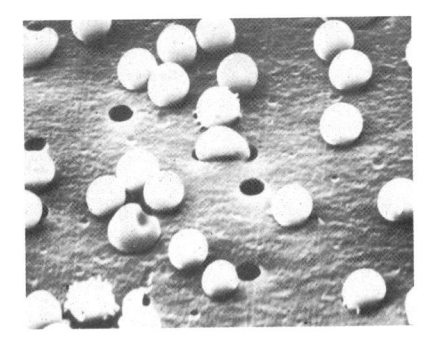

Electron micrographs showing the passage of sickle (left) and normal (right) red blood cells through filters. Left from S. Chien, 1981 in Publication A156; Right modified from S. Chien, 1971 in Publication A60.

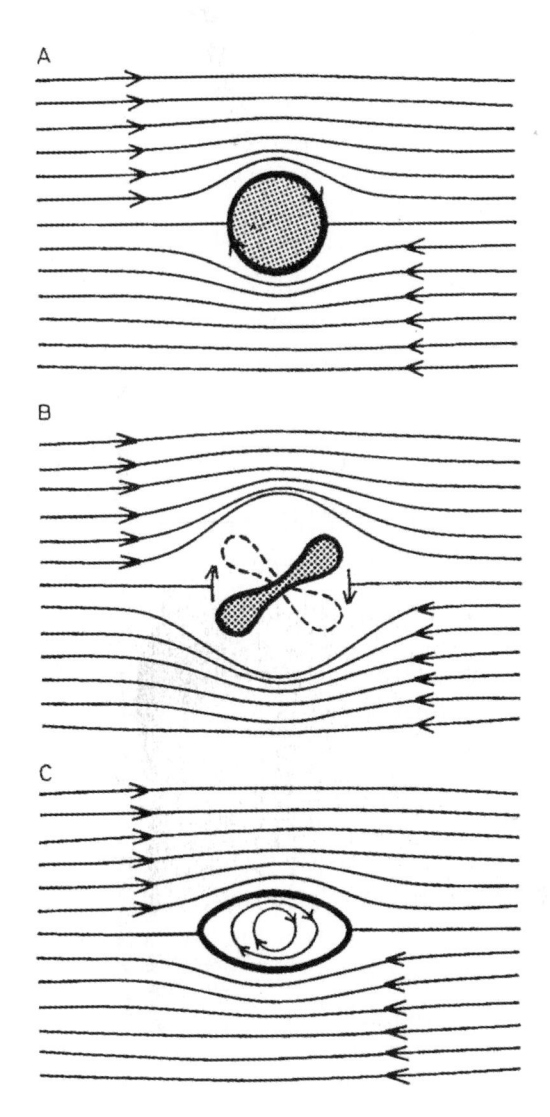

Effects of rigid sphere, ridge discoid and deformable particles on streamlines. From S. Chien, 1972 in Publication C15.

The World is Better Because of You

Lanping Amy Sung, Ph.D.[*]

There are major events in one's life that have tremendous impacts on how we think and what we do. Based on some of these major events, my life can be divided into pre- and post-American, pre- and post-martial, pre- and post-doctoral, pre- and post-motherhood, or pre- and post-California periods. Getting to know Professor and Mrs. Shu Chien and being influenced by them for many years is also a significant event in my life.

It was in the fall of 1974 that my husband Paul and I first met the Chiens. Earlier that year Professor Chien had been kind enough to offer Paul a scholarship to study Physiology at Columbia, and I, as a new wife, happily came along to explore the wonderful world of New York.

Paul was quickly immersed in the curriculum, and I found myself in need of a job. Professor Chien was kind enough to offer me a laboratory position. (I interviewed for two positions and landed one, so the success rate was 50%.) With a B.S. degree from NTU (Taita) in Taiwan and a M.A. degree from the College of William and Mary, I worked hard among a group of M.D.s and Ph.D.s. This was the time when I donated many times my own blood (and begged many others for theirs) and carried out hundreds of red cell aggregation and flow channel studies using dextran as bridging molecules.

---

[*]Associate Professor, Department of Bioengineering, UC San Diego.

With Elvin Kabat and Amy Sung, 1983.

Dr. Usami showed me how to put the flow channel together, and Professor Chien taught me the ways of analyzing the data. I was so impressed by Professor Chien. Often an experimentally based graph or calculation that took me so long to complete took him only a few seconds to interpret and to come out with an interesting new experiment or a meaningful new number.

Later as I entered the Ph.D. program of the Department of Human Genetics and Development, I chose Professor Elvin A. Kabat as my thesis advisor and began working on the interaction between blood group-specific lectins and carbohydrate receptors on erythrocytes (red blood cells). I did not leave Chien's lab behind, however. On the contrary, I integrated the training of these two internationally renowned laboratories into my thesis. (For people who are not in the field of immunology: Dr. Kabat was a recipient of the Presidential National Medal of Science, who had trained Dr. Baruj Benacerraf, a Nobel laureate of 1985.) Often my weekly progress report at 8 o'clock in the morning consisted of three of us: Elvin Kabat, Shu Chien and me, in Kabat's office in the Hammer Health Science Building across the street diagonaly from the P&S and Black Buildings.

When I was about to finish my Ph.D. work, I thought seriously about what to do with my knowledge and skills in immunochemistry from Kabat's lab and biorheology from Chien's lab. This was also an exciting time for biomedical scientists as several revolutionary recombinant DNA technologies had just been invented. After reading, reporting and defending three journal articles every Saturday (8 a.m. to 11 a.m.) for five years in Kabat's lab, I knew I wanted to include molecular biology in my future research. I thought I could do one or more of the following three new projects: (1) A vaccine against influenza virus based on the molecular interaction between hemagglutinin on the virus and receptors on the red blood cell surface; (2) the molecular basis of red blood cell membrane mechanics; and (3) the molecular recognition and interaction energy between cytotoxic T cells and tumor target cells.

Knowing that Professor Chien would be more interested in my second project, I decided to bounce that idea off of him first. So one day I walked into his office on the 11th floor of P&S and laid out a plan for cloning the membrane skeletal proteins, identifying mutations in their genes, and testing the mechanical properties of mutated cells and membranes. To this date I remember vividly his excitement. With his fist tightened and his arm striking in the air, he said, "Go for it!"

With Professor Chien's strong support, I later set up a molecular biology lab from scratch and began trying every protocol I could find from various sources to clone red cell membrane skeletal proteins. [Editor's note: She basically duplicated the molecular biology lab in IBMS in 1987, mentioned by Tang Tang (p. 327).] Now as I work closely with students and fellows in my Molecular Bioengineering Laboratory at UCSD on the creation of knockout mouse models and mechanical analysis of genetically engineered cells and tissues, I realize that the 15-minute proposal has become the theme of a lifelong pursuit. It was his enthusiastic support that helped shape the direction of my research. I worked briefly on cytotoxic T lymphocyte and published a paper in *Science* (p. 99), but have not had a chance to work on influenza.

With Amy Sung, John Conboy, Helen Yin, Mohandas Narla and Saul Yedgar. Symposium on Molecular Biorheology, Vancouver, 1986.

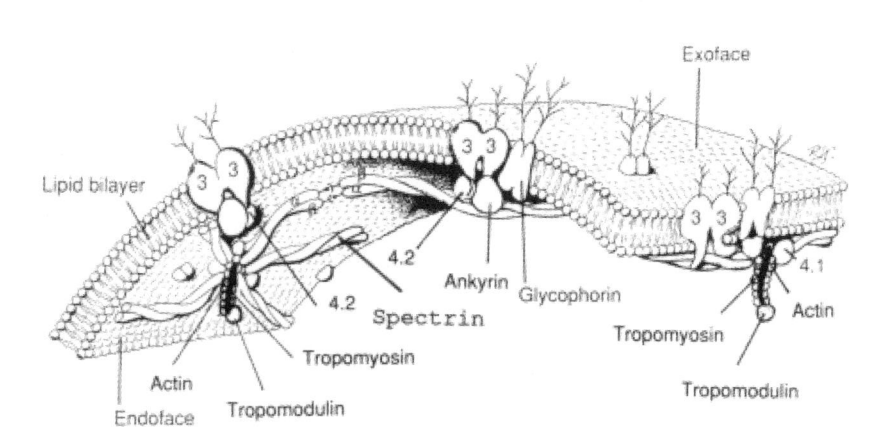

The red cell membrane, modified from S. Chien and L.A. Sung, 1990 in Publication A302. [The cloning of protein 4.2 was reported by L.A. Sung et al., 1990 in Publication A295; and that of tropomodulin by L.A. Sung et al., 1992 in Publication A313.]

As a woman scientist, balancing a life between family and career has been a challenge. Mrs. K.C. Hu Chien, a pediatrician herself, who has raised two wonderful daughters, lent me her wisdom on numerous occasions. Her thorough examinations of my son Eric and daughter Wendy after they were born, and much professional advice throughout their youth have become part of the sweet memories of the Sung family.

Dear Dr. Chien and Mrs. Chien (Shih-Mu, that's how I call her, meaning the wife of the professor or teacher): On this very happy occasion, I would like to thank both of you for being my mentors for so many years, 27 years to be exact. I am counting on you to do the same for many years to come.

Truly, the world is better because of you.

Eric Sung, almost 3, in Demarest, New Jersey, 1981; and Wendy Sung, almost 3, in Tappan, New York, 1985. Eric was one of Dr. K.C. Chien's "little patients" for 3 years until the family moved further north to Tappan in 1981.

To Professor Shu Chien for His 70th Birthday Celebration

Richard Y.C. Chen, M.D.[*]

Dear Dr. Chien,

It gives me great pleasure to participate in your 70th Birthday Celebration. You are not only the greatest mentor, but also my best friend. I still remember that day during the last year of my clinical training, I asked you a question about fluid management and its rheological implications on those infants undergoing induced hypothermia for surgical repair of congenital heart lesions. You encouraged me to find answers through research protocols instead of guessing. I was so gratified that you accepted me to work as a Research Fellow in the Hemorheology Laboratory on the 17th floor of P&S Building, where I very much enjoyed the learning experiences you provided me. I learned not only about basic science research, but also clinical skills of physiological monitoring as well. Your guidance and advice inspired me to change from an intended clinical practitioner to become more involved in scientific research. I have gratefully devoted my professional life to the academic anesthesiology ever since.

Shu, your support to me during all these years have been invaluable! You have always been a role model for both my professional as well as personal lives. Your lessons and teachings are still helping

---

[*]Chair, Department of Anesthesiology, VA Greater Los Angeles Healthcare System, Professor, Department of Anesthesiology, UCLA School of Medicine.

Trudy Lipowsky, Lee Laufer, Shu, Foun-Chung Fan, Shunichi Usami, Shlomoh Simchon and Richard Chen (left to right). The Pink Flamingo, New Orleans, 1983.

me whenever it is needed. Your wisdom and words continue to sway and enrich my life so deeply and profoundly.

With unfathomable heartfelt appreciation for all you've given me, I sincerely present to you my warmest wishes of love and thanks on your 70th Birthday!

Affectionately yours,

Richard Chen

Culture Clash

Herbert H. Lipowsky, Ph.D.[*]

When I reflect upon my 25-year association with Shu Chien, it is clear that there were obvious occasions when he had a great influence on my life, such as when providing a word of inspiration or a recommendation for a job. There were also many subtleties of our relationship that had a much greater effect. These were the events that contributed the wisdom I needed to be successful, both in life and academia, and arose in part from the diversity of our backgrounds.

At the beginning of our association, I had a very limited knowledge of Asians. In high school, I had one Chinese classmate out of a thousand, and I didn't know him too well. I, like every other Jewish person loved Chinese food, but we always greeted the bottle of soy sauce on the table in our restaurant with suspicion because it was affectionately referred to as "pigeon blood." My knowledge of the orient was derived almost entirely from Hollywood representations, such as Charlie Chan, Fu Man Chu and Kung Fu. At least there was a lot of diversity among these stereotypes. Fortunately, I was prepared for my association with Shu by my exposure to Y.C. Fung, Mike Yen and a few others I had met during my stay in La Jolla, such as my fellow transplantees from New York City, Dr. and Mrs. Ben Zweifach, who first introduced me to Shu Chien.

Shu came to La Jolla in the early 1970s, when I was a grad student with Ben, and gave a seminar on blood rheology. I vaguely

---

[*]Professor and Chair, Department of Bioengineering, Pennsylvania State University, University Park, Pennsylvania.

recall attending his seminar although I thought he was much more rotund at the time, probably from eating all that good dim sum in New York. Unfortunately, we never got to meet at that time because I was sequestered in the darkest bowels of Urey Hall on the UCSD campus. To be honest, I don't recall any details of his visit, but I do remember Shunichi Usami coming to give a seminar and wowing us all with beautiful pictures of Amphiuma red cells that were so large one didn't need a microscope to see them.

Dr. Usami spent some time in Ben's lab, so I am told; I never met him either because I was just starting as a student and did not have any experiments going on in the lab at that time. A few years later, he reminded me that he was there and learned to do micropuncture on small blood vessels. "I never miss," he was to say with pride some years later. Usami was known in La Jolla as "Lefty." Those of us who grew up with TV cops and robbers movies knew what that meant. He was Shu Chien's right hand man. Certainly when it came to micropipettes he was a hitman. I suppose it was similar to labeling the left and right sides of the heart. In actuality, it was because he always carried his drink in his left hand. At least that is what the fellows in the lab told Usami. The truth was closer to referring to a right hand man who had difficulty in distinguishing left from right. In spite of such disorder, Chien and Usami were a dynamic duo in La Jolla long before their exodus from Columbia, and as you can see, it is always difficult to discuss one without the other. Together they represented a conduit to the clinical world. After all, the early 1970s was the age in which engineers were going to unravel all mysteries of the human body, and Chien and Usami symbolized a focal point for studies on hemorheology. Ben Zweifach would say some years later that the engineers were the one bright hope in physiology, until the molecular biologists came along and ruined it for everyone.

In 1975, I gave a paper at the First World Congress on the Microcirculation where I talked about my thesis work with Ben on *in vivo* studies of blood rheology in the microcirculation. Dr. Usami was in the audience and he returned to Columbia to tell Shu that there was an engineer making measurements of apparent viscosity in the living animal and we had to have him. Shu said we had already gone after the white cell kid from La Jolla and we couldn't afford another expensive engineer. After some careful budget analysis, Shu found some additional funds and called me in early spring of 1976 to talk me into coming to Columbia. And by early, I mean early. At the time I thought that he forgot about the three-hour time difference and he called me at 6 am, San Diego time. That early in the morning I would have agreed to almost anything, and so I was recruited to Columbia. It wasn't long after my arrival in New York that I realized that Shu had not neglected the three-hour time difference. I quickly learned that Shu did not have a normal circadian rhythm. It seemed as though he never slept. For the next 12 years, he would continue to call at all sorts of weird hours and deprive me of my precious sleep.

Nonetheless, we worked together for the next decade plus, and I believe that it was a learning experience for both of us. I taught Shu everything I learned growing up in Brooklyn about how to say something nasty to someone and make them walk away thinking it was a compliment. I will never forget the look in his eyes when I had a friendly exchange with a fellow native New Yorker in the elevator at Columbia which prompted Shu to advise me to be more careful about my remarks to people. I had forgotten about that incident until about 20 years later, when I was sitting at a dinner listening to Penn State football coach, Joe Paterno, another transplantee from Brooklyn, "rank out" a fellow member of the audience. It brought to mind the culture gap of my early years with Shu and how our association had enhanced my appreciation for the diversity of our backgrounds. Shu taught

"Who says I don't sleep. I slept through every one of Herb's seminars."

Looking at Herb Lipowsky, 1984.

me his secret of success, which I share with you now: "Make no judgments." When he told me that, I questioned how could one go through academic life without ever being critical of someone. It is a seemingly impossible task in a scientific system that has peer review as a foundation. In spite of the difficulty, I managed to try my best to adhere to his philosophy, although he was indeed the master of telling people their work was not very good while making them feel as if they received a compliment.

One of the most important lessons I learned from Shu during our association at Columbia was the value of thorough planning. During my first nine years at Columbia, Shu had orchestrated no fewer than 12 NIH site visits, during which he charmed people into giving away lots of NIH dollars. His grant applications were so well written that they became the accepted model for a program project application. He was a master at preparing for these annual events. He also put just as much effort into planning our Division's annual picnic. I remember that for the first picnic, he even held a private practice session for barbequing hamburgers. The week before, he and K.C. spent countless hours grilling hamburgers until they came out picture perfect. Best of all, however, were the gatherings he hosted in his home in Englewood. On several occasions he set up a large round table in his family room, just like in a Chinese restaurant, complete with that ever-present threatening bottle of pigeon blood. One year, Shu hosted a dinner party for Y.C. Fung and put out an elaborate spread. By this time I felt so at home with Shu that I blended right in with the oriental theme of the dinner. Everything was going very well that evening until, at one point, Shu, Dr. Fung and I were conversing and Dr. Fung looked at me and asked a question. In response, I grunted uh? Shu looked at me and said, "Dr. Fung just asked where are you living, in Chinese." It was indeed a nice feeling of acceptance.

After that experience I did try to make an effort to learn some Chinese. In 1987, the group had gone to Taiwan to help set up the Institute of Biomedical Sciences in Taipei. Trudy, Josh and I considered ourselves very fortunate to be able to spend a few weeks there, helping to set up a microcirculation laboratory. At the end of our stay, we had a going away party and I attempted to say a few words in Chinese, in order to demonstrate my affection for our hosts and show off my very limited vocabulary. I uttered about two words before Paul Sung jumped up, and insisted that I refrain from butchering the Chinese language. I was really startled by this, since for over ten years I had spent countless hours listening to his assassination of the English language. I think that this was Shu's way of teaching me how hard it is for an immigrant to become acclimated to a foreign country. That is one thing the Chiens did very well.

Of course, it was Mrs. Chien who was responsible for their assimilation into American culture. K.C. was an essential ingredient of our family at Columbia. I was amazed at how much like a Jewish mother she was. She knew the Jewish secret of marriage: always let your husband think he is the boss. I am sure it was she that ferreted out the hamburger cooking instructions for those picnic rehearsals in their Englewood backyard. She also let it be known how difficult it was for Shu and her when they first came to live in New York. They had so little money that when they went to lunch at a dim sum restaurant in Chinatown, K.C. would stash the plates in her pocketbook. Since the final bill was computed by the waitress who counted all of the used plates at the end of the meal, they managed to save a few dollars.

It is indeed interesting how our social lives revolve around the act of eating. Historians suggest that the Chinese invented the restaurant, and that it is one of the foundations of modern civilization. Clearly, the restaurant business depended heavily on our group at Columbia, so much so that when Shu left in 1987, our favorite restaurant on Broadway, Chun Cha Fu, went out of business.

Although the westward migration of the Chien Lab was a great loss to the New York area, many positive things came out of it. Shu has done an outstanding job of becoming a bioengineer and leading UCSD into prominence in the field. Along the way he has contributed to the growth of many careers in academia, medicine and bioengineering. He has truly made a difference in the world around him. What more could one ask for, besides a little sleep?

Herbert H. Lipowsky
June 2001

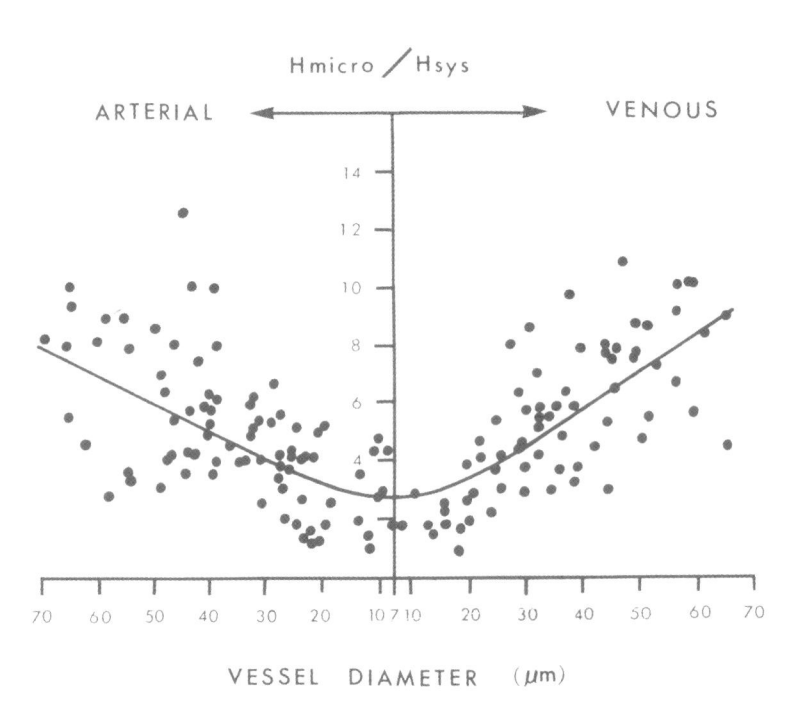

Arteriovenous distribution of micro vessel red cell concentration. From H.H. Lipowsky et al., 1980, in Publication A132.

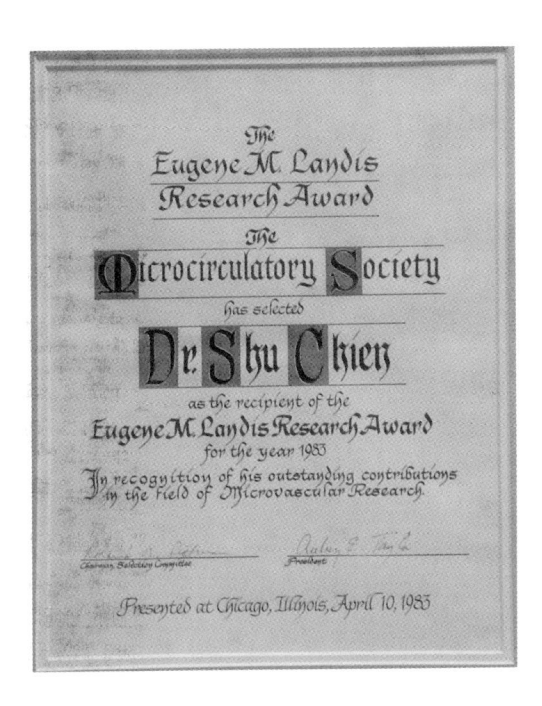

With Eugene Landis in his home, 1983.

Shu Chien
A Leader and Perfectionist in Physiology and Bioengineering

Geert W. Schmid-Schönbein, Ph.D.[*]

Any historian will recognize Shu as a man of many outstanding talents. He is a model citizen, an excellent and most productive scientist, an admired and respected colleague, a wise counselor, a beloved teacher, and a caring individual. But there is one talent that stands out: Shu is a superb diplomat who knows how to communicate with a diverse range of people, from card-carrying civil engineer to frontier molecular biologist. Shu's presence makes them feel at ease as he inspires them to work with each other. This is Shu, the initiator, the facilitator, at the center of the activity.

In the process he created, some 30 years ago, the Columbia University's outstanding brand of Bioengineering. Columbia University is still proud of his pioneering effort of collaboration between medicine and engineering. Shu gave it its own flavor and created a vibrant activity that eventually spanned from physiological controls to blood rheology, cell biophysics, membrane mechanics, microcirculation, atherosclerosis and mechano-transduction.

Shu made an especially strong impact on blood rheology and created a way of thinking about blood cells and viscosity that was

---

[*]Professor, Department of Bioengineering, UC San Diego.

embraced by medical doctors and biologists as readily as by engineers. His thinking about Blood Rheology has become standard in physiology and bioengineering textbooks. His theme of integration in physiology has become the motto of the American Physiological Society. Shu made fundamental contributions to the development of Physiology and Bioengineering in the United States and in the world.

After a first brief encounter at the International Union of Physiological Sciences in Munich in 1971, I had the opportunity to meet you, Shu, in 1975 at the 1st World Congress for Microcirculation in Toronto. Shortly afterwards you visited UC San Diego and invited Herb Lipowsky and myself as postdoctoral fellows to the Hemorheology Laboratory at Columbia University. As graduate students, Herb and I enthusiastically accepted your offer, having just read your pioneering work on blood rheology and the cellular basis of the non-Newtonian properties of blood. There was no greater attraction than Shu Chien and his team at Columbia University.

What followed then in 1976 were three exciting and almost carefree years as postdoctoral fellows, devoted to microcirculation and to bioengineering. Kuang-Chung and you instantly made Renate and me feel at home. You provided a rich environment, which formed the basis for many new ideas and created lasting friendships. Your interest and contagious enthusiasm for studies in microcirculation and in leukocyte rheology was an inspiration. You served as a sounding board for all ideas, you saw the opportunities, you shared the excitement of discoveries, and you patiently offered your wisdom and skills. You gave us unlimited freedom and had endless patience. We are forever grateful to you for these marvelous years. It was only afterwards that we fully recognized how great the opportunity was which you had provided.

From there on our ties remained unbroken even though Renate and I moved across the continent back to La Jolla. After a number of years of regular visits and continued consultation, you again assumed a leadership role and invited us to the opening summer semester at the Institute of Biomedical Sciences of the Academia Sinica. It was your vision to carry this newfound understanding in bioengineering to the young people at the Academia Sinica in Taiwan and to start the study of blood flow at the level of molecular biology. As one of the Institute's first directors you wrote the first Program Project application in China, which was funded, and we started many projects with the talented students of the National Taiwan University.

When in 1987, UCSD was about to start a new phase in the development of Bioengineering with a major expansion, we could convince you to join us together with Richard Skalak, Amy and Paul Sung, Jerry Norwich as well as Shunichi Usami a short while afterwards. This was a tremendous gain for UCSD. In a period of less than ten years, you created a new Institute for Biomedical Engineering and a new Department of Bioengineering. You started the planning for a new building devoted entirely to Bioengineering and funded by private sources, especially the generous Whitaker Foundation and Powell Foundation. This was a first at UCSD and in Bioengineering and was the fruit of many years of pioneering work in bioengineering. You moved mountains and it was such joy for all involved.

These achievements were made while you maintained a most active research program and taught us about signal transduction

Dick Skalak, Geert Schmid-Schönbein, Y.C. Fung, Shu and Sid Sobin, 1994. Department picnic at La Jolla Kellogg Park celebrating the Whitaker Foundation Development Award.

pathways under fluid shear stress and gene expression. Besides all these activities you have maintained a most active life of service to national organizations, a record that, for most workers, is in itself the equivalent of a lifelong achievement. You were President of the American Physiological Society and the first to serve both as President of FASEB and as President of AIMBE. The officials at the National Institute of Health admire you and seek your council. We know you enjoyed these opportunities and on the way you touched many lives such that they will always be grateful to you. We have nothing but admiration for your excellence and the effectiveness of your approach.

Dear Shu, Happy Birthday and many returns. With deep gratitude we thank you for your incredible service and your leadership. We know you still have many plans. Bioengineering has many ways to go. We wish you the best possible success.

Tribute to Dr. Shu Chien

Anne Chabanel, Sc.D.[*]

When I arrived in New York with one suitcase and one backpack in April 1980, I had planned to stay for six months.

As it turned out, I remained six years and a half.

What I remember most and what is most unforgettable is how warmly and open the welcome was when I dropped into your laboratory, Dr. Chien.

I also remember that, as I continued working in your lab, everyday was a feast for me (at least this is how I see it today, 20 years later): I learned a new language, a new way of life, a new way of studying, and a new way of working with thousands of new concepts and experimental settings. Learning all of that while I worked in your lab, Dr. Chien, was such a great pleasure! During all my time in your laboratory, you were such a close guide, generous and always ready to help, correcting and encouraging me.

Your laboratory was like a wonderful busy hive: full of activities with open-minded people, and where I had the opportunity to meet all these famous authors of my favorite scientific papers.

Of course some days (or nights) were more difficult than others: when the deadlines for grants were approaching… when at 10 p.m. I had to write my proposal all over again! But, I learned so much on the importance of the choice of words, on the necessity of precise writing of scientific papers in English. Your efforts were not wasted on me.

---

[*]Scientist at the Medical and Scientific Direction, French National Blood Service, Paris, France.

What I have learned from you in English then, I can use it in French now.

"Excellent!" Besides red cells, elastic modulus, hemorheology or microcirculation, I think this is the most frequent word I have heard from you in my days at Columbia. I have learned to strive for the same. It is through your guidance that I was able to receive the first Al Copley Young Investigator Award in Bordeaux in 1987 (shortly after I had left Columbia). Once again, I need to thank you for being such an excellent teacher, if I may use your own word "excellent." How lucky I was to have been able to work with you. How honored I feel!

I shall also never forget the fun part around work: laboratory picnics with baseball games, Chinese meals… and my first American Christmas with the Shu and Kuang-Chung family.

For all your teachings, your enthusiasm and your kindness, I would like to express my warmest gratitude to you.

Fondly,

Anne

A Great Scientist Taught Me That...

Steven D. House, Ph.D.[*]

Dear Dr. Chien,

The four years (1983–1987) I spent at Columbia University with you, Dr. Usami and Herb were some of the most stimulating, rewarding and enriching years of my life. The three months we worked together in Taiwan during the summer of 1987 were unbelievably exciting and

With C.D. Kuo and Steven House, 1987.

---

[*]Dean, College of Arts and Sciences, Elon University, Elon, North Carolina.

educational. I will never forget the intensity of that experience or the camaraderie that it produced.

Through all of these experiences you taught me many, many things, but most importantly you taught me that being a great scientist is more than doing great science. You taught me that being a great scientist includes being a creative and stimulating teacher. You taught me that being a great scientist includes being a caring and empowering leader. You taught me that being a great scientist includes being a passionate and ethical colleague. You taught me that being a great scientist includes nurturing great scientists.

Everyday I strive to follow in your footsteps. I am very proud that you helped nurture me to be a great scientist.

Happy 70th Birthday and thank you for all that you have done for science and for so many scientists.

Your student and friend,

Steven

From Columbia to UCSD

K.-L. Paul Sung, Ph.D.[*]

At the end of 1985, on the 11th floor in the southwest corner of the College of Physicians and Surgeons building at Columbia University, Dr. Chien called my wife Amy and me to meet with him in his office. I thought he was going to discuss the abstracts for the Sixth International Congress of Biorheology. But to my greatest surprise, he told us that the University of California at San Diego had approached him with an offer.

He went on to say that the bioengineering group was interested in him and his research, in particular, its cellular and molecular aspects. At that time, I had been working for several years in his Laboratory of Hemorheology on the biorheology of single cells, e.g. red blood cells, neutrophils and lymphocytes, by using the micropipette technique, and had just extended my work to study the adhesion properties between cytotoxic T lymphocytes and cancer cells. The latter work, which was supported by a Young Investigator Award from the National Cancer Institute, was later published in *Science* in 1986 (Fig. 1 and Ref. 1). The Laboratory of Hemorheology at Columbia was one of the few laboratories in the world with the sophisticated biophysical technology to measure the force of interaction between two specific cells by using dual micropipettes. This work quantified for the first time the interaction force between individual cytotoxic T lymphocytes and their target cells.

---

[*]Professor, Departments of Orthopaedics and Bioengineering, UC San Diego.

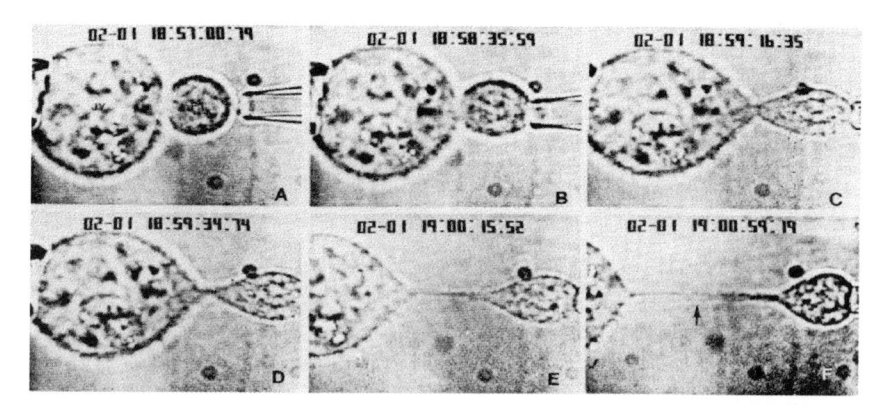

Fig. 1. Determination of the interaction force between a cancer cell (JY) and a cloned cytotoxic T lymphocyte (F1) by using two micropipettes.

At the same time, Amy had been busy setting up a molecular biology laboratory in Shu's lab and began the molecular cloning of membrane skeletal proteins of erythrocytes. Shu said these areas of research would be the main direction of his laboratory for the next decade, and asked if we had any interest in accompanying him to San Diego. If both of us proved interested, he would pursue an offer from UCSD. We gave him an initial positive response.

The following year in August, after attending the 6th International Congress of Biorheology in Vancouver, Amy and I flew south to visit the La Jolla campus and the city of San Diego in general. The University was constructing the new Engineering Building I at that time. We have very nice memories of that trip and reaffirmed that it would be a positive move. After all there are few places in the world that would allow one to both pursue scientific excellence and raise a family in a positive environment.

From January 1987 to the summer of 1988, Shu was busy at the Institute of Biomedical Sciences (IBMS) in Taiwan while we stayed at Columbia with the anxiety of not knowing whether Dr. Chien was going to move or not (How many of us shared this feeling?). During the summer of 1987, Amy and I assisted Shu at IBMS in setting up a molecular biology laboratory and a cellular biophysics laboratory, respectively. In the following summer (August 1988), we moved to San Diego. Dick and Anna Skalak had arrived in San Diego two months earlier, along with Dick's two postdoctoral fellows, Cheng Zhu (now at Georgia Tech) and Cheng Dong (now at PSU). These two young men actually drove Skalak's Dodge Aspen all the way across the continent from New York to San Diego in 14 days. Shu and K.C. arrived in September. At last, the two groups from New York with a long history of collaboration were reunited in San Diego.

Amy and I assisted Shu in setting up the research facility at UCSD with one lab in the Basic Science Building and two in the new EBU 1. This included recruiting and training technical personnel, postdoctoral fellows and graduate students, developing research projects, teaching new courses, and writing PPG and other grant proposals. The class we taught was Cell and Molecular Biology. Shu and I taught the class at first, and Amy joined us two years later. In the first few years we had three textbooks, switching from one textbook to another, and then returning to a revised edition of the first. Each change meant more reading, new summaries and revised handouts, 50 or more pages per chapter. The team discussed the content of each chapter, handout, and quiz before giving them to students. It was a tremendous experience and involved a challenging amount of work.

No matter how busy Shu was, he always made student learning and research his first priority. I remember studying harder than many of the students, often until midnight, and then waking up early in the morning to review the notes for the lecture that started at 8 a.m. Shu had a tradition of testing students frequently to be sure that they stayed on top of the coursework. At the beginning of each class we would administer a ten-minute quiz on the material from the previous lecture. Sometimes I thought about giving students a break from taking them (and myself from preparing comprehensive, multiple choice quizzes designed to test a large amount of materials in a short period of ten minutes). But Shu constantly reminded me that it was important to emphasize to the students the need to prepare early for the cumulative midterm and final exams and not to leave preparation until the last minute. Though difficult, he told me that our hard work was essential because teaching in front of a class is like an actor

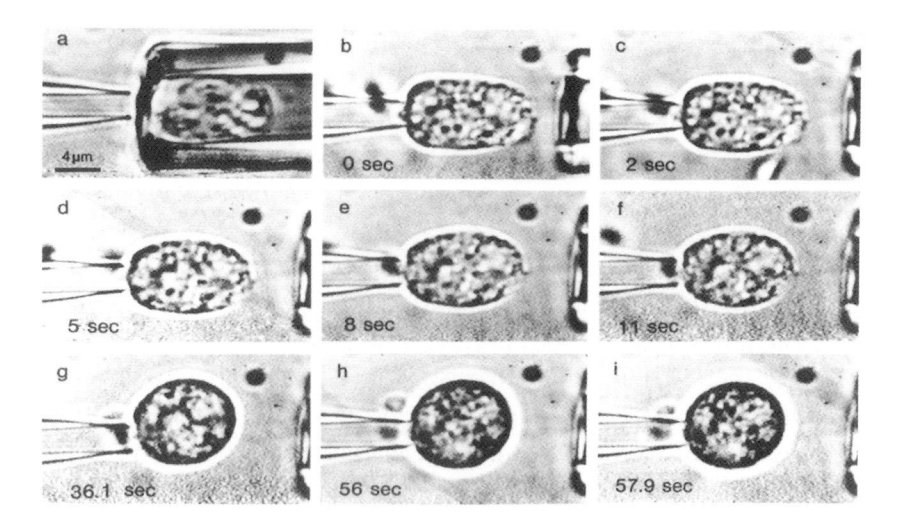

Fig. 2. Neutrophil viscoelasticity properties determined from the shape recovery after its release from a micropipette.

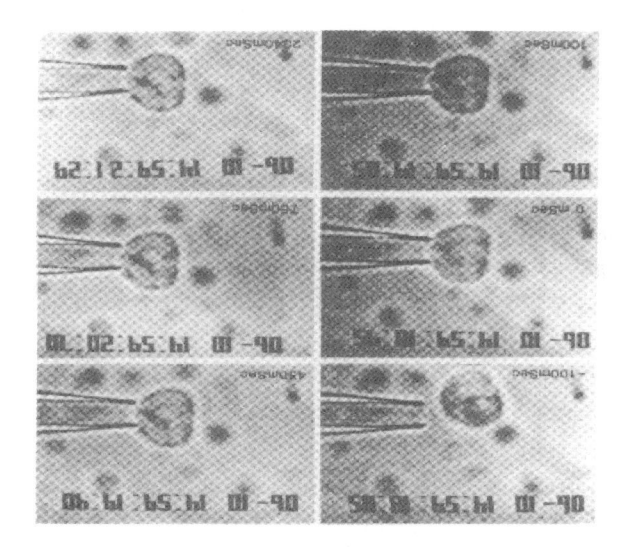

Fig. 3. Determination of the mechanical properties of lymphocyte nucleus by micropipette aspiration shows that the rigidity of the nucleus is about 20 times higher than that of the cytoplasm.

performing his/her art on a stage, and that everybody watches the teacher and the actor alike. Therefore, he told me, when you teach you should prepare yourself as much and completely as possible. An unprepared teacher should not teach. Shu told me that no matter how busy we were, it was of utmost importance to prepare for the students. This impressed me greatly and I admire him for his teaching philosophy.

Dr. Chien gave me and all of his students the "7 C's" as guidance during our studies and work: Compassion, Commitment, Comprehension, Creativity, Cooperation, Communication and Consummation. His scholarly spirit and teaching philosophy were greatly appreciated by his students and co-workers. Under Shu's guidance, I received the Whitaker Research Award from the Whitaker Foundation in 1989 and together we received the Best Journal paper award in 1989 and the Melville Medal in 1990 from the American Society of Mechanical Engineers, along with Cheng Dong, Geert Schmid-Schönbein and Richard Skalak (Figs. 2 and 3, picture, and Refs. 2–4).

After the UCSD Institute for Biomedical Engineering received the Development Award from the Whitaker Foundation in 1993, a new Bioengineering Department was built upon the existing outstanding program and won the Whitaker Foundation's Leadership Award. This Award affirmed that the molecular and cellular aspects of biomechanics had become integrated with that at tissue and organ levels to become key components of bioengineering. Shu's vision and appreciation of molecular and cellular biomechanics had been a key factor that allowed our department to continue developing as one of the top Bioengineering programs in the United States.

I have been very fortunate to have had the chance to work with Shu for these 27 years both at Columbia and at UCSD. I have distinct and cherished memories of the development of the bioengineering program and department since 1988 and I am proud of participating in its evolution and growth. In 1993, I received an appointment from the Department of Orthopaedics, a distinguished program in a field with traditionally strong ties with bioengineering. I believe that it was a great and invaluable experience in my life to work with Shu, and to be involved in his hard work, experiences and achievements. He is my

The Gang of Five. Left to right: Shu, Cheng, Geert, Dick and Paul (photograph taken at the President's luncheon after the Melville Medal Award Ceremony at the 1990 ASME Winter Annual Meeting in Dallas, Texas).

San Diego Chapter of Phi Lambda Fraternity, 2001. Anna and Sawyer Hsu, Michael and Bertha Wang (seated), Ken Lau, Paul Sung, Fah Seong and Polly Liew, Fang Hui Chou, Theodora Lau, K.C., guests Wendy and Cheng H. Lee, and Shu (standing). Sawyer, Ken, and Shu are the founding members of this young chapter.
Not pictured: Amy Sung.

Ph.D. mentor (Columbia-Rutgers combined program), colleague, friend and brother.

Shu is a "giant" in Bioengineering among his colleagues, friends and associates. These years are some of my happiest and most precious memories, ones that I will surely never forget. I wish him and K.C. many happy years, and we are expecting more exciting things from you in the years to come.

Congratulations and many happy returns!

Shu — your name, accomplishments and compassion have been branded in our minds forever!

REFERENCES

1. Sung, K.-L.P., Sung, L.A., Crimmins, M., Burakoff, S.J. and Chien, S. Determination of junction avidity of cytolytic T-cell and target cell. *Science* **234**: 1405–1408, 1986.
2. Sung, K.-L.P., Dong, C., Schmid-Schönbein, G.W., Chien, S. and Skalak, R. Leukocyte relaxation properties. *Biophys. J.* **54**: 331–336, 1988.
3. Dong, C., Skalak, R., Sung, K.-L.P., Schmid-Schönbein, G.W. and Chien, S. Passive deformation analysis of human leukocytes. *J. Biomech. Eng.* **110**: 27–36, 1988.
4. Dong, C. Skalak, R. and Sung, K.-L.P. Cytoplasmic rheology of passive neutrophils. *Biorheology.* **28**: 557–567, 1991.

A Mentor and a Friend

Cheng Zhu, Ph.D.[*]

Growing up as a graduate student in Dick Skalak's group, there was no escape from Shu's influence. At Columbia, I was at the Engineering School at 120th Street, 40 some blocks south of Shu's laboratory. There were numerous occasions when I and other students were sent uptown to attend seminars given by visitors of Shu's lab. During my meetings with Dick, he often took out his notebook to look up notes of a discussion he had with Shu or a set of notes that Shu passed to him. Here, the influence took the form of expectation, not only experimental data that I as a theoretician had to explain, but also the high standard with which I had to measure the quality of my work. In fact, such influence reached me even before I became Dick's student. That was when Dick interviewed me as a prospective graduate student in his first visit to China. I was then a student of Peking University. I had little idea about what he described as work done in collaboration with Shu, but I understood that Dick was also describing what he expected from me at the same time. From then on, I knew I had to live up to such high expectations.

But it was not until we all moved to UCSD — Shu and Dick joined the Bioengineering Faculty, and I went along for my postdoctoral training — that I had much more direct interactions with Shu. Although still a theoretician, I had the opportunity to spend time in Shu's lab to observe experiments and to talk to those who did experiments at a

---

[*]Professor, Schools of Mechanical Engineering and Biomedical Engineering, Georgia Institute of Technology, Atlanta, Georgia.

106 *Tributes to Shu Chien*

regular basis. We also had joint lab meetings in which both experiments and theories were discussed. It was during this period that Shu had the most significant influence on my career. I was looking for a new project at the time, and a paper published by Shu (with Paul Sung and Amy Sung) in *Science* sparked my interest. Two key observations described were the time course of increase of contact area and the force required to separate the adhesion. I started working on theoretical modeling of the cell adhesion processes, which turned out to be a project that I have been doing since.

In addition to being a great mentor of my scientific career, Shu is a wonderful family friend. In fact, the Zhu family began when Shu first served as the Master of Ceremony at my wedding banquet on 12 August 1989 in San Diego. My wife Frieda and I have many fond memories of that event, some of which were captured in the accompanying photos. These will stay with us for the rest of our life. Happy Birthday to you, Shu.

Happy Birthday, Most Esteemed One

Anna L. Skalak[*]

For 29 years Shu Chien was a close friend and valued colleague of my husband, Dick Skalak. Dick is the one who should be writing this. Unfortunately, in God's great plan, that was not to be. I am unqualified to speak about their work, but it was plain to see that Shu was an ideal collaborator — absolutely trustworthy, intellectually stimulating, reliable, truthful, generous with his ideas and his time, inspiring, and always pleasant to be with. Over time, Dick and I came to know his wife, K.C., and to meet his daughters, May and Ann, and their families too. The same attributes present in the academic scientist are also present in Shu, the family man, husband, father and grandfather. On one of our trips to Asia we were privileged to visit the Academia Sinica in Taipei of which Shu is an honored member, and where he founded the Institute of Biomedical Sciences. His influence is obviously far-flung, and his work on behalf of UCSD, as well as many Societies, is unceasing and greatly appreciated. K.C. is his constant, willing helpmate and also sees to it that he relaxes at the movies once in a while. I value their continued friendship very much.

---

[*]Anna L. Skalak is Mrs. Richard Skalak. The late Dr. Skalak was James Kip Finch Professor of Engineering Mechanics and Director of the Bioengineering Institute at Columbia University, Professor of Bioengineering at UCSD, and Shu's close collaborator for 30 years.

Skalaks and Chiens at Dick's 70th birthday celebration, 1993. "The Great Tie Exchange!" Dick was wearing Shu's tie and Shu was wearing Dick's bowtie.

Shu once shared with Dick this Chinese saying: "When a man reaches 70 years of age, he will no longer be confused; he can do whatever he wishes because by then he knows the proper limits." So, Shu, it's your turn now! I wish you a most enjoyable 70th birthday. Given your intellect and energy, I know there will be many more productive years, and more celebrations to come.

Top: Theoretical modeling of RBC rouleau formation, showing that an increase in aggregation energy can lead to a change in curvature of the end cells from concavity to convexity (R. Skalak and S. Chien). Bottom: Experimental results. From S. Chien et al., 1984, in Publication A192.

Sweet Memories

Sheldon Weinbaum, Ph.D.[*]

It was 32 years ago, in 1969, that I first met Shu. I had been at CCNY only two years at the time and Bert Fung was visiting CCNY for the 50th anniversary of the School of Engineering. Before Bert's first lecture, he said to me there were two people that he had specially invited and wanted me to meet. One was Dick Skalak and the other was Shu. I still have many colleagues from that period, but there is only one who has been my partner and collaborator, my co-author and co-PI till this very day, my dear friend, Shu Chien. One of our joint grants from NIH on the role of the endothelium in atherogenesis was continuously renewed for a total of 27 years. There were also numerous NSF grants over this same period including an NSF "Special Creativity Grant Award."

More important than these grants have been the lives of all the graduate students whose careers have been initially shaped by this collaboration. Shu has been on the doctoral examining committee of ten CCNY students and has served as a co-mentor for five of these students, Leslie Arminski, Ghebre Tzeghai, Fan Yuan, Yaqi Huang and Peter Butler. One was from Columbia, Daniel Lemons. Dan's thesis advisor left Columbia, just at the point when we had received a major grant from NIH in bioheat transfer, and Shu stepped in and assumed the role as Dan's mentor, although he had no previous experience in microvascular heat transfer. I speak on behalf of all of these former students in saying how enriched and fortunate they have been to have

---

[*]CUNY Distinguished Professor of Engineering, City College of New York.

experienced Shu's wisdom, patience, encouragement, scientific insight and largesse.

I think when you are part of a great team, everyone senses it and feels a great pride in its success. Shu has been the nucleus of several such teams and the celebration today is also a celebration of the many individuals involved in these successes. One painful absence is Dick Skalak. I think it is fitting on this occasion to give a brief history of the team which might be called Shu's endothelial-atherogenesis group. It had many central players over the years in addition to Shu, Dick and myself, some for just short periods and others for the entire lifetime of the group. A partial list in loose chronological order includes: Robert Pfeffer, Kung-ming Jan, Shunichi Usami, Colin Caro, George Palade, Maia and Nicolae Simionescu, Ann Baldwin, Tom Carew, David Rumshitzki, John Shyy and many invaluable supporting staff, technicians and graduate students. There were also many others from Shu's other groups that interfaced with us, such as Herb Lipowsky, Amy and Paul Sung and George Schuessler.

The beginnings of the group started shortly after Bert's visit in 1969. In the summer of 1970, Bob Pfeffer and I asked Shu if he would arrange for us to take the Physiology course with the Columbia medical students. Bob then went off to Imperial College of Science and Technology for a year and I followed a year later, joining Colin Caro's Physiological Flow Studies Unit. Colin and I had made our first venture into vesicular transport in arterial endothelium. Nearly all work on vesicular transport at the time was done on capillary endothelium, and George Palade was a central figure. Unknown to Shu, Colin and I had gone through a whole list of potential collaborators in the Eastern U.S. Shu was at the top of our list, although his group had no major experience at the time in large blood vessels. Persuading Shu was far easier than I thought. He really loved the idea of branching into a new area, much like he did 15 years later when he jumped into molecular biology. He was, however, very leery about breaking into a new NIH funding community, the large blood vessel-atherosclerosis community. In 1975, Shu, Bob Pfeffer and I jointly approached George Palade at the New York Heart Association Meeting to be a consultant for our first proposal. Palade then introduced us to the Simionescus. Kung-ming went up to Yale to learn the freeze fracture technique from the Simionescus and our initial group was born.

I will briefly relate just a few of the highlights that have occurred since then. Most of the next decade was spent believing that macromolecules were transported by vesicles across vascular endothelium or through sites of denuding endothelial injury. By the early 1980s, there was a growing body of evidence that neither was the case. There appeared to be no overt endothelial injury in atherogenesis, and work from our own group, Ann Baldwin, suggested that vesicles were not capable of transporting ferritin across arterial endothelium in animal models and similar results were observed for colloidal gold. In 1985, a new cellular level hypothesis, the leaky junction-cell turnover hypothesis, was proposed, namely that only a few endothelial cells in 10,000 would have junctions that were leaky to LDL at any given time. However, to find such leakage was like looking for a needle in a haystack. This captured people's imagination and to our surprise we were informed that NSF was considering us for an unsolicited "Special Creativity Grant," one of only four that were awarded nationally that year. Three years later in 1988, Shu and Kung-ming devised a convincing experiment to demonstrate that cell junctions were, indeed, leaky during both mitosis and cell death.

By the early 1990s, the group was faced with a very paradoxical observation. The leakage spots for HRP and LDL observed by our own and other groups showed an amazingly rapid spread, a spread that was 100 times faster than could be explained by diffusion. This observation led to a new model for transport in the arterial intima, in which it was proposed that the matrix structure of the intima could differ dramatically from that of the media and that the growth of the leakage spots was convection, not diffusion dominated. The paper explaining this hypothesis received the 1996 Melville Medal, the highest literature award of the ASME (see picture attached). This was the second time that Shu had received the Melville Medal, the first was with Cheng Dong and Dick Skalak on leukocyte mechanics, just a few years earlier. To the best of my knowledge, no engineer has co-authored two Melville Medal papers, or for that matter anyone but Shu.

114 *Tributes to Shu Chien*

ASME Melville Medal Ceremony, 1996. Ernest L. Daman, Chair, ASME International Committee on Honors; Richard J. Goldstein, President, ASME International; Sheldon Weinbaum, Melville Medallist; Shu Chien, Melville Medallist; Yagi Huang, Melville Medallist; Belden, Executive Director, ASME International (left to right).

In the last ten years, Shu's interests have turned much more heavily into molecular biology. This is an undertaking that he has taught himself. He told me that the best way to learn something difficult and new is to teach it and then write a book about it. This would be a most nerve-racking experience for most of us. Not all of Shu's advice should be taken too literally. He is credited with the discovery of the shear stress sensitive element in the promoter region of the monocyte chemotactic protein and the elucidation of the signaling pathways in the activation of the endothelial cell by fluid shear stress and the signaling pathways in naturally occurring cell death.

The story of just this group would be sufficient to fill one's plate, but when one realizes that this is just one story among so many others that one starts to see the full kaleidoscope of Shu's far-reaching research and professional involvement. It has been such a pleasure for me to be part of just one story, to have shared such sweet memories with such a special individual. When you have such a special partner it is a treat to, in some way, let others know. This did happen for me a few years ago. Bert asked me if I would nominate Shu for the NAE. I couldn't have asked for a nicer present.

I Shall Always Look Up to Him

David H. Cheng, Ph.D.[*]

In the early 1970s, the School of Engineering (SOE) at The City College (CCNY) was about to embark on initiating Ph.D. programs. Since SOE is the only engineering school in the City University of New York (CUNY), it could expect no assistance from other units within CUNY. Concerned faculty members were genuinely apprehensive about the prospect. By consensus, we decided to invite highly regarded advisors to come to evaluate us and offer guidance. My suggestion was Professor Y.C. Fung (Now well-known as the Father of Bioengineering) who at the time just left CalTech to join UC San Diego. At Dr. Fung's first lecture, a young man whom I did not know was sitting on the first row with undivided attention to the lecture. At the end of it, I went over and introduced myself and found that the young man was none other than Chien Shu.

At lunch, we found out that Shu knew Y.C. only through reputation. We were told that Shu was a professor of physiology at the Columbia medical school and had received a basic M.D. training in Taiwan. Only after 1973, I got to know him and his lovely family better; his better half, Kuang-Chung is also an M.D. and they have two lovely daughters.

---

[*]Dean-Emeritus of Engineering of the City of New York. Visiting Scholar, Department of Philosophy, The William Paterson University of New Jersey.

David Cheng, T.C. Tsao, Thomas Chang, Shu and Fred Yu at Phi Lambda Twin State Gathering, 1985.

It makes no sense for me to dwell on his achievements which are still going higher. He maintains humility at success and humbleness in receiving praises, a rare quality among high achievers. Once he and Kuang-Chung sang a duet at a performance for a group of friends, wearing very colorful costumes and performing with unreserved enthusiasm. The audience loved it and went wild. This shows another aspect of his personality: Shu takes his research, obligations, responsibility and every other person's business entrusted to him seriously but himself.

According to Chinese custom, we believe in predestined affinity among people. I believe Shu and I have it. Despite the separation of three thousand miles, he is the person I shall always look up to!

Shu — An Appreciation of His Achievements and Friendship

Colin G. Caro, BSC, M.D., FRCPE[*]

Sending this note gives me great pleasure, but also causes me much regret.

The regret derives from being unable to be present in person to celebrate Shu's 70th birthday. So much had I intended to come that I had made an airline reservation. Frustratingly, something has come up in London on the very day — June 23rd — which compels me to be here. I so wish it were otherwise.

But to the pleasure. I can look back on 25 years of happy stimulating contact and friendship with Shu. As I recall it, it began when Shelly Weinbaum spent a sabbatical year with us in the Physiological Flow Studies Unit (PFSU) at Imperial College, working in the area of arterial physiology/pathology. Prior to his return to CUNY, he spoke of the hope that he, Dick Skalak and Shu had of building an atheroma research program in New York City and, clearly misguidedly, they asked me to be an advisor.

Nothing changes. There we were, driving ourselves into the night and on weekends on proposals to NSF and NIH. However, it soon became clear (and this comment in no way diminishes Dick or Shelly)

---

[*]Emeritus Professor of Physiological Mechanics, Department of Biological and Medical Systems, Imperial College, London.

With Colin and Rachel Caro in London, 2003.

that there was only one maestro in "proposalism" — the ever calm, ever analytical, ever reasonable, ever perceptive Shu.

Support was attracted and I remember visits to them at P&S, visits with them to workshops at Ohio State University, arranged by Bob Nerem, and visits to George Palade and the Simionescus at Yale. I also vividly remember, I wonder if Shu does, that when he and his colleagues expanded work begun at the PFSU, they obtained the novel finding that cyclical mechanical extension of excised canine common carotid arteries caused them to dilate. By consent, as I remember, this result was attributed to the removal of spasm. Could it have been due to the release of a vaso-relaxant material, say the then undiscovered EDRF? I think we shall not know!

Though I was later to be less frequently in contact with Shu, I felt that the closeness of our relationship was fully retained, indeed enhanced by our periodic comings together at meetings in the U.S. and elsewhere. One of the "elsewheres" was when he honored me by attending my Festschrift at Imperial College.

From a little distance, I could also watch with amazement Shu's meteoric rise and transformation. The rise included prizes and medals, presidency of the American Physiological Society and high office in AIMBE. The transformation was from physiologist/bioengineer to "accredited" physiologist/cellular-molecular biologist/bioengineer and was associated with leading discoveries and, from time-to-time, the most lucid of keynote lectures.

I have omitted a translocation. Shu and Dick accepted Bert's invitation to join him at UCSD. We all know what achievements there have been at La Jolla — attributable to many, but leading among them Bert, Dick and Shu. Everyone here, including Rachel, sends love and congratulations to Shu and K.C.

A Tribute

Samuel C. Silverstein, M.D.[*]

Dear Shu,

I deeply regret that I cannot be at UCSD to join in celebrating your 70th birthday. I trust that reaching this venerable age does not presage retirement. Your mind and spirit are as youthful and spry as when you were at Columbia. I am confident they will continue in this vein for many years to come.

You have been a wonderful friend and colleague at Columbia and elsewhere. Your clear thinking and leadership helped save FASEB at a critical juncture in its development. On behalf of all of your friends and colleagues in the Department of Physiology and Cellular Biophysics, I wish you a happy birthday and many more years of good health.

Very best personal regards.

Yours sincerely,

Samuel C. Silverstein, M.D.

---

[*]John C. Dalton Professor and Chairman, Department of Physiology and Cell Biophysics, College of Physicians and Surgeons, Columbia University, NYC.

A Tribute to Professor Shu Chien

Van C. Mow, Ph.D.[*]

As I sit here this bright and beautiful spring morning on graduation day at Columbia University, anticipating the festivities to come this afternoon, when at Convocation four of my Ph.D. students will receive their degrees from this venerable Institution. With this degree, each will begin to face life's challenges as they commence his/her professional career with exciting and limitless possibilities. At such moments, one cannot help but reflect on one's own career, or the career of someone else who has been a dear and close friend for a long time.

Shu Chien is a Columbian. As with a number of us of Shu's age, he came to Columbia University by a long and circuitous route. He received his Ph.D. degree in physiology from Columbia in 1957 at the College of Physicians and Surgeons. He received his M.D. degree from National Taiwan University College of Medicine in 1953, thus making him one of the first M.D./Ph.D.s in the nation. Shu was always precocious, receiving his Premed education from National Peking University in 1948 at the age of 17. This is remarkable when one thinks about the turmoil that embroiled China, particularly after 1937 when war began officially between Japan and China that finally

---

[*]Stanley Dicker Professor of Biomedical Engineering and Orthopaedic Bioengineering Chairman, Department of Biomedical Engineering; Director, New York Orthopaedic Hospital Research Laboratory, Columbia University, NYC.

ended in 1945. Immediately following the end of World War II, the revolution began that led to the establishment of the People's Republic of China in 1949. Yet throughout these turbulent times, Shu managed to attain the scholastic achievements that eventually brought him to America, first at Columbia University, and then in 1988 when he arrived at the University of California, San Diego. In that fateful year, Richard Skalak of the School of Engineering and Applied Science (a Columbian for 48 years) and Shu Chien left Columbia and New York City for the sunny shores of San Diego.

This was a great loss for Columbia, and a major gain for UCSD. Shu and Dick had formed a quintessential team for multidisciplinary research in bioengineering at Columbia. They developed and kept the spirit of bioengineering going from the early 1970s. To a large degree, they defined the best in bioengineering and biomechanics research and education. Today, this spirit of multidisciplinary bioengineering activity is alive and well at Columbia University. Indeed, a little more than ten years after their departure, and nearly 40 years after the beginnings of bioengineering activity at Columbia, on 9 December 1999, the Trustee of the University formally established a Biomedical Engineering Department in the Engineering School. This new Department will be provided with 18 new biomedical engineering faculty members, all of whom will be pursuing important multidisciplinary bioengineering research with colleagues from the College of Physicians and Surgeons.

While Shu's intellectual abilities are undeniable, his charm, dedication and energy are also most compelling. I first met Shu in the late 1970s when he and Dick Skalak invited me to contribute to the Bioengineering Handbook that they were editing. This book was pivotal when a large segment of bioengineering needed to be defined. It launched a number of new directions for the field that would not have occurred otherwise. To this day, Shu's dedication to bioengineering is everywhere evident. For Columbia, we are fortunate to have him as the chairman of our External Advisory Committee for our successful Whitaker Biomedical Engineering efforts thus far; to this date, he continues providing sage advice that has been invaluable in the building of our Biomedical Engineering Department. As Isaac Newton once

K.C. with Barbara and Van Mow, 1997.

observed, an apple does not fall far from its tree; so Shu comes back to Columbia University to help us build an outstanding program worthy of the legacy in bioengineering that he and Dick Skalak left.

A quick perusal of Shu's resume shows that many have appreciated his immense qualities. To list just a few awards of note, he is member of the National Academy of Engineering and Institute of Medicine of the National Academy of Sciences. Only 14 individuals in the nation concurrently hold membership in these two prestigious National Academies. He is also a member of the Academia Sinica in Taiwan. These elected memberships are based on many years of research where Shu won numerous outstanding research awards, including the prestigious ASME Melville Medal (2), ASME Bioengineering Best Paper Awards (2), and others.

So, after a long and distinguished career, Shu has clearly and uniquely established an outstanding model to whom young Ph.D. bioengineers can admire and aspire. With his quality of incisive intellect, perseverance, duty, honor and warmth, it is no wonder Shu has gained the highest level of academic achievement possible in bioengineering, and has the genuine heartfelt admiration from all his friends and colleagues around the world.

My wife Barbara and I wish Shu a most joyous 70th birthday, and to both Shu and K.C. many more happy and fruitful years to come in sunny La Jolla, California.

With warmest regards,

Van C. Mow

The Blooming of Taiwan Biomedical Sciences — A Tribute to Dr. Shu Chien

Cheng-Wen Wu, M.D., Ph.D.[*]

It is a great pleasure for me to write something about Professor Shu Chien with whom I share a belief that scientific research is a process of exploring the unknown and a pursuit of excellence.

Sometimes life is a series of surprising co-incidents. Both Professor Chien and I graduated from National Taiwan University, College of Medicine. Both of us devoted most of our lives to teaching and research in biomedical sciences as professors in U.S. universities, and both were elected as members of the Academia Sinica. In addition, both of us served as Directors of the Institute of Biomedical Sciences, Academia Sinica.

Professor Chien is one of the founding fathers of the Institute of Biomedical Sciences (IBMS). With his hard work, the first group of faculty joined the budding institute in 1986, a year when Professor Chien served as Director of IBMS. In the following one and a half years, Professor Chien paved the way for IBMS to grow from scratch to a well-established institute among the Institutes of the Academia Sinica. He also helped build up the reputation of the Life Sciences in the Academy. One successful program launched for the entire Taiwan biomedical scientific community during his tenure as the

---

[*]President, National Health Research Institutes, Taiwan.

Director was the Oncologist Training Program. Through a multi-centered collaboration, the Oncologist Training Program attracted the support from both the National Science Council and the Department of Health of the Taiwan Government. The Training Program eventually led to the founding of the Taiwan Cooperative Oncology Group (TCOG) two years later, when I succeeded him as Director of IBMS. TCOG is the first multi-institutional clinical trial organization consisting of 22 major hospitals in Taiwan. In fact, Professor Chien will never get tired of teaching the young and nurturing the future scientists. There were lots of other examples in IBMS during his directorship at IBMS.

I joined IBMS along with Professor Chien in July 1987 when there were already 30 plus Principal Investigators and more than 200 staff members at the Institute. Needless to say, all the faculty and staff at IBMS were heart-broken to see Professor Chien leave. I was no exception. But I was assured then that he would continue to help IBMS as the Chair of its Advisory Committee and that he would eventually return to Taiwan to serve the country.

With Cheng-wen Wu, 1995.

Though Professor Chien did not return to Taiwan in the subsequent years to take up a long-term position, he made various efforts to strengthen biomedical research in his home country. In 1988, with the same vigor and drive that he showed in establishing the IBMS as a first-rate research institute in Taiwan, Professor Chien and former president Ta-You Wu of Academia Sinica were instrumental in fulfilling a dream in Taiwan's academic history: the breakthrough in a special salary system for Distinguished Medical Scientists at IBMS. This unprecedented change of personnel system later led to the birth of the Distinguished Research Fellow system of the Academia Sinica and consequently equipped the Academy with more competitive power in the international scientific community.

His magic power did not end here. Professor Chien continued his influence to help launch another milestone project for Taiwan — the establishment of the National Health Research Institutes (NHRI), a national research organization similar to the National Institutes of Health in the U.S. and the Medical Research Council in England, in January 1996. The mission of NHRI has been dedicated to the coordination and promotion of medical and health-related research in Taiwan. Professor Chien now serves as a member of the Board of Directors, and as the chairman of the Advisory Committee at NHRI.

It is hard to describe how important Professor Shu Chien is to the development of Taiwan's biomedical sciences over the past two decades. I admire his great enthusiasm in promoting biomedical sciences and in encouraging and inspiring the young scientists. No matter how busy he is, Professor Chien is always generous in offering his time and advice to IBMS and NHRI. He is one of the invisible hands that help shape Taiwan's biomedical sciences to meet the challenges of vigorous competition at the international level.

It is a real blessing for me to be a friend and colleague of Professor Chien. I share the great delight in his 70th birthday.

A Tribute to Professor Shu Chien

Yuan T. Lee, Ph.D.[*]
Nobel Laureate

Professor Shu Chien is a scientist, an educator and a true gentleman. He is affectionately respected by his peers and deeply loved by the people who are fortunate enough to share his life. He is indeed a wonderful friend who can be counted on whenever needed.

Receiving award from Yuan Tseh Lee, 1998.

---

[*]President, Academia Sinica, Taiwan, R.O.C.

Professor Chien played a key role in the establishment of the Institute of Biomedical Sciences at the Academia Sinica. He was, in fact, among the handful of our Academicians who took the initiative to lay the groundwork for the future development of the Institute. For more than two decades, Professor Chien has also made extremely important contributions to the advancement of science and education in Taiwan.

As President of the Academia Sinica, I would like to thank him, both sincerely and profusely, for what he has done for the Academia Sinica and the scientific community of Taiwan. I heartily wish him many happy returns on the occasion of his 70th birthday.

"Li De; Li Yen; Li Gong"

Paul O.P. Ts'o, Ph.D.[*] *and Muriel Ts'o*

To: Dr. Shu Chien for his 70th Birthday

    Following your beloved father,

        Build Virtue
        Create Knowledge
        Accomplish Achievement

                      Paul and Muriel Ts'o

With Paul Ts'o at NHRI Advisory Committee Meeting, 1996.

---

[*] Professor, Department of Biochemistry and Molecular Biology, Bloomberg School of Public Health, Johns Hopkins University.
Chairman and CEO, Cell Works, Inc., Baltimore, Maryland.

錢煦學兄七十壽辰誌慶：

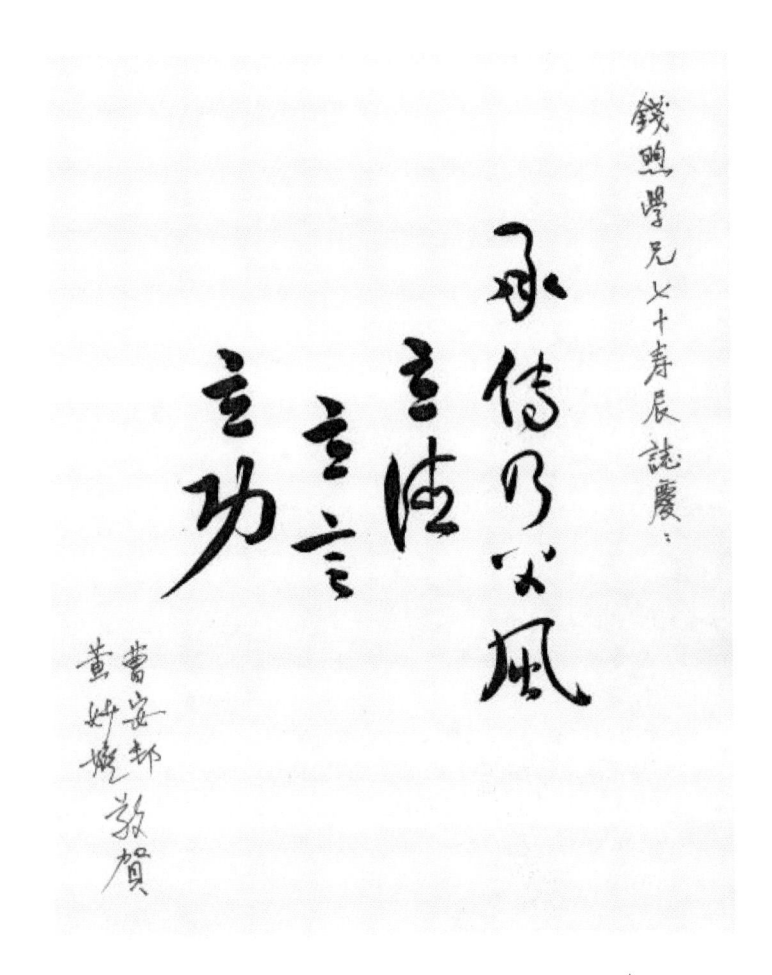

承傳乃父風
立德立言
立功

曹安邦
董妙嫻 敬賀

Working Together with Shu

Monto Ho, M.D.[*]

Dear Shu,

The first time we met was during the 1980 biennial meeting of Academia Sinica. We were often together during the post-meeting tour and I remember our playing tennis early in the morning. The meeting was notable in passing two important resolutions that laid the foundations of the Institute of Molecular Biology and the Institute of Biomedical Sciences of Academia Sinica. From then on until this very day, both of us have been engaged in helping the development of biomedical sciences in Taiwan. Your efforts in this area have commanded my deepest appreciation and respect. I believe both of us, as expatriates, have been similarly motivated to help Taiwan. But you always were a notch ahead of me in the depth of your dedication, in your persistence and your know-how. Your style of getting things done is by an unsurpassed combination of diplomacy and efficiency. The result is that you have accomplished much more than me. Among other things, I have the greatest admiration for your fluency and mastery of spoken and written Chinese and English. Among the hundreds of Chinese I know, there are only a handful who come close to you.

---

[*]Member, Academia Sinica, Taiwan, R.O.C. Director, Division of Clinical Research, National Health Research Institutes, Taiwan, R.O.C. Emeritus Chair, Department of Infectious Diseases and Microbiology, Graduate School of Public Health, University of Pittsburgh, Pittsburgh, Pennsylvania.

On this joyous occasion of your 70th birthday, let me congratulate you on your many achievements of a lifetime, both personal and professional. It has been an honor and privilege for me to be associated with you in one of them.

With best regards,

Sincerely yours,

Monto

National Health Research Institutes Advisory Committee Meeting, Taipei, 1996 (Monto Ho is at far left).

Monto Ho (second from left) and Shu posing before tennis game with colleagues at National Cheng Kung University, Tainan, Taiwan, 1982.

Some Small Role

M. Lea Rudee, Ph.D.[*]

Dear Shu,

During the nearly 12 years I served as Dean, the School of Engineering made more than 80 appointments. This was a spectacular group of faculty. I hope I don't hurt anyone's feelings, but I felt at the time your appointment was the top in which I was involved. Your subsequent performance at UCSD has exceeded my high expectations.

You have provided leadership in every way — as a scholar, as a teacher, as a fund raiser, and as the creator of the separate department of Bioengineering. What is truly amazing is that you have accomplished so much while making friends, not enemies, and helping develop the careers of so many younger faculty.

Shu, UCSD is much the better place for your presence, and I feel very proud that I had some small role in your coming to La Jolla.

All the best in your future endeavors.

Sincerely,

M.L. Rudee

---

[*]Professor Emeritus, Department of Electrical and Computer Engineering, University of California, San Diego. The Founding Dean of the School of Engineering, which was later named Ivwin and Joan Jacobs School of Engineering in 1997.

Lea Rudee speaking at Whitaker Foundation Development Award Inaugural Ceremony, UCSD Price Center, 1993.

Counting Precious Pearls in the Family: What Makes Shu the Leader in So Many Fronts?

Yuan-Cheng Fung, Ph.D.[*]
National Science Medalist

Our lives are shaped often by a few lucky breaks. In my life, one of my breaks was meeting you, Shu and Kuang-Chung, in 1966, through our common interest in the behavior of red blood cells. Twenty years later, I was lucky again. Your acceptance of the UCSD offer, and your bringing Dick and Anna with you to UCSD was an event that I consider the highest achievement of my career and the best contribution I made to the University of California. With you, success of the bioengineering enterprise at UCSD was assured. History proved me correct. I am pleased that this was a success that everybody in our field can enjoy.

Luna and I have three and a half decades of memories of our interactions with you. Between our families, we have anecdotes galore.

---

[*] Professor Emeritus, Department of Bioengineering, UC San Diego. Professor Yuan-Cheng Fung is a founding member of the Bioengineering Group at UCSD in 1966. He is widely regarded as the father of Biomechanics.

Shine like a rainbow.

I remember many interesting occasions. I think of our lives often. With time, many stories gather some philosophical shading. Writing about anecdotes is easy. However, picking one against the others to write about is difficult.

With a piece of nice white paper in front of me and a black pen in my hand, I let all anecdotes flow through my thought. Then I looked out of my window. I saw a rainbow in the sky. It was beautiful, as always, but especially that day. Looking at it, I was struck by a thought. Here is a beautiful person, who shines like a rainbow. A rainbow is the result of an organization of optics of all those particles of water. A person is represented by all the decisions and actions he/she has taken. Only when all the events interact cohesively in a certain way can a personality be revealed as a rainbow.

Therefore, a theoretical problem is: "What are the qualities that make Shu shine like a rainbow?" To solve the problem, one has to count all the marbles in the bag, or as we Chinese say, to count all the precious pearls in the family chest. Here is a try:

Outstanding Teaching: Shu is a natural teacher. He is always well organized for his classes. He gives inspiring lectures. He gives clear penetrating answers to students' questions. He is always clear and consistent in his assignments.

Shu regularly teaches Cell and Molecular Biology and Respiration and Renal Physiology, two core courses in the bioengineering graduate program. He also teaches a part of the School of Medicine's core course Organ Physiology, and other graduate and undergraduate courses. He gave individual instructions and supervision to an average number of seven graduate students per quarter. The rate at which his students received their Ph.D. is 7/5 or 1.4 per year over the years. He always participates at the Friday afternoon TGIF parties and welcomes any opportunity to get involved with student issues and functions.

Extraordinary Research: Shu has published 382 full-length papers in professional scientific journals, and edited nine books. In 1999, he published five journal papers and in 2000, seven. He submitted research proposals to NIH, NSF, NIST, the Whitaker Foundation, the Alliance Pharmaceuticals, etc. at a rate of about five per year. He received direct cost research funding at a level of $2.5 million per year.

Extraordinary Service to the Bioengineering Department at UCSD: Shu arrived at UCSD in late 1988. He immediately took charge of the bioengineering program at UCSD and served as the coordinator of the bioengineering group within the Department of Applied Mechanics and Engineering Science (AMES). UCSD began offering bachelor of science, masters of science, and Ph.D. programs in September 1966, and was believed to be the first in the nation to offer such a complete program in bioengineering. Our program functioned under the umbrella of the Department of AMES. Our AMES Department was and is very distinguished. Our UCSD's Bioengineering Program within AMES, however, had little visibility nationwide. For example, the U.S. News and World Report did not list UCSD in Bioengineering before 1994. To gain visibility and national recognition was a real need.

Shu took the initiative to organize the Institute for Biomedical Engineering at UCSD, which was approved as an Organized Research Unit by the University of California in November 1991. In 1992, Shu led the UCSD group to apply for a Development Award from the

Whitaker Foundation. After strong competition, the funding was granted in early 1993. The Institute for Biomedical Engineering (now named the Whitaker Institute of Biomedical Engineering) has over 100 members coming from UCSD faculty and research staff in the School of Medicine and the Departments of Engineering (AMES, CSE and ECE), Biology, Chemistry, Physics and Mathematics. Also joining us are scientists in our neighboring institutions Salk, Scripps and Burnham, as well as industry. All these people united to advance the interdisciplinary area of Biomedical Engineering. Its establishment was pivotal to our development. It selected the area of "tissue engineering" as an area of concentration. With Tissue Engineering, which means scientific understanding of natural tissues and creation of tissue substitutes by engineering, the nature of bioengineering becomes transparent.

Many people say that bioengineering is a bridge between engineering and biology. But I do not think so. I think that it is a new entity grown out of the two old fields, like a baby created by two parents. As the baby grows, the parents are enriched by the new baby. Thus, the fields of biology and engineering are enriched by bioengineering while bioengineering develops its own character.

After the establishment of the Institute, Shu and the Bioengineering faculty members, with the strong support of UCSD administration, proposed to form a Department of Bioengineering. This was accomplished in 1994, with the support of Dean Conn and the Whitaker Foundation. Following the establishment of the Institute and the Department, our visibility to the outside world increased rapidly. In 1995, the publication of "Evaluation of Doctorate Programs in the United States" by the National Research Council of the NAS, NAE and IOM listed the UCSD Bioengineering as number one in "Ph.D. Programs" in the United States, and number two in "Bioengineering Faculty" in the United States (MIT being number one in that category). From 1995 to 2000, the U.S. News and World Report has given UCSD Bioengineering high ranking year after year, improving annually, from number 7 to number 3.

Following the establishment of the department status, Shu, as the chair of the department, took the initiative to write a Leadership Award Application to the Whitaker Foundation for funding of a new building to house the department and to hire additional faculty members. In a nationwide competition, we won. Much of the credit should go to the University administration, but Shu should be given a major share of it. Deciding the scientific directions of the proposal, filling in the details of the plan, unifying the input and ideas of the faculty members, getting the collaboration of other departments of UCSD and neighboring institutions such as Salk, Scripps, Burnham, etc. was Shu Chien's responsibility. Shu really did a superb job! Our asserted plan was to create the Nation's best research institute. I feel we have succeeded in our goal.

The reputation of UCSD's Bioengineering Department is enhanced, undoubtedly, also by Shu's participation and collaborations with various U.S. Institutions and Societies such as the IOM, NAE, AIMBE, BMES, ASME, APS, and Universities in the U.S. and other countries. Undoubtedly, the awards and recognitions that Shu received from these institutions do help also.

One of the most gratifying recognition came from the Powell Foundation. In 1999, the Powell Foundation gave UCSD a grant of $8 million for the new building, which is sponsored by the Whitaker

Shu, K.C., Luna and Bert Fung, San Diego, 2000. Fung was the honorea at the Union of Pacific Asian Communities.

Foundation Leadership Award with a funding of $18.2 million. The new building will be called the Powell-Focht Building. Ground has been broken and we are anxious to move in when it is completed.

Extraordinary Service to UCSD and the UC System: Besides chairing the Department of Bioengineering from 1994 through 1999, Shu served also as the Director of the Whitaker Institute of Biomedical Engineering from 1991 to the present. After retiring from the Bioengineering Department chairmanship in 1999, he continues to serve as the Director of the Whitaker Institute and he devotes an even greater attention to the Institute's affairs to promote collaboration among our leaders on campus. The biomedical research on this campus has added a new focus on bioinformatics to tissue engineering, both trying to understand the functions of genes. The campus has been recharged with new energy and new financial support. The Whitaker Institute of Biomedical Engineering is strategically situated for this new task. Shu is a natural catalyst.

Shu participates in the campus-wide affairs by accepting membership in many committees. For the UC System, Shu serves as the chair of the Steering Committee of the UC System-wide Bioengineering Programs, and on the Advisory Committees on Bioengineering of UCB, UCSF, UCI, UCLA and UCR.

Extraordinary Service to the Profession: Most professors serve professional societies in some capacity. Shu serves a number of important large professional societies at the very top. For example, he served as the President of the largest organization of American biologists — the Federation of American Societies for Experimental Biology. For the American Physiological Society, Shu was President in 1990–1991 and Chairs of the Public Affairs and Long Range Planning Committees in subsequent years. The most important worldwide organization in physiology is the International Union of Physiological Sciences: Shu is its Treasurer since 1998. In the prestigious U.S. National Academy of Engineering, Shu serves as the chair of Peer Committee of Section 2, Bioengineering. In the American Institute for Medical and Biological Engineering, Dr. Chien was chair of the College of Fellows in 1998 through 1999, and is President in 2000 through 2001.

Shu is respected in every organization he joins. For example, he initiated and served as the first chairman of the North American Society of Biorheology. He served as the Director of the Institute of Biomedical Sciences in Academia Sinica (analogous to our NAS, IOM and NAE) in Taiwan. He is the chairman of the Advisory Committee of the National Health Research Institute of Taiwan (analogous to our NIH).

He is on the Advisory Committees of Columbia University, Duke University, Georgia Institute of Technology, Emory University, Stanford University, Keck Graduate Institute, National Institute for Advanced Interdisciplinary Research, Tsukuba, Japan, Tsinghua University in China, and the National University of Singapore. His participation in so many countries and organizations is an example of his extraordinary leadership.

Honors Received: The honors and awards won by Shu are too numerous to be named here. His CV lists 50 entries. Here are a few that stand out: He is one of the 14 distinguished Americans who hold memberships in both the National Academy of Engineering (elected in 1997) and the Institute of Medicine of NAS (elected in 1994). This is quite an honor. He is also a member of Academia Sinica, the highest Academy in Taiwan. He won the Nanci Medal in blood rheology, Fahraeus Medal in Clinical Hemorheology, the Landis Award and the Zweifach Award of the Microcirculatory Society, and the ALZA Award of Biomedical Engineering. He was UCSD's Distinguished Faculty Lecturer in 1994 and Columbia University's American Society of Mechanical Engineer's (ASME's) Awardee for authoring the best paper in the *Journal of Biomedical Engineering* twice, once in 1989 and again in 1995. He won ASME's Melville Medal for the best contribution to permanent literature also twice, once in 1990 and again in 1996. Truly, this is an extraordinary feat.

What is the secret for Shu's success? People who have worked with Shu will probably say that it is his phenomenal memory, congenial personality, power of concentration, superior communication ability, clarity of thought, diligence and persistence, among other qualities. But this is only the counting of jewels. Jewels alone, no matter how many, do not a rainbow make. For the rainbow effect, a greater chemistry is needed. I feel it. I see it. But I do not know the secret.

THE WIND UNDER HIS WINGS — THE K.C. FACTOR

K.C. is the one who enables Shu to fly. When I met K.C. and Shu in Atlantic City on an Easter Sunday attending a FASEB meeting, they were carrying their toddler daughters to listen to a lecture. FASEB met regularly in Atlantic City in old days. I have a photograph with the Chien family in Captain Starn's Lobster House in Atlantic City. I treasure that photo because it exhibits Shu and K.C.'s devotion to their family and profession. K.C. and Shu were always together. They were together when Shu was installed as the President of the American Physiological Society. They were together in Taiwan when Shu initiated the Institute of Biomedical Science at the Academia Sinica. To K.C. and Shu, science and family is one enterprise to be pursued together.

Kuang-Chung Hu and Shu were schoolmates at the College of Medicine of National Taiwan University. K.C. got her M.D. degree in 1956, and took postgraduate courses at the New York Polyclinic Postgraduate School in 1969. She is a pediatrician and holds medical licenses in New York, New Jersey and California. She practiced medicine full time in New York and New Jersey, and later part time in California.

Y.C. Fung and the Chiens, Atlantic City, NJ, 1966.

Now, she pursues many other interests in her free time. In fine arts, she has become a serious watercolor painter. She paints flowers not only for their beauty, but also for their character. In one you see the roses are just happy. In another they seem crying. She has also become a truly fine Chinese brush painter. Her "four seasons" indeed show successively the cool mist, the hot sun, the cutting wind, and the quiet waiting. Her paintings have been exhibited in shows and won high praises. She participates in volunteer groups of physicians treating and teaching old and young patients, minorities and disadvantaged people. She teaches Chinese to English-speaking kids. All kids love her. She took dance and singing lessons and teaches children dancing and singing. She is active in San Diego Chinese Historical Society and Museum. She is enthusiastic to help people. A many splendoured lady she is. Dear Kuang-Chung, in your future painting, please paint more rainbows. Someday it will reveal the secret of Shu's glow.

A Tribute

Beatrice Zweifach[*]

Dear Professor Shu Chien,

I am so pleased to be part of your birthday celebration. You have always made me feel welcome on many occasions. I remember attending the parties that you hosted for the Bioengineering students — always with warmth and caring.

You have been a wonderful asset to UCSD and the Bioengineering Department in particular. It is through your efforts that the department

With Ben and Bea Zweifach, 1996.

---

[*]Beatrice Zweifach is Mrs. Benjamin Zweifach. The late Professor Zweifach was a founding member of the Bioengineering Group at UCSD in 1966. Professor Zweifach made fundamental contributions to physiology, microcirculation and blood rheology.

will have a building of its own. Ben was so pleased that you made the change from Columbia to UCSD.

All I can say is do have a very Happy Birthday and all the days that follow.

Sincerely,

Bea Zweifach

Powell-Focht Bioengineering Hall Ground-breaking, 2000. (Left to right) David Gough, Wendy Baldwin, Portia Whitaker Shumaker, Shu Chien, Robert Dynes, Joel Holiday, Minerva Kunzel, Duane Roth, Peter Katona, Robert Conn.

Powell-Focht Bioengineering Hall (completion in December 2002).

148   Tributes to Shu Chien

Bob Dynes, Irwin and Joan Jacobs (seated). Shu, Bob Conn and K.C. (standing). At the ceremony to celebrate the establishment of the Whitaker Institute of Biomedical Engineering, the initiation of the Powell-Focht Bioengineering Hall, and Y.C. Fung's 80th Birthday, UCSD Faculty Club, 1999.

With Drs. Burtt and Ruth Whitaker Holmes at the same celebration.

Shu Chien: A Diamond with Many Facets

David Gough, Ph.D.[*]

Diamond is one of Nature's amazing materials. In its purest form, it is composed only of carbon atoms and has only covalent, carbon-carbon bonds, four to each atom. The atoms are aligned in a perfect crystalline array in three dimensions. Photons pass easily through the matrix, which explains diamond's unusual brightness. The strong covalent bonds result in diamond's extreme hardness and insulating properties. The stone can be cut to numerous facets, which reflect light internally in a myriad of ways, giving rise to its brilliance and fire.

Shu Chien is like a fine, many-faceted diamond. His radiant contributions can be seen from many angles. As Shu's successor in the department chair, I've come to appreciate his enormous intellectual energy, his breadth of interests and concerns, and his range of activities and service. Shu is truly brilliant.

Just to mention a few of his broad range of contributions:

Shu is an **outstanding teacher**. His lectures are always carefully prepared and delivered, and he has received highly favorable ratings from the students at every level. Regardless of his many other service and research activities, Shu has always maintained an above-average teaching load and the strongest commitment to teaching.

Shu is an **excellent researcher**. He has published over 400 research articles, many of which have been described as "ground-breaking" by

---

[*]Professor and Chair, Department of Bioengineering, UC San Diego.

his peers in the field. Indeed, a recent paper won a second Melville Medal for Shu from the American Society of Mechanical Engineers for being the "Best Paper of Permanent Value" published in ASME Transactions in 1995. These are no small achievements.

Beyond the above achievements in teaching and research, Shu has contributed **unique leadership** by leading UCSD's successful application for a Whitaker Foundation Leadership Award to construct a building for the Bioengineering Department. Although the application had the enthusiastic support of UCSD's administration, Shu organized the overall plan, coordinated the contributions of the many contributors, wrote the proposal, and orchestrated the site visit. The overall result was astonishing: not only did the Whitaker Foundation make one of then only two Leadership Awards to UCSD to fund half of the construction, but this success, together with the national first-place ranking of the Department, encouraged the Charles Lee Powell Foundation to commit additional substantial funding for the building.

UC System-wide Bioengineering Steering Committee Meeting, UCSD Faculty Club, 1999. Front (kneeling): Geert Schmid-Schönbein, Scott Simon, Chih-Ming Ho, Thomas Budinger, Shu Chien, Kathleen Ferrara, Andrew McCulloch. Back (standing): David Gough, John Shyy, Ashok Mulchandani, Stanley Berger, Maury Hull, Scott Simon.

The new Powell-Focht Building, now under construction, will represent a new vision for the future for Bioengineering at UCSD in educational, research and technology transfer initiatives. This successful bid to the Whitaker Foundation is a model of leadership, teamwork and interdisciplinary cooperation.

Shu has also contributed **outstanding service** outside UCSD. He initiated UC System wide activities in fostering development of Bioengineering on other UC campuses and closer interactions among researchers and students. He has served as: President of the American Physiological Society, President of the Federation of American Societies of Experimental Biology (FASEB), Chair of the First World Congress on Biomechanics, Invited Speaker to the Second and Third World Congresses on Biomechanics, Organizer and Chair of the Biomedical Engineering Society Meeting in San Diego in 1997, Member of the Executive Committee of the Academia Sinica, Chair of the College of Fellows of the American Institute for Medical and Biological Engineering, and many others. Shu has also served as the president of the American Institute of Medical and Biological Engineers, and was key in the establishment of the National Institute for Biomedical Imaging and Bioengineering, the newest institute of the National Institutes of Health, created under the Clinton administration. Shu has made outstanding contributions to the campus, to students, to the UC System, at the highest levels within the NIH, and to the public at large.

The **honors and awards** won by Shu are too numerous to list completely, but only the most important will be mentioned here. Shu is an elected member of the prestigious National Academy of Engineering and the Institute of Medicine of the National Academy of Sciences. He was elected to leadership positions in the American Institute for Medical and Biological Engineers, including Chair of the College of Fellows and President.

Shu is one of the true giants in the development of the biomedical engineering profession. In spite of his modest and gentlemanly manner, his contributions are as brilliant and multi-faceted as a fine diamond.

I wish him the very best on his 70th Birthday and many more.

Constraining Shu: Mission Impossible?

Bernhard Ø. Palsson, Ph.D.*

I became to know Shu in early 1995 during my move to UCSD. He came across as an affable, accommodating, and pleasant individual. He engineered my recruitment and move with great administrative skill.

UCSD Bioengineering Faculty, 1999. Bernhard Palsson, John Frangos, Sangeeta Bhatia (p. 244), Y.C. Fung, Amy Sung, Marcos Intaglietta, Shu Chien, Andrew McCulloch, David Gough and Geert Schmid-Schönbein (left to right). Not pictured: Bob Sah (out of town).

---

*Professor, Department of Bioengineering, UC San Diego.

After I joined UCSD I began to realize what a great human being and builder Shu is. His consensus-building leadership skills transformed the bioengineering department over a ten-year period. He kept a good social agenda for the faculty, staff and students and thereby created a great team spirit and enthusiasm for the bioengineering department, its growth and its well-being. He is always gracious in a social setting, quick with words and perceptive observations, and very comfortable in front of a crowd. Shu is a natural leader and is truly a man for all seasons.

My research is in part focused on determining the constraints that govern the function of living systems. We have been quite successful in determining these constraints for lower organisms and discovering how they guide cellular function and their evolution. We have been interested in determining if similar analysis can be performed with higher organisms. Unfortunately, what I know of Shu as a person is greatly discouraging in this regard. There seem to be no constraints that limit what Shu can do. Whatever problem Shu faces, he solves and moves along in life with great speed, efficiency and agility. He seems to know no bounds.

When I told an acquaintance that I was going to Shu's 70th birthday party, the answer was "Shu 70? Can't be. He looks just over 50." So Shu, you even defy the constraints associated with aging.

My best wishes for a productive future. I will always appreciate the positive influence that you have been on me personally and my career development. Congratulations on a truly outstanding and influential career, and best of luck!

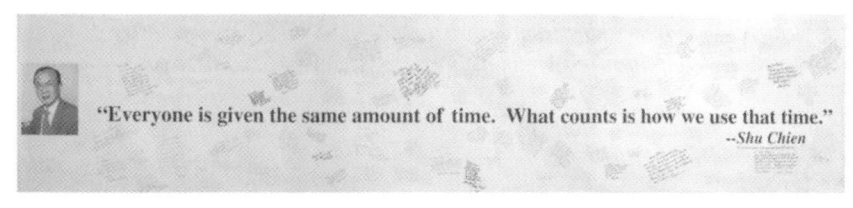

"Everyone is given the same amount of time. What counts is how we use that time."
--Shu Chien

A banner given by the Department of Bioengineering for Shu's oustanding contributions as the Chairman, 1999.

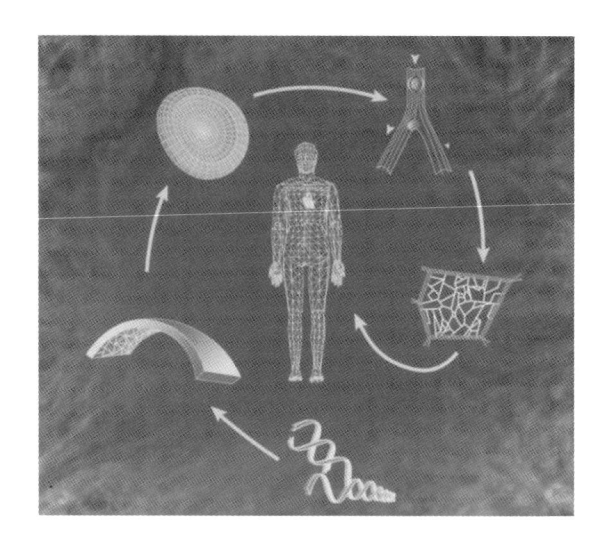

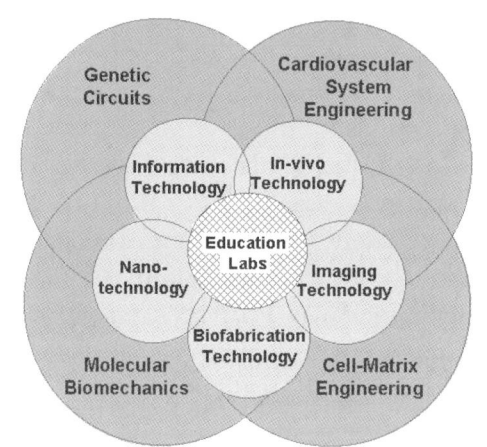

Integrative Bioengineering. Top: Illustrating biological hierarchy from genes to molecules (membrane), cell, blood vessels, vascular network, to organs and body, with theoretical analysis complementing experimental studies. Bottom: Showing the four thrust areas in UCSD Bioengineering (Genetic Circuits, Molecular Biomechanics, Cell-Matrix Engineering, and Cardiovascular System Engineering), which are supported by five core technologies, and all of these have an education focus, as well as research functions.

Shu: You Can Really Cook!

Robert L. Sah, M.D., Sc.D.*

Dear Shu,

I always marvel at what you have done and continue to do for the field of biomedical engineering, as well as your advances in cardiovascular bioengineering research. As a leader in the field of biomedical engineering, and as a scientist and engineer, you have touched the lives of many people, not only in San Diego but around

Bob Sah (first from left), undergraduate and graduate students in the Cartilage Tissue Engineering Lab wish Dr. Chien a Happy Birthday!

---

*Associate Professor, Department of Bioengineering, UC San Diego.

the world. You have made the UCSD Department of Bioengineering a place that people want to come to, to work on Bioengineering problems both individually and in collaborations, and to stay forever at.

On the more personal level, I cannot thank you enough for being my mentor, collaborator, colleague and friend. You have provided me with many precious memories and also pearls of wisdom that I will

Bioengineering seniors graduation party at Shu's backyard in La Jolla Farms (p. 555) (June of every year).

always treasure. I will always remember my very first visit to UCSD, when you spent so much time with me and made me feel so very much at home.

Enclosed are a few pictures of some happy and proud undergraduates, enjoying your company at the 1998 and 1999 graduation parties at your house. Shu, whether you are standing by the grill or working at Bioengineering, you can really cook!

With cheerful wishes for the future,

Bob Sah

A Tribute

Shankar Subramaniam, Ph.D.[*]

Dear Shu,

Greetings! It is a great privilege for me to be a colleague to felicitate you on your 70th birthday. In the short time of our association, I have begun to deeply appreciate and value your kindness, generosity and intellectual stamina. I look forward to coming years towards the opportunity to learn from you and share in your voyage of discovery in science and engineering.

Best regards.

Sincerely,

Shankar

With Gary Huber (p. 159), Michael Heller, Andrew McCulloch, Berhnard Palsson, and Shankar Subramaniam following a Bioengineering faculty meeting, 2001.

---

[*]Professor, Departments of Bioengineering, Chemistry and Biochemistry; Director, Bioinformatics Graduate Program; Senior Fellow, San Diego Supercomputer Center.

A Tribute

Gary Huber, Ph.D.[*]

Although I have been in this department only a short time, I have enjoyed my interactions with Shu very much. As department chair, he was very friendly and hospitable to me when I was interviewing for my faculty position. Since I've started, it has been fascinating to learn about his groundbreaking research that is being done on the link between the forces exerted on the outside of a cell and the biochemical events taking place inside. Also, I have been privileged to behold and enjoy the fruits of his labor as department head, when our department came into being. Finally, it is truly an inspiration to see someone five years past traditional retirement age still going strong in such a rigorous field; I was very surprised to learn that his chronological age is already 70! My hopes and prayers are for continued health and joy in Shu's eighth decade of life.

---

[*]Assistant Professor, Department of Bioengineering, UC San Diego.

The Making of a Bioengineering Dynasty: A Tribute to Dr. Shu Chien

Andrew McCulloch, Ph.D.[*] and Deidre MacKenna, Ph.D.[†]

When I arrived at UCSD, a naïve 26-year-old, in 1987, the small core of faculty in what would become the nation's top Bioengineering department, had already hatched their plan. It would take another year or so, for all to be put in place, but the excitement only grew from then leading up to the day when Shu and K.C., joined by Dick and Anna, Amy and Paul, Jerry and their associates, arrived in La Jolla. It was an event that was bigger and more worthy of celebration than moving into EBU1. In 2001, we are looking forward to moving into a new building. Can you imagine an event that would eclipse that day? The arrival of Shu Chien to join the Bioengineering faculty at UCSD was such an occasion.

I didn't fully appreciate the significance of this move (in professional sports they would call it an "acquisition") at the time. But it didn't take me long to figure it out. Shu Chien almost single-handedly built our Department and Institute to their present heights. He set the standard of excellence that we all aspire to. And he did it

---

[*]Professor, Department of Bioengineering, UC San Diego.
[†]Research Scientist at Tanabe Research Laboratories, La Jolla, and Adjunct Assistant Professor of Bioengineering at UC San Diego.

while energetically seeking the perspective of everyone involved and assiduously assigning them all the credit.

Deidre and I have been privileged and delighted to have Shu as a friend and colleague, teacher and mentor. Perhaps more than any other leader we know, he inspires everyone around him by his fine example of wisdom, intellect, insight, compassion and tireless energy. No job is too large for him to take on; no detail is too small to be unworthy of his attention; no responsibility is too large for him to accept; no student is too young to lend an ear to; no effort is too great to spare on behalf of his department; no task is beneath him.

After all these years, Shu, we still don't know how you do it; but we continue to admire you. We congratulate you on your unparalleled accomplishments and well deserved honors, we thank you for your friendship, leadership and tireless pursuit of excellence on behalf of us all, and most of all we wish you a very Happy Birthday. With fondest wishes to you and K.C. for countless more years of happiness and fulfillment,

Andrew and Deidre

With Deidre and Andrew McCulloch at their home, 1993.

Reminiscences from a Physiological Perspective

Paul C. Johnson, Ph.D.[*]

Dear Shu,

At this most important milestone in your life, may I congratulate you for all that you have contributed to the lives of others through your friendship and many kindness, as well as to the development of biomedical science as a researcher, scholar and mentor and as a spokesperson and leader in the field. It is difficult to say which of these many contributions is most important but it is safe to say that the depth and breadth of your contributions is quite unusual among scientists.

You and I have known each other for many years, beginning in the 1960s and I recall well your visit to the Physiology Department at the University of Arizona to lecture on blood rheology. While you have moved on to other areas, that work still forms the basis of our current understanding of the fundamental properties of blood, and my colleagues and I in the area of peripheral circulation and microcirculation consult it regularly for its many insights.

It has been a special pleasure for me to join this department as an Adjunct Professor and have the opportunity to continue to teach and do research. It has also been a pleasure to see first-hand the department grow under your leadership and take its place as one of the premier

---

[*]Adjunct Professor, Department of Bioengineering, UC San Diego.

departments in the field of bioengineering and in the Jacobs School of Engineering here at UCSD.

Genny and I have appreciated very much the warm hospitality that we have received from you and K.C. since coming to San Diego. It has helped to make our transition to a new university and new community quite pleasant and we are very grateful to both of you for your thoughtfulness.

Again, our congratulations and best wishes. You deserve all the recognition possible at this time for your many achievements.

Sincerely,

Paul Johnson

Gathering with Marcos Intaglietta, and Genny and Paul Johnson at Chiens' home, 1990. Background: the La Jolla Shore with a view of palm trees and the Pacific Ocean.

On the Occasion of Dr. Chien's 70th Birthday

Peter C.Y. Chen, Ph.D.[*]

In 1983, Geert and I were working on a blood cell separation project with a biomedical manufacturer. There was a problem and I needed to go back to New York for a technical meeting. Geert suggested that I discuss with Dr. Chien to seek some advice and guidance before talking to the company. I thought to myself, "Dr. Chien is a very famous and busy person, he does not know me and he will not see me." Nevertheless, as I wanted to see my good friends Herb, Paul and Amy, I decided to go to Columbia. I walked by Dr. Chien's office, noticed he was working very hard, as usual, and knocked. I introduced myself and before I knew it, Dr. Chien spent more than an hour explaining the red cell aggregation process. Then I thought to myself, "Dr. Chien is a very famous and busy person, and he is also very nice."

At one of the Microcirculation meetings, I decided to sneak out to do some sightseeing on the local tram. I stepped on the tram and noticed Dr. and Mrs. Chien sitting at one corner. I immediately moved to the other end and hoped they did not see me. I was totally embarrassed when I saw Dr. Chien walk slowly towards me. It was a relief when he handed me a camera and asked me to take a picture of him and Mrs. Chien. He even showed me what angle to use and what scene he wanted to include. I did not ask him what he was doing on the tram.

---

[*]Project Scientist, Department of Bioengineering, UC San Diego.

With Peter Chen, 2001.

We all know Dr. Chien's passion for photography. Each time I see him at a meeting or party he or Mrs. Chien will be holding a new camera. I am into new photo equipment myself and I try to keep up. At the last Department's party celebrating Dr. Fung's award (National Model of Science), I was taping his speech with a brand new digital camcorder. Once again I noticed Dr. Chien slowly moving towards me. This time I thought to myself, "He is coming over to admire my camcorder." But I was wrong. Dr. Chien came by and said, "Peter, I have one of those. Its small size is great for traveling, right?"

Dr. Chien, I plan to use a D-30 to take some pictures at your 70th birthday celebration and I hope you will be impressed, with the camera of course. I am pretty sure when you read this album again at your 80th birthday party you will say, "Peter was using a D-30? That's a dinosaur."

Dr. Chien, I am not qualified to comment on your achievements, your work and your dedication to work. But I like your hobby and I will continue to keep up.

Happy Birthday!

Peter

Leader of Bioengineering, Friend to Many

Thomas C. Skalak, Ph.D.[*]

Happy Birthday Shu! I wish you the time to do what you dream of, and to dream of ever new things.

Since we've known each other, our relationship has certainly evolved tremendously, and has always been a true pleasure and privilege for me. I think my memories go back to the time in the early 1970s when our families both lived in New Jersey, and you were a trusted friend and colleague of my father at Columbia University in New York. At that time, I was in high school, and I remember sometimes carrying drafts of NIH proposals and other materials by bicycle from our home in Leonia to your house in Englewood. That was a bit of a thrill, but not only for the reason that I was a courier carrying urgent material to a prominent professor — it was because Englewood had some excellent basketball players who were our dreaded opponents on the court. One in particular, Bill Willoughby, was a tall forward and the best player in the area, who in fact was the very first player in history drafted to the NBA directly out of high school. So, for me it was like riding through foreign territory, to arrive at your home as a safe house behind the enemy lines.

Later, my connections to you became stronger as I worked on my Ph.D. dissertation in microvascular biomechanics at UCSD in the early

---

[*]Professor of Biomedical Engineering, University of Virginia, Charlottesville, Virginia.

1980s with Geert Schmid-Schönbein, who had recently studied with you as a post-doc at Columbia. After I had worked as a faculty member for some time at the University of Virginia, your visits as a member of our advisory board in the 1990s were tremendously helpful to my career development and to that of our department at Virginia. Your remarks to our faculty group after just a single day's discussions were always remarkably incisive and helpful. One of your major observations that has helped our program to excel was how the engineering school and medical school cultures were different, and how we, our Deans, and other community members could unite the two schools into a great strength of the institution, which we have done. Thank you for that advice.

More recently, we had the fortune to serve together on several national society committees and boards, most notably this year as you were President of AIMBE and I was President of BMES. Your leadership of the entire discipline of bioengineering and your guidance of AIMBE during the initiation of the new NIH institute in 2000/2001 were historic services to the discipline, and will never be forgotten. What is amazing to me is that, in the midst of all that other critical activity, you were also willing to volunteer to remain as Chair of the BMES Publications Board past the end of your term, until your search

Barbara, Diane, Richard, Martha, Tom, Anna, and Steve Skalak, 1993.

committee identified a new editor for the BMES journal. Beyond simply being willing to serve, as always you were efficient and reliable, and produced the highest quality results. We are grateful to you for enabling BMES to consider several outstanding candidates and appoint Larry McIntire as our new journal editor. We hope that the journal will rise to new heights in this growing age of bioengineering, adding new cutting-edge focus areas and representing the discipline at the level it deserves, and you will have played a very key role in that evolution.

I have often turned to you for advice on the simultaneous pursuit of so many threads of a personal career, program teamwork, national service and family life. Because our work has been directed at somewhat related subjects, including recent work on shear- and stretch-mediated mechanotransduction events in vascular endothelial and smooth muscle cells, I have found your advice scientifically useful, and applicable in some detail to my own attempts to find time for many things. But the best and most profound advice I remember was not specific to a technical area or society commitment — in fact it was completely general. I'll call it Shu's advice: "Just do more." Like all useful advice, it is short and every word tells. Each word has real meaning: "just" means that we ought to simply get on with whatever is next, and proceed apace with both our dreams and responsibilities; "do" means that we should not mistake motion for action, but rather act with effect whenever possible; and "more" means that when faced with many worthwhile causes and commitments, one should try to make room for many of them — an area in which you, Shu, have set a very high example for all of us.

Doing more includes time for family and friends. You have always been such a gracious friend to my family, and now to my wife Susan (although you've not yet met our 4-week-old baby boy, Scott!). I wish you the time to spend with your own family, and to feel the breeze on the links of Torrey Pines a bit more often now, as you hone some new shots in your golf game. Finally, I still remember your frequent calls at the time of my father's illness, to check on his condition. To me, that sums up Shu Chien — because I sensed then the depth of your feeling, and your desire to still remain closely connected to the life of a long-time friend, while you knew that your dear friend was becoming too weak for visitors or even to speak. Thank you for that.

Shu, you have deeply touched the lives of many. Your contributions to the discipline of bioengineering will be enduring. God bless you and Happy Birthday!

Dr. Chien's Impact on My Career

Jeffrey H. Price, M.D., Ph.D.[*]

Dr. Chien has made and continues to make an important impact on my career development in Bioengineering and WIBE at UCSD. There were many occasions when I needed his advice, letters of collaboration or departmental support and recommendations critical for my career advancement. I cannot count the number of times I realized (often at

With Jeff Price, 2001.

---

[*]Assistant Research Scientist and Lecturer, Department of Bioengineering, UC San Diego.

the last minute) that my proposal needed a letter of support from him as chair of the department or director of WIBE. In spite of his mindlessly busy schedule, he somehow not only found the time to write these letters, but also added positive comments and suggestions for improving my chances of funding. It's difficult to convey how much of both an emotional and substantive lift this gave to my grant writing and research efforts. His positive outlook and energetic encouragement have been remarkable and he is an important reason UCSD Bioengineering has been such an overwhelmingly positive experience for me.

A Tribute to Dr. Shu Chien

Richard Lieber, Ph.D.*

I arrived at UCSD in 1982 during the "reign" of Y.C. Fung. Could anyone follow Dr. Fung in terms of stature, influence and determination? I thought not. However, I was naïve, not knowing the person of Shu Chien. Who was this man being recruited to La Jolla and how would he impact our precious program in Bioengineering? Those of us who were "raised" here felt very protective and, I sheepishly admit, skeptical.

With Richard Lieber at the celebration symposium, 2001.

---

*Professor, Departments of Orthopaedics and Bioengineering, UC San Diego.

I could not have been more pleased as I watched the new legacy unfold. I observed tremendous vision, compassion and inclusion in this man's leadership style. A quiet confidence. A kind word. A skillful analysis. Such traits are the traits of Kings, not Professors. Leadership. Style. Grace. Command. However, in spite of these skills, it was the man himself that everyone seemed to follow. The character of the man. As the architect of the famous Gulf War stated so eloquently, "Leadership is a combination of strategy and character. If you must be without one, be without the strategy." (Gen. H. Norman Schwarzkopf). Now I know how Bioengineering has flourished at UCSD for the past 35 years. These two men — Fung and Chien — poured from the same mold — have raised the bar to a new standard. We have watched them lead and we have been blessed to live in their reflected glory.

At the Top of My List

Michael W. Berns, Ph.D.[*]

Shu, I have known you for only a short two years. But in this time, I have experienced a person who ranks at the top of my list. I see not only your fierce commitment to excellence in research, but also in two other key areas — mentoring students and colleagues, and in dedicating yourself to establishing biomedical engineering as a discipline of high stature and impact.

Your research has literally shaped the field of cellular biomechanics. The insightful research on cellular shear forces has provided the seminal foundation for understanding key elements of vascular physiology and cell behavior.

But even as important as the far-reaching research results that have emanated from your lab over your long and still enormously productive career, is what I have personally observed in your mentoring style. Shu, you have graciously allowed me to have an office and lab in the midst of the high-energy activity of your research group. Thus, I am both an "outsider" as well as a person on the "inside" who really sees what happens on a daily basis. What impresses me is that you always take the time to meet both as a group and individually with each of your team. The one-on-one time that you provide is impressive — especially to those of us who really know how busy you are, and how many other "projects" you are involved in. The reward, of course, is that your colleagues are totally dedicated to both your high

---

[*]Arnold and Mabel Beckman Professor; Chairman and CEO, Beckman Laser Institute; University of California, Irvine.

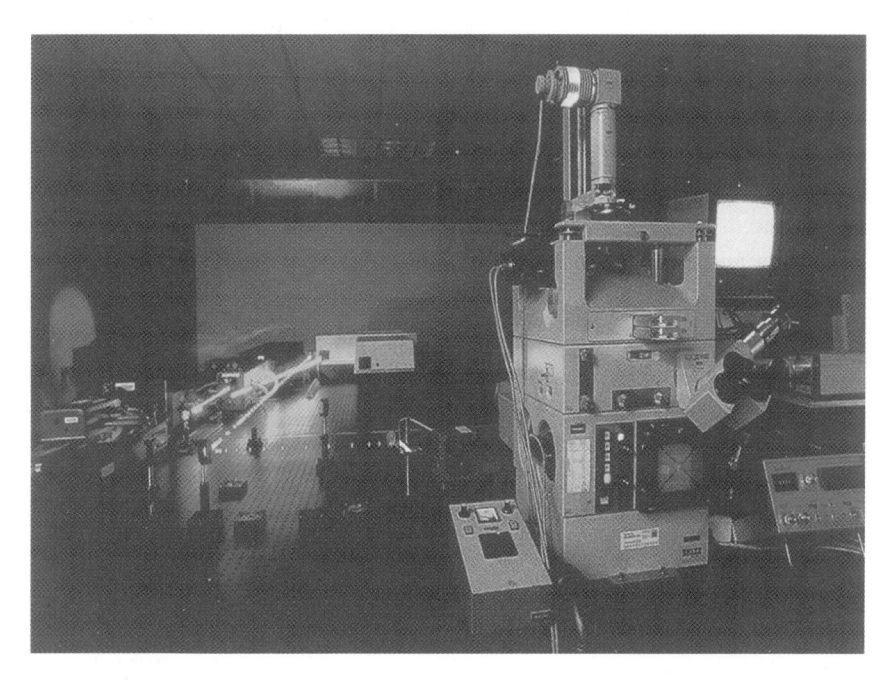

Laser trap and scissors system in Michael Bern's laboratory.

With Michael Berns, 2001.

standards and ideas, and to you personally. Shu, this is rare today, and it is a virtue that we all appreciate and try to emulate.

Of course, your efforts on behalf of bioengineering are legendary. You built the top academic department and Whitaker Institute in the country, in the exciting and highly competitive environment of UCSD. But you also spearheaded the national effort to bring bioengineering to the "next level" of prominence and credibility. Your creative energy, perseverance and "bio-political" talents have ideally positioned bioengineering on the doorstep of the 21st century. For that, as well as for being a role model for all of us — we are eternally grateful. Shu, I am proud to be even a "peripheral" member of your team.

Bringing Life to Bioengineering

John Watson, Ph.D.[*]

Critical times in world history have produced men and women of astonishing vision and energy. The last ten years have witnessed Bioengineering as a rapidly growing field of immense opportunity without a national research support infrastructure to optimize its potential. My friend, Shu Chien, is one of these rare individuals of vision and energy who sensed that bioengineering was in "the" crucial period of its young history. He accepted the leadership baton (as necessary) to benefit bioengineering, running the gauntlet without any thought of personal recognition or the impact on his personal time. Shu stepped into the breech, beginning to bring bioengineering to "life" and longevity using his keen understanding of existing research support processes and a vision for integrating bioengineering as a full partner within the federal research system.

The series of his extraordinary roles began with the 1993 Congressional legislation requesting a comprehensive national study of "Support for Bioengineering Research (1)." He served on the External Consultants Committee, charged to study the issues raised in the Congressional language and to make recommendations to the National Institutes of Health (NIH). As a result of Consultants' report and the

---

[*]Director, Clinical and Molecular Medicine Program, Division of Heart and Vascular Diseases, National Institutes of Health, Bethesda, Maryland. Adjunct Professor of Bioengineering, UC San Diego.

NIH report to the Congress (2), the Bioengineering Consortium (BECON) was established by the then NIH Director, Dr. Harold Varmus (3). This was done because of the recognition of the potential benefits to human health that could be realized from applying and advancing the field of bioengineering. While the scientific and engineering communities universally endorsed BECON, investigators proposing design research and technology development remained concerned that they did not have access to a peer-review system.

Shu was then called to serve in two different national advisory roles addressing these peer-review questions. He began service on the Huntsman's Committee providing bioengineering peer-review recommendations to the NIH Center for Scientific Review (CSR) (4). He quickly became a key member and the writing engine for this committee. The Secretary of Department of Health and Human Services also recognized Shu's talent and appointed him to the CSR National Advisory Committee. On this Committee, he is providing leadership on peer-review policies generally and bioengineering specifically. Of particular importance is the Board's oversight of the implementation of the Report of the "Panel on Scientific Boundaries for Review," chaired by Bruce Alberts, President of the National Academy of Sciences

With John Watson after his seminar at the San Diego Supercomputer Center, 1999.

(5). The implementation of this report holds the keys for impedance-matching peer-review with bioengineering research interests.

BECON quickly became the National focus for encouraging bioengineering research. The first three BECON years have witnessed dramatic growth for bioengineering research, increasing at twice the rate of the percentage growth of the NIH budget. However, Congressional leaders noted that BECON lacked permanency and resources to optimally foster bioengineering research, although under the able leadership of Dr. Wendy Baldwin, NIH Deputy Director for Extramural Research. Legislation developed over the last two years, was signed by former President Clinton, 29 December 2000, as his last official act calling for establishing the National Institute of Biomedical Imaging and Bioengineering (NIBIB) at the NIH (6). Health and Human Service's Secretary Thompson officially authorized NIBIB, 20 April 2001.

For the good fortune of Bioengineering and all of this legislative activity, Shu had just ascended to the Presidency of the American Institute for Medical and Biological Engineering (AIMBE). AIMBE worked with the Academy of Radiology Research (ARR) to assist in the passage and signing of the House Bill (H.R. 1795), which also passed the Senate. In early January 2001, AIMBE was again asked to attend high-level meetings including the Acting NIH Director, Dr. Ruth Kirschstein, and the ARR to discuss the mission and future of the New Institute.

As AIMBE President, Shu was the perfect representative and negotiator for Bioengineering. His reputation as an outstanding scientist and engineer, natural leader and courteous negotiator was well known and respected by all parties. As usual, he was fully prepared with a complete command of the issues and options for success. On behalf of AIMBE, he drafted a letter for submission to *Science Magazine* that was also supported by the NIH and the ARR. The letter had inputs from the AIMBE Board and was co-signed by Dr. Doug Maynard of the ARR and appeared in Science last March (7). The discussion by all parties surrounding the formulation of this letter in part guided the NIH in designating the NIBIB mission.

Yes, Shu is a man for all the bioengineering seasons. He was in the right place, at the right time to bring "life" to bioengineering history. Shu, thanks for all your efforts, for all of us.

REFERENCES

1. Nerem, R.M. (Chair), Taylor, K.D. (Vice-Chair), Arnold, F., Chien, S., Katona, P.G., Littell, C. and Young, W.D. *Support for Bioengineering Research*. Prepared for the National Institutes of Health by the External Consultants Committee, 1994.
2. Watson, J.T. (Chair), Cassman, M., Coleburn, T.R., Hausman, S.J., Huerta, M.F., Medgyesi-Mitschang, S., Postow, E. and Vaitukaitis, J. *Support for Bioengineering Research*. Department of Health and Human Services, Public Health Service, National Institutes of Health, 1994.
3. Varmus, H. *Bioengineering Consortium (BECON) Charter*. February 1997.
4. Huntsman, L.L. (Chair), Chien, S., Davis, R.W., Griffith, L.G., Hendee, W.R., Henry, S.A., Hubbell, J.A., Koonin, S., Phillips, W.M. and Engel, L.W. *Expanding Opportunities*. Report of The Working Group on Review of Bioengineering and Technology and Instrumentation Development Research, 1999.
5. Alberts, B.M. (Chair), *et al. Recommendations for Change at the NIH's Center for Scientific Review*. Phase 1 Report Panel on Scientific Boundaries for Review, 1999.
6. Public Law 106–580. *National Institute of Biomedical Imaging and Bioengineering Establishment Act*, 2000.
7. Chien, S. and Maynard, C.D. Newest member of the NIH family. *Science* 2: 291(5509): 1701–1702, March 2001.

Please note Items 1 to 6 are available on the NIH BECON website!

An Appreciation

Larry V. McIntire, Ph.D.[*]

Dear Shu,

Best wishes on your 70th birthday! We all look forward to celebrating the next 70 years with you. Your work has been a continuing inspiration. I remember first reading your papers on the importance of blood rheology in human physiology and sickle cell anemia pathophysiology. The important series of papers with Dick Skalak on red cell membrane mechanical properties and then progressing to leukocyte individual cell viscoelastic properties are a lasting contribution to our basic understanding of the importance of individual blood cell rheological properties in vascular physiology. Your continuing collaboration with Shelly Weinbaum has greatly extended our understanding of the mechanistic basis of the disease processes of atherosclerosis. Your recent research on the mechanical regulation of gene expression in endothelial cells, particularly elucidating the molecular signal transduction pathways involved, is a model for combining bioengineering and modern molecular biology to answer important fundamental questions in vascular biology.

While doing all of this research, your contributions to the greater development of bioengineering and biosciences at the national and international level have been truly amazing. Serving as president of the American Physiological Society, the Federation of American Societies

---

[*]E.D. Butcher Professor and Chair, Department of Bioengineering, Institute of Biosciences and Bioengineering, Rice University, Houston, Texas.

182 *Tributes to Shu Chien*

Kroc Foundation Meeting, 1983 (Larry McIntire is at the center of the last row).

for Experimental Biology, forming a new Institute of Biomedical Sciences in the Academia Sinica in Taiwan, and most recently guiding AIMBE in pushing to a successful conclusion in establishing a new Institute for Biomedical Imaging and Bioengineering within the NIH are examples of your unselfish willingness to contribute your enormous talents and efforts to the greater good of our collective advancement in bioengineering and biosciences.

Knowing you has been a joy and an inspiration. I look forward to your continued active involvement in all of our causes, both scientific and political. Wish best wishes.

Sincerely yours,

Larry

A Tribute

Don P. Giddens, Ph.D.[*]

A hearty congratulations on your 70th birthday, and best wishes for many, many more ahead! You have been a major figure in bioengineering, especially with helping to move the field more solidly into the biological arena. On the "political" front, your tireless efforts in working to create the new NIH Institute are also greatly appreciated. We all owe you a great debt, and we look forward to your continued leadership.

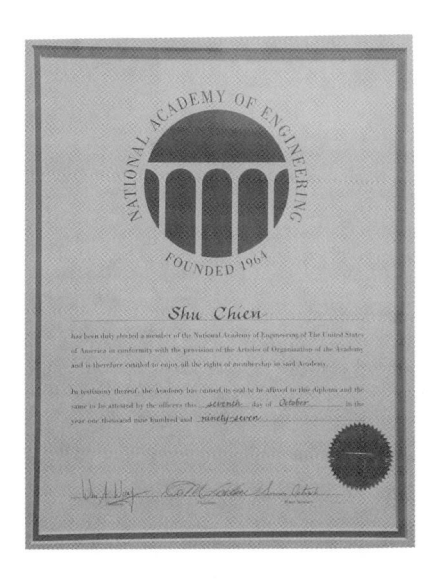

---

[*]Lawrence L. Gellerstedt, Jr. Chair in Bioengineering, Wallace H. Coulter Department of Biomedical Engineering, Georgia Institute of Technology and Emory University School of Medicine.

Redwoods are Tall, Mt. Everest is Big, and Sunsets are Beautiful

Peter G. Katona, Sc.D.[*]

Dear Shu,

If it had not been for certain indirect signs, such as your completing your term of chairing your department, I would have never known that you were approaching your 70th birthday. You can out-think, out-produce, out-travel, and out-stand almost anyone decades younger.

Congratulations!

---

[*]President, The Whitaker Foundation, Rossyln, Virginia.

So, I have two wishes for your 70th birthday. The first is for you. In this, Jane joins me in wishing you a great birthday, with hopes and expectations of very many happy returns. The other wish is for me. May I be as active and vigorous at 65 as you are while turning 70.

With congratulations and warmest personal regards.

P.S. Some time ago I was asked to write a letter concerning you. The attached pictures illustrate my comment: "Writing such a letter is like arguing that redwoods are tall, Mt. Everest is big, and sunsets are beautiful."

Visit by Whitaker Foundation to Powell-Focht Bioengineering Hall at UCSD, 2003. From left: Bernhard Palsson, Geert Schmid-Schönbein, Peter Katona, Portia Whitaker Shumaker, Ruth Whitaker Holmes, Burtt Holmes, Jack Linehan, and Shu Chien.

The Microcirculatory Worker's Microcirculationist!

Aubrey E. Taylor, Ph.D.[*]

Dear Shu,

I have been told that you will celebrate your "70th" birthday on 23 June 2001. On that day you will host a fantastic, modern symposia entitled "Integration of Biorheology and Molecular Biology into Physiology and Bioengineering." Unfortunately, I will be in Europe at that time and cannot be there to help you celebrate such an outstanding milestone in your life, i.e. living to age 70. But more important than age is the legacy of your outstanding research accomplishments in microcirculation and bioengineering, your training of numerous students who are still active in your disciplines and your service as President of almost all organizations to which you have belonged!

I, of course, first became aware of you through John Prather, one of my students in Jackson, who did a postdoctoral fellowship at San Diego with Drs. Zweifach and Intaglietta. I also lost the election in San Diego for President Elect of the Microcirculatory Society to you many years ago by one vote! Later, you became President of the APS, and also served as FASEB's President in 1993. In each instance you were instrumental in substantially changing each society by altering their scientific emphasis, which allowed them to develop more modern

---

[*]Professor and Chairman, Department of Physiology, University of South Alabama, Mobile, Alabama.

APS Executive Committee, 1991 (Shu, Aubrey Taylor, Vernon Bishop and Martin Frank).

approaches and to adapt their science to the rapidly changing scientific world. As you know, I worked with you on the MCS, APS and FASEB as these societies were totally reorganized and redirected to enter the third millennium allowing them to grow in the process and lead our disciplines to greater excellence within the worldwide scientific community. I also remember as President of MCS I was supporting a motion to increase the yearly membership dues. I allowed a discussion to occur, even though Dr. Johnson had made another motion to not discuss this item. Well, as you know, the discussion went on for hours, or at least it seemed so. Finally, we voted and the motion passed and dues went from $15 to $25 dollars a year. Afterwards you told me something very special. You said "Aubrey, when you are in a leadership position and someone of Paul Johnson's stature recommends something, pay attention, as their idea will very often simplify a difficult situation." I've used this advice throughout my career and have many times used my lesson from the awkward situation above to emphasize how to not do things.

I know of no one who has produced such a positive impact in Microcirculation, Physiology and Bioengineering in the world as you. I congratulate you on your 70th birthday but most of all wish to thank you for being such an outstanding role model for me and especially for all of my many students and collaborators. You are certainly a scientific treasure and I wish you Godspeed and good health as you continue your outstanding career into your 70s in the years of the snake!

Finally, I know you could not have accomplished so much without your supporting wife, so give K.C. my regards and thank her for allowing you to give so much of your life and effort to our sciences.

Life in Physiology

Martin Frank, Ph.D.[*]

Dear Shu,

Sorry I can't join you today on the celebration of your "Life in Bioengineering." As a man devoted to family, you can understand why I could not attend.

Recognizing your contributions to the bioengineering program at UCSD, I can understand why the celebration of your 70th birthday might focus on bioengineering. However, for your colleagues at APS, the celebration should focus on your "Life in Physiology." After all, it was through your leadership that the Society made the transition into molecular biology and subsequently into physiological genomics. You provided the vision that helped to shape the future of the Society.

As a result of your presidential leadership, many of us learned to appreciate the beauty of Taiwan and the many scientific contributions that were being made by physiologists in that country. The visit helped to start many scientific collaborations and friendships.

Shu, it has been a pleasure working with you in my role as APS Executive Director. You have been a friend and mentor, demonstrating to me that the science of physiology cannot be stopped by international borders. Through your efforts at APS and IUPS, you have promoted international collaborations and interactions for the benefit of humankind.

---

[*]Executive Director, American Physiological Society, Bethesda, Maryland.

190 Tributes to Shu Chien

My parents taught me that age is relative, the actual number is immaterial, it is how you feel and the contributions you are making to society. While we celebrate your 70th birthday as a milepost along the journey we call life, it must be noted that based on your vitality and

中美生理學聯合研討會
APS-CPS JOINT MEETING
November 2-5, 1990, Taipei, Taiwan, R.O.C.

American and Chinese Physiological Societies Joint Meeting, IBMS, Taipei, 1990. APS and CPS Council Members (Left to right) First row: Tsai-Hsin Yin, Vernon Bishop, Shu Chien, Tze-Kong Yang, Eminy H.Y. Lee. Second row: Hsheng-Kai Lee, Chau-Fong Chen, Chok Yung Chai, Beverly Bishop, Hsing-I Chen. Third row: H.S. Fang, Mao-Tzun Lin, Shou-Teh Chiang, Martin Frank, Yun-Lai Chan, David Ramsey. Fourth row: Hung-Jung Liu, Jon-Son Kuo, Wann-Chu Huang, Danny Ling Wang.

scientific contributions, it is a milepost early in your journey. There is still much to be done and contributions to be made to society. Consequently, I look forward to future opportunities to celebrate your journey with stops at 80, 90, 100 and beyond.

If I were present, I would lift my glass of wine and say "L'chiam, to life." May you and K.C. have many more happy, healthy and productive years together.

With friendship and admiration,

Marty

Establishing a Standard

Michael J. Jackson, Ph.D.[*]

Dear Shu,

I would like to extend to you my congratulations and best wishes on the occasion of the Symposium celebrating your 70th birthday. I regret that I am unable to be present for this landmark event. I know your colleagues and students attending the symposium will appropriately describe your work in bioengineering, but I would like to take advantage of this opportunity to highlight your contributions more broadly to science and the scientific community. I refer, of course, to your service as a Board Member and President of the Federation of American Societies for Experimental Biology.

Following the reorganization of FASEB, you were the first Society representative to become President in an open election, a position you assumed in 1992. One of your colleagues referred to this period of FASEB history as a perilous time, in which the Federation sought to define its new priorities with much reduced resources and among conflicting expectations of the members. Accepting the position at this time was an act of considerable personal courage, and we were fortunate in being able to call on your diplomatic skills. I recall particularly well, one meeting in which you skillfully steered Board members from threatened resignation to unanimous acceptance of the annual operating budget. Your term as President established a

---

[*] Executive Director, Federation of American Societies of Experimental Biology (1990–1999).

With Michael Jackson at Experimental Biology Meeting in New Orleans, 1993.

standard for civility and collegial conduct at Board meetings that carried over and flourished in the terms of your successors.

The reorganization plan called for enhanced activity by FASEB in the areas of advocacy and representation for biomedical research. Recognizing that effective advocacy must be built on a solid foundation of information; you were successful in persuading the Board of the need for a well-founded program of research and analysis. This was embodied in the Office of Policy Analysis and Research, and activities in this area remain a hallmark of the FASEB advocacy effort. I think it may be characteristic of your work that you not only made the program feasible, but you participated in some of the early work. Indeed you co-authored one of the first studies resulted from this program, a study to determine if the benefits of biomedical research could be evaluated quantitatively.

Another expectation of the reorganization was the recruitment of new societies to membership in the Federation, and the presence of the Biophysical Society and the American Association of Anatomists in the FASEB coalition can be directly attributed to your efforts.

In summary, the impressive growth and development of the FASEB organization in the last decade has roots that were established under your leadership, and the scientific community owes you a debt of gratitude for undertaking this important work.

Sincerely,

Michael J. Jackson, Ph.D.

Shu Chien's Contributions to FASEB and Public Affairs

Howard H. Garrison, Ph.D.[*]

Shu Chien joined the Board of Directors of the Federation of American Societies for Experimental Biology (FASEB) at a very crucial period in the organization's history. In 1989, the FASEB Board held a retreat in Williamsburg, Va. to assess its activities and structure. Decisions made at this historic meeting changed the Federation and its mission. As a new board member, and subsequently as president (1992–1993), Dr. Chien helped to turn FASEB into a dynamic and effective public affairs organization. His leadership helped shape the climate for medical research in the United States, and his contributions to science policy helped ensure that the views of bench scientists would be influential in decision-making circles.

As one of the first "Post-Williamsburg" presidents, Dr. Chien helped to build a volunteer-led public-policy program that represented the individual scientist. He had the perfect leadership style for an evolving organization. He led with firmness, yet was patient and tolerant of diverse points of view. Under his guidance, new structures were established. To help implement the new policy of service to member societies, Dr. Chien traveled to meet with eight of the nine FASEB society councils. As president, he established the Executive Committee to expedite decision-making and initiated FASEB's program of policy

---

[*]Director, Office of Public Affairs, Federation of American Societies for Experimental Biology, Bethesda, Maryland.

"S. Chien, FASEB President, 1992–1993."

research by commissioning an economic analysis of the benefits of basic biomedical research.

In the public affairs arena, Dr. Chien helped to create the activist tradition that subsequent FASEB presidents have adopted. In meetings with advocacy partners, members of Congress and administrative officials, he was a tireless and effective advocate for bench scientists and for research. He launched a successful petition drive to boost funding for the National Institutes of Health, collecting more than 20,000 signatures and delivering them to the petitioners' congressional representatives. Dozens of letters were sent to members of Congress and key administration officials in support of research funding.

Those of us who were fortunate enough to work with Dr. Chien were inspired by his dedication and devotion to FASEB, to science, and to medicine. He was always available to respond to an emergency, no matter where he was or when it happened. Even when he was in Taiwan he never missed a beat, chairing conference calls in the middle of the night and responding to an endless stream of faxes.

His warm and inspiring relations with the FASEB staff helped to foster a climate of camaraderie and respect. Dr. Chien's thoughtful and enlightened leadership built the foundation for FASEB's growth as an organization devoted to public affairs. An organization on the verge of dissolving became a strong and coherent force. The game plan that was established for nine societies and 40,000 members became the foundation for a thriving coalition of 21 societies with more than 60,000 members.

Dr. Chien, we are grateful for all of your contributions to FASEB and your service to the research community. Your friends and colleagues at FASEB congratulate you on your 70th birthday.

What You May Not Know About Shu Chien

Kevin W. O'Connor[*]

What a great honor it is to salute my mentor and friend Shu Chien on the occasion of his 70th birthday! As Executive Director of the American Institute for Medical and Biological Engineering (AIMBE), I learned much from Shu during his tenure as our President and member of our Board of Directors. One of the many highlights of Shu's career was his role in founding AIMBE, which in its ten-year history has become recognized as the umbrella organization for medical and biological engineering. AIMBE now represents 650 premier engineers and scientists in its College of Fellows, 15 societies with more than 32,000 individual members, 70 academic institutions (including UCSD), and an industry council. Shu is a Founding Fellows of AIMBE, a list which includes many of the pioneers in the medical and biological engineering field. During his AIMBE presidency (2000–2001), Shu played a leading role in advocating the creation of the National Institute for Biomedical Imaging and Bioengineering at the National Institutes of Health. Shu's leadership was crucial to the creation of the new Institute, which was signed into law this past December by President Clinton.

We all know that Shu Chien is a great man, a superb scientist, a true leader, and a gracious soul. I suspect others writing tributes here

---

[*]Executive Director, American Institute for Medical and Biological Engineering, Washington, D.C.

Shu Chien can dance, and so does K.C.

will describe in more detail these aspects of Shu Chien. On the occasion of his birthday, however, let me tell you something about Shu Chien that you might not know.

Shu Chien can dance.

At the World Congress on Medical Physics and Biomedical Engineering last summer in Chicago, one of the major social events was a dinner cruise on Lake Michigan. After a delightful meal while cruising along the sparkling Chicago skyline, the band began to play. Absolutely nobody got up to dance, until Shu Chien jumped up, grabbed his lovely wife K.C. by the hand and declared, "Let's dance!" Within two minutes, scores of others on deck followed, and the dance floor was jammed. Let the record show that Shu danced every dance. The night was a hit. Even on the dance floor, the man knows how to lead!

My life has been enriched by knowing Shu Chien. Happy birthday, Shu. Many happy returns, and many more dances, I wish for you!

Tribute to Shu Chien

Wendy Baldwin, Ph.D.[*]

We at the National Institutes of Health are greatly indebted to the scientists and engineers who have willingly given their time and talent to develop and sustain new programs in bioengineering. A great institution can thrive only when high caliber people are willing to think beyond the confines of their own work and seek a larger vision.

Shu Chien has given generously of his time to help us see the future of bioengineering and define the role that the NIH can play. But the development of a national agenda requires more than just an understanding of the science. It requires an ability to communicate that vision to many different audiences; to work with many varied types of constituencies; and to refocus continuously on the bigger picture. Shu has demonstrated over and over his generous willingness to help the NIH, the scientific community and the field of bioengineering. I applaud and thank him.

---

[*]Deputy Director for Extramural Research, National Institutes of Health, Bethesda, Maryland.

Wendy Baldwin speaking at groundbreaking ceremony of Powell-Focht Bioengineering Hall at UCSD, 2000.

Working with Dr. Chien

Song Li, Ph.D.[*]

While I was packing my books and files in my office, preparing for my move to Berkeley in December; I found several papers given to me by Dr. Chien on my first day in UCSD seven years ago. I still remember some of our conversations in Dr. Chien's office and in the lab on that day. Everything seems to have happened yesterday.

For more than seven years, Dr. Chien has been my Ph.D, postdoc and research advisor, and I matured academically with his guidance. He always encouraged us to be creative, and is committed to educating the next generation of bioengineers. He has a strict scientific approach, and we often learned from his rigor and scrutiny in our group meetings. Sometimes Dr. Chien spent hours helping us to go through a presentation, from scientific aspects to presentation skills. His corrections and comments on our manuscripts and theses were very detailed, and he usually gave us speedy feedback. Frequent traveling did not prevent him from communicating with us. On the contrary, he somehow managed to get more things done. He worked on manuscripts at airports and on the plane, and we often received his faxes and e-mails from hotel rooms. Dr. Chien is such a cheerful and youthful person. I remember that in my graduation party, we were playing soccer and Dr. Chien joined in the game. His skill, flexibility and energy made him a challenging player. Dr. Chien has also been recognized as the "Most Valuable Player" at work in our department. As the founding chair, he put his heart and efforts into the organization and development of our

---

[*]Assistant Professor, Department of Bioengineering, UC Berkeley.

With Song Li at his Ph.D. commencement ceremony, UCSD, 1997.

department. With his leadership, we experienced and enjoyed the birth and growth of our new department in the past few years.

Commitment, collaboration, communication... them Dr. Chien practices these principles in his daily work, and we learned them when working with him. It is so fortunate that I have him as a mentor, a role model and a friend. From the colorful San Francisco bay area, I send my best wishes to Dr. Shu Chien.

Happy Birthday, Dr. Chien!

Http://www.chienlab.com

Julie Y.-S. Li, Ph.D.[*]

Dr. Chien,

It is such an honor for me to be in the chienlab.com. The market has shown a recent influx dot com companies, amongst which chienlab.com has demonstrated its outstanding performance in the hi-biotech field. Not only are there significant achievements gained in the present and past; there is also a strong potential for the future. With its rapid expansion rate, chienlab.com is and will be the giant in the field.

With Shunichi Usami and Julie Li, 2000.

---

[*]Assistant Project Scientist, Department of Bioengineering, UC San Diego.

The reason behind the growth of the group is the BOSS, Dr. Chien. You are a true leader, teacher, scholar, visionary, executor, and most importantly, a caring person. You possess the vision of the bioengineering research direction, have the ability to guide it into the area of integration of all scientific disciplines, and parade into the challenge of the new millennium. However, as busy as you are playing the role of superman, you have showed extraordinary caring spirit to all the people around you. You respect all their differences, you honor the collaborations, and you cherish education. During the years of my association with you, you constantly amaze me with your energy and generosity. I have learned so much from you, there are no words that can describe my appreciation. I can only simply say thank you very much. Your legend will be passed onto the next Chien generation.

On this great day for celebration, Happy Birthday to You.

Happy 70th Birthday

Gerard Norwich[*]

Dear Dr. Chien,

Words are inadequate in expressing my thoughts and the kind wishes I would like to extend to you on your 70th Birthday celebration. Please accept my deepest thanks for your kindness and the gracious hospitality that you and your family have shown me.

Best Wishes and Joyous Birthday!

Jerry

With Jerry Norwich, 2001.

---

[*]Staff Research Associate, Department of Bioengineering, UC San Diego.

A Tribute

Ellen Tseng[*]

Dear Dr. Chien,

After all the moving around between the United States and Taiwan, we have settled in a city one hundred miles away from San Diego. Of all the years living in the U.S., 1993 is the most memorable. That was the year that Ham-min worked under a fellowship in your lab. Having

With Ellen, Joshua, Ham-min and Wendy Tseng, 2000.

---

[*]Ellen Tseng is Mrs. Ham-min Tseng. Ham-min Tseng, M.D., was a postdoctoral fellow in Shu Chien's laboratory and his father was Shu Chien's classmate at National Taiwan University College of Medicine.

never lived in America before, the first year was very difficult. However, you and Mrs. Chien made those days very special.

I appreciate those days in which you looked after us and made us feel at home. The first day we arrived in the U.S., Mrs. Chien helped us to find an apartment and open bank accounts. After the endless paperwork, she brought us a bag of food. The aroma still remains in my heart. You lent us your furniture, which relieved quite a bit of our financial burden. Even though many years have past, I can still feel the warmth in my heart.

I'll never forget your friendship. When we visited you, you were always smiling and welcoming us. We enjoyed relaxing in your backyard and seeing pictures of your family. The parties at your house were interesting. I always feel that a Chinese old saying suits you: The humbler you are, the nobler you appear.

We idolize your wisdom and your intellect. Your dedication to molecular biology benefits generations, including Ham-min. I also enjoy your elegant command of English. I am glad to know you.

Because it is your birthday, it is the right time to express our appreciation. We appreciate the things you've done for us, and we respect the career you have built and your wisdom.

Happy Birthday!

Ellen Tseng

A Tribute to Mark Dr. Chien's 70th Birthday

Gang Jin, Ph.D.[*]

It is my honor and privilege to write a birthday tribute to Dr. Shu Chien, who is and always will be one of the most respected men in my life. People told me that a Ph.D. advisor would follow you throughout your career; I now realize that this is so true. I hope my entry in the tribute book for Dr. Chien's 70th birthday tells Dr. Chien, my family, and the world how much he means to me, to other students, and to the academic world.

I know I speak on behalf of many when I present the following survey of his extraordinary scientific achievements and honors leading up to his 70th birthday.

(1) 43 years of professorship at Columbia University (since 1958) and University of California, San Diego (since 1988).
(2) 378 original scientific articles in peer-reviewed journals, and the editor of nine books.
(3) Ten professorship and directorship appointments. He was the Director of Circulatory Physiology and Biophysics at Columbia University, Chair of the Department of Bioengineering at University of California, San Diego, and is currently the Director of Whitaker Institute of Biomedical Engineering at the University of California, San Diego.

---

[*]Director and Principal Scientist, Purdue Pharma Ltd., Irvine, California.

(4) President and chairman of 18 national and international well-known scientific societies and organizations. He was President of the Federation of American Societies for Experimental Biology with 40,000 members, and President of The American Institute for Medical and Biological Engineering.
(5) Over 50 high level honors and awards such as a Member of the Institute of Medicine, the National Academy of Sciences since 1994, and a Member of the National Academy of Engineering since 1997.
(6) Many areas of research interest. His well-known interests are blood rheology and microcirculatory dynamics in health and disease, and effects of mechanical forces on gene expression and signal transduction. His current major interest is microarray/bioinformatics applications in bioengineering.
(7) He has given both his personal and professional guidance to numerous national and international scholars and outstanding students.

Dr. Chien is a very versatile man and is a prolific contributor in many areas. His national and international reputation is not limited to his professional achievements; he is also highly respected by everyone for his kind personality. He is so much more than an academic giant. He is a man with such an outgoing and generous personality that it comes as no surprise that he is well liked by all. He never pushes his people. With his personal charisma, he impressively inspires his students, scholars and colleagues with his scientific concerns. Although he is always busy, he can still find time to provide valuable advice, from project design to completion. He always revises our manuscripts, abstracts and theses over and over with his careful and thoughtful corrections. His revisions are so perfect and enlightening that I have kept all the manuscripts with his original corrections. He not only oversees all of the scientific work of his people, but is also highly supportive in helping to build their careers. He is an excellent teacher and professor and also a keen student of his disciplines with a strong attitude towards learning new things. He never turns people away, although he is continually telling himself he will learn to say "no."

He is a great professor in his disciplines, a great leader for his organizations, a great friend to his friends, a great father to his daughters, a great husband to his wife, and an outstanding advisor to his students. It has been my great fortune to be his Ph.D. student and postdoctoral fellow for many years. Even though I have left his lab, he still provides me with scientific advice and personal support in my career. Below is a list of the many examples of Dr. Chien's constant support for his students such as myself, which I will remember throughout my life.

(1) The first time I received Dr. Chien's direct guidance was in the Fall Quarter of 1993 when I was taking his course in cell and molecular biology at the University of California, San Diego. I was amazed by his excellent teaching presentation, and clear lecture notes.

(2) I started doing my thesis project in his lab in the Summer Quarter of 1995 after I passed my Ph.D. qualifying exam. I really appreciated Dr. Chien in assigning me the cutting-edge project, gene therapy to prevent cardiovascular diseases such as restenosis. By working on this gene therapy project, I had the chance to learn a series of lab skills from *in vitro* to *in vivo*, from non-viral to viral, and from cellular to molecular.

(3) I became his official Ph.D. student in 1997 after we lost Dr. Richard Skalak, my former Ph.D. advisor and also a world famous professor in bioengineering, who collaborated with Dr. Chien for three decades. Dr. Chien was kind enough to take care of my Ph.D. senate exam on behalf of Dr. Skalak at the time when Dr. Skalak was suffering from cancer. After I passed the senate exam, Dr. Chien immediately went to Dr. Skalak's home to bring him the good news and also asked for his approval signature on my exam.

(4) After Dr. Skalak passed away, Dr. Chien made a smooth transition for me to continue my thesis project in his lab until I completed my research for the thesis at the beginning of 1999.

(5) I have always benefited from Dr. Chien's dynamic interests in cutting-edge studies. Although he has established many research fields in his life, he still keeps capturing new research topics

from conferences and meetings. The most important issue that has directly influenced my job career is that he asked me to undertake the initial development of the microarray technology in his lab in 1998. He encouraged me to take the course of chip fabrication in the Department of Electrical and Computer Engineering. He always encouraged me to attend many workshops and conferences for training in microarray and bioinformatics.

(6) After I received my Ph.D. study in bioengineering, I was most fortunate to become Dr. Chien's postdoctoral fellow in March of 1999. I greatly appreciated the microarray project that he assigned to me. Since then I have had the great opportunity to be involved in every aspect of microarray technology, including array fabrication, array hybridization, array image scanning, and array data analysis. Mastering the integrative skills in microarray made it possible for me to obtain my first job position as the Director and staff scientist of the Functional Genomics Center (Microarray Center)

Professor Shu Chien hooded Gang Jin during the Ph.D. commencement ceremony at UC San Diego, 1999.

at the Salk Institute in San Diego. Without Dr. Chien's strong recommendation to the Salk's faculty committee, it would have been impossible for me to acquire such a tremendous job opportunity. I understand it was difficult for Dr. Chien to make this recommendation at that particular time due to the consequence of losing a well-trained array person in his own lab. When I finished setting up the microarray facility at the Salk Institute and made it fully functional for the Salk's investigators, Dr. Chien extended his kind support by helping me to obtain my current position as the Director and Principal Scientist of the Genomics Lab at the well-established pharmaceutical company, Purdue Pharma Ltd. in Irvine, which gives me a great opportunity to discover new drug targets using microarray technology.

Happy 70th Birthday Dr. Chien, with my greatest respect, admiration and appreciation.

Professor Shu Chien — A Person with Great Wisdom and Kindness

Yingli Hu[*]

Dr. Shu Chien has made remarkable contributions to bioengineering and physiology and has received numerous awards and honors.

After joining his laboratory in 1991, I became immediately impressed by the kindness of his demeanor, breadth of his knowledge, and tirelessness of his energy. Dr. Chien's tremendous potential must have been recognized very early on by his parents, teachers and fellow students, and certainly by Mrs. Chien.

I remember that when I came to UCSD, I could hardly converse in English. When Dr. Chien found out that I had taken a whole year of English language training in China, he said "One year of concentrated study should yield much better results. One should be able to learn any new language in a whole year." I did not dare to tell him that actually I had many additional years of English courses in high school and college. From his remarks, I appreciate that talent results from hard work and that learning requires focused concentration.

Dr. Chien's marvelous talent and tremendous diligence are important factors in his success. During my years of working with him, however, I learned that intelligence and hard work are essential, but not sufficient, for success. One must focus one's effort and make best

---

[*]Assistant Scientist, Department of Bioengineering, UC San Diego.

With Yingli and Huiping Hu, 2001.

use of one's time. Dr. Chien often says "Everyone is given the same amount of time. What counts is how we use that time." (This quotation from him was printed on the banner (on p. 153) given by faculty, students and staff to Dr. Chien upon his completion of his five-year term as Chair of Bioengineering. Dr. Chien pursues his tasks with unusual levels of energy and enthusiasm. He almost works non-stop, but he works so efficiently that he has always been able to squeeze out time to do other things. Then, whatever he does, he puts in his entire heart and mind. He cares about his family. His daughters say that he gave them quality time when they grew up and that he was fully focused and attentive when he was with them. Dr. Chien is always so very busy and he makes use of every moment of time, and I rarely see him resting. It seems that the time for relaxation is when he is with people, for example at parties. In caring for Dr. Chien's health, Mrs. Chien helps to arrange time for him to play tennis or golf. Dr. and Mrs. Chien both watch their diet and weight. Dr. Chien told me that he is able not to overeat at banquets because he can anticipate the pleasure when he weighs himself. This shows his ability of motivating and disciplining himself.

According to Confucius, "In learning it is essential to have interest; and it is best to be able to enjoy it with zest and rapture." In whatever Dr. Chien does, he aims at achieving excellence and enjoys it with total immersion.

As a result of his zestful learning and hard work, Dr. Chien has accumulated an extremely rich body of knowledge and experience and he has made outstanding accomplishments and contributions. It is said that "The wise person is happy, and the kind person has longevity." Dr. Chien is a person of great wisdom and kindness. I sincerely wish him all the health, happiness and longevity on the occasion of his 70th birthday.

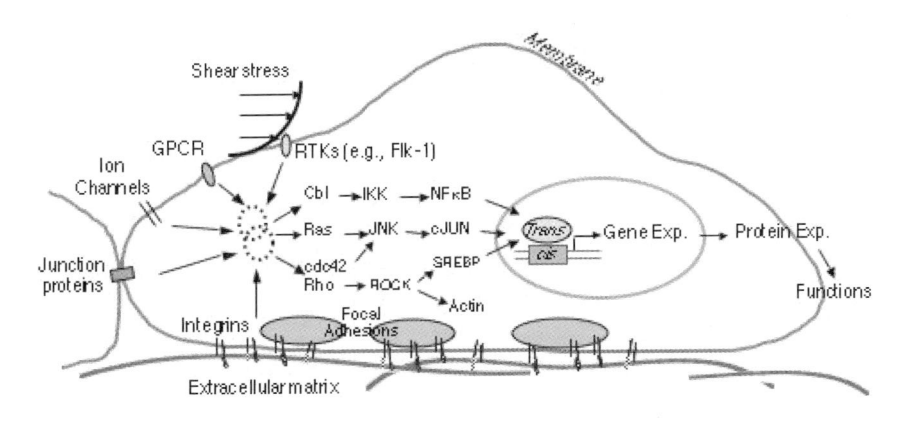

Schematic diagram showing the mechano-transduction of shear stress to gene expressions and endothelial cell functions. From S. Chien, Progr. Biophys. Molec. Biol. 83: 138, 2003.

The Family of Dr. Chien

Peter J. Butler, Ph.D.[*]

Dr. Chien, you have engendered quite a large scientific family with each offspring exhibiting traits of such consistency and persistence that the inheritance is analogous to gene transfer. The genes I'm referring to are not *MCP-1* nor *c-fos* but rather the genes for perseverance, understanding, caring, accessibility, clarity of direction, and dedication to excellence shared by many of your former students and collaborators. My experiences with you, both direct (as a mentor and friend) and indirect (through your alumni family) are greatly treasured. My friendship with you will be nurtured as you have nurtured my career. And through me your influence will be passed on to future (scientific) generations.

---

[*]Postdoctoral Fellow, Department of Bioengineering, UC San Diego.

Happy Birthday, Dr. Chien!

Yan-Jing Elizabeth Shiu, Ph.D.[*]

I joined Dr. Shu Chien's Vascular Bioengineering Laboratory at UCSD in July, 1999. It has been a wonderful experience working with Dr. Chien. He is not only a terrific supervisor in research, but is also a mentor to me in many ways: he is very modest and amiable in spite of his extraordinary academic achievement; he is a true gentleman and always treats people with respect and honesty; he likes to collaborate with people and to help them succeed; he aims at excellence in handling tasks, no matter how little/big they are... etc. It is a privilege to have the chance to work with him and learn from the best.

Happy Birthday, Dr. Chien!

With Elizabeth Shiu, 2001.

---

[*]Postdoctoral Fellow, Department of Bioengineering, UC San Diego.

Happy Birthday to Dr. Chien

Yingxiao Wang[*]

I feel the urge to write something at this moment — the 70th birthday of Dr. Shu Chien.

I came to UCSD as a graduate student in 1996, under the direction of Dr. Shu Chien. Homework and exams were the main theme of the first year, which passed quickly. Then my wife came to America, enrolled in the University of Michigan at Ann Arbor. After half a year of difficult separation, I decided to move to Michigan. It was a big mistake, which I realized soon after my departure from UCSD. Dr. Chien and his laboratory is really the best place where I belong, although I did not realize this until after I left. Feeling my unhappiness and helplessness in Michigan, my wife said to me "Based on what you told me about Dr. Chien, maybe you want to call him to see whether there is a possibility for you to go back to his laboratory. Since he is such a nice guy, probably you could return and be happy again. And if he agrees, I shall go with you." So I called, and he agreed, with kindness and generosity, as always. I am now enjoying the time in his laboratory, and when I sometimes think of the past, I know I owe him quite a lot.

As time goes by, I had the chance to know Dr. Chien better. It is one of my greatest pleasures to be his student. Like the excellent advisor that he is, he never urges me to do what he intends to do. He would usually listen to my proposals or thoughts carefully and comment

---

[*]Ph.D. Candidate, Department of Bioengineering, UC San Diego.

With Yingxiao Wang, 2001.

only when he foresees problems. Even when he had strong feelings for some interesting scientific explorations, he would persuade me with reason and logic. Therefore, all his students, including me, enjoyed the ultimate freedom in research, which we truly appreciate.

He is also fully committed to all his responsibilities, no matter how important or trivial. For example, despite his extremely busy teaching and research commitments, he would take the time to modify any of my drafts over many times, correcting any mistakes including typos until they are perfect to him. He works like an engine that knows no rest. I was very surprised that he can continue doing this for so many years.

One day, as I was walking along the Pacific Beach, watching the tide push back and forth incessantly, I realized that there must be an endless ocean of enthusiasm and love of life in his heart.

As many people have mentioned, Dr. Chien is a real gentleman with malice toward none, with charity for all, and with firmness in the right. I have heard so many times that he has no enemies in his life, but friends. As one of his students, I am proud of him and wish him a wonderful time.

Beach party at La Jolla Kellogg Park, 1992.

A Tribute

Roland Kaunas[*]

Dr. Chien,

I am happy to join everyone in celebrating your birthday. I would just like to share with you how enjoyable it has been to study under your guidance. I recall, as I was researching the literature from potential researchers with whom I might choose to perform my graduate studies,

With Chris and Roland Kaunas and Tsui-Chun Tsao, 1999.

---

[*]Ph.D. Candidate, Department of Bioengineering, UC San Diego.

that among your multitude of technical papers there was an article you wrote in which you called upon the bioengineering community to rally to support the development of their field. It was quite inspiring and reflected my own excitement about joining this new and growing discipline. After reading this article and learning of your past accomplishments, I knew that working in your lab would be an enriching experience in many levels.

I look forward to hearing stories from your past students and colleagues as they share more details of your colorful and productive life.

Best wishes,

Roland Kaunas

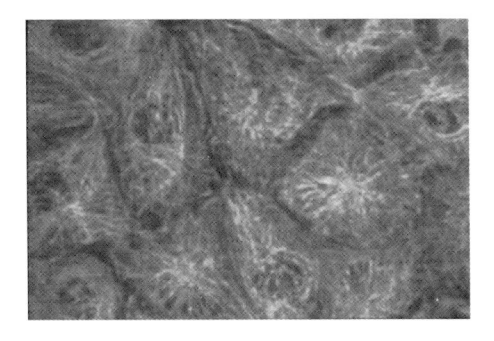

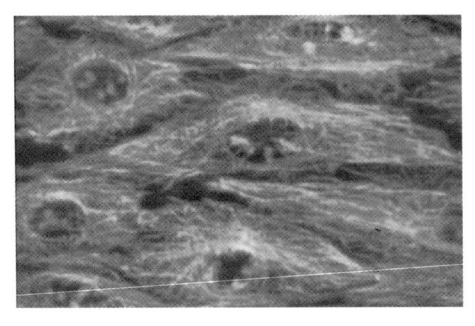

Effects of flow on the reorganization of cytoskeletal fibers and orientation of endothelial cells. Micrographs taken during the study for Publication A364.

Dr. Shu Chien, My Teacher and Collaborator

Xiong Wang, Ph.D., CR/CNRS[*]

The first time that I met Dr. Shu Chien was in 1989 in Nancy, France. Even then I knew very well his extraordinary contributions to hemorheology. In fact, Dr. Chien came to Nancy that year to attend the 7th International Congress of Biorheology. While still a Ph.D. student, I tried, with some feeling of a Nancéien,[a] to invite Dr. Chien and some other scientists from mainland China, Taiwan and Japan to a gathering at my tiny apartment. To my surprise, everyone happily accepted my invitation.

Five years later, the Second World Congress gave me another chance to meet Dr. Chien, this time in Amsterdam. By then I had been a researcher at the French National Scientific Research Center (CNRS) for almost four years. I talked to Dr. Chien for a long time regarding my sabbatical project in his lab — to learn the fast developing field of Mechanobiology. Again, Dr. Chien accepted my request and a research program was discussed and defined.

Thus, with financial support from the French Ministry of National Education, Research and Technology and from CNRS, we (my wife Li, our daughter Valentine-Lanting, and myself) arrived in San Diego on 3 July 1995 and began an unforgettable year there.

---

[*]Chargé de Recherche, French National Scientific Research Center, Laboratory of Energetics and Theoretical and Applied Mechanics, Vandoeuvre, France.
[a]In French it means resident of Nancy.

UCSD Chien lab gathering, in front of EBU1, 1994 (Xiong Wang is at the right front).

Before going to San Diego, my main research interest was in hemorheology and hemodynamics. However, the early 1990s was a period of transition for scientists doing basic research in these fields, because on one hand, from the point of view of mechanics, powerful methods and tools had been developed to approach mechanical properties of blood and blood vessels, and a kind of "stagnation" drove people to look for new research themes. On the other hand, the effects of mechanical forces (or mechanical environment) on tissues and cells drew more and more attention from researchers. Pioneers, like Dr. Chien among others, already concentrated their efforts to elucidate phenomena of mechanoactivation and mechanotransduction. My research program at San Diego was based on the following hypothesis: "Disturbed flow (or secondary) induces pathogenesis of atherosclerosis." This hypothesis became the first motivation of my research work upon my return to Nancy in July 1996. We invested completely in this new research area, and in April 2001, Professor Jean-François Stoltz and myself successfully co-organized the Euromech Symposium N°420 "Mechanobiology of Cells and Tissues" at Nancy. Dr. Chien's arrival was eagerly anticipated, and as expected, he gave a marvelous plenary lecture on the mechanobiology of endothelial cells.

Since 1998, Dr. Chien and I have collaborated on a project investigating the effects of flow on endothelial cells, jointly supported by the French CNRS and the American NSF. Annual exchanges between researchers from the two groups are very beneficial. It must be mentioned that Dr. Chien kindly conducted a wonderful course for our masters students of bioengineering at Nancy in 1999, and my friend and colleague Dr. Sylvaine Muller had the chance to spend one month at Dr. Chien's lab, undertaking experiments on gene expression of endothelial cells under flows.

My stay at San Diego constitutes one of the most important transitions in my life as a researcher. Today I do not hesitate to say that Dr. Chien is the teacher who has most influenced my opinion of scientific research and who has given me crucial help in my career. So, once again, from the bottom of my heart: Thank you, Dr. Chien!

A Salute to a Renaissance Man

Koung-Ping Cheng, Ph.D.[*] and Lilly Cheng, Ph.D.[†]

On this very auspicious day, we celebrate the 70th birthday of Dr. Shu Chien.
His numerous accomplishments make us proud.
He should be very proud and yet he is always humble.

With Lilly Cheng, 1999.

---

[*]President, Intelligent Modeler Company, San Diego, CA.
[†]Professor, San Diego State University. Director, Asian Pacific Affairs, Global Partnership Development, CSU, Chancellor's Office.

His leadership and vision have made a difference in the lives of many.
His dedication to community service is impressive.
He has inspired us all to do our best.
He is a role model and is well respected.
He is the embodiment of a scholar, leader and visionary.
We celebrate his lifelong achievement, and share the joy of this grand occasion.
A salute to our beloved Dr. Chien.
Happy Birthday!

Koung-Ping and Lilly Cheng

The Golf Connection

Ernest Chun-Ming Huang, Ph.D., Huei-Jen Su Huang, Ph.D. and Natalie Huang[*]

It is a pleasure to pay tribute to someone like Shu Chien because it is so easy to find qualities that are easy to admire and to describe. However, it is also a challenge because it is difficult to fully capture how to describe him — he is so multifaceted and multidimensional and has so many varied interests.

Drs. Shu and K.C. Chien moved from Columbia to UCSD shortly after Su and I moved to San Diego. Although we did not know them personally at that time, we knew of him because he was from a very famous family in Taiwan, Republic of China, and Shu was a very reputable scholar in his own right. Shu's father was a very well known scholar, and the family raised three successful and famous sons. One son is very accomplished in finance, another in diplomatic and foreign affairs, and yet another, Shu, became very successful in academia, sciences and biomedical engineering.

After each of us arrived in the San Diego area, Su and I met Shu and K.C. at several social occasions. We were impressed by their caring and gentle demeanors. We were also amazed by the level of energy they both possessed. But we never really got to know them

---

[*]The Huang Foundation, Rancho Santa Fe, California. Dr. Ernest Chun-Ming Huang, President Emeritus and Founder of BD PharMingen, La Jolla, CA; Dr. Huei-Jen Su Huang, Formerly Head of Molecular Oncology, The Ludwig Institute for Cancer Research, and Adjunct Associate Professor, UC San Diego (retired 31 March 2001); Natalie Huang, their daughter.

Ernie, Natalie and Su Huang, Rancho Santa Fe, X'mas, 1997.

simply because at that time Su and I were both busy building our careers.

One day, I was flying back from a business trip and I saw that Shu was sitting in another seat on the plane reading a golf magazine. At that time, I just picked up the game and had fallen in love with it. So, I went over and re-introduced myself to Shu and asked whether he played golf. He modestly replied that he just started playing, and like me, loved it. So, I invited Shu to play a round of golf with me. Because of the golf connection, Su and I have had numerous opportunities to interact with Shu and K.C.

I would like to describe an example of Shu's passion for the golf. Several colleagues have told me that they have seen Shu at the Torrey Pines municipal course waiting for a tee time at 5:30 a.m. He would play nine holes and then go back into the office and "no one would be the wiser." Later we found that the Chiens' passions, apart from work, include not only golf, but also so many other aspects of life such as singing, dancing and table tennis.

Over the years and since those early times, Shu also had numerous opportunities to interact (sometimes through golf) with my colleagues at PharMingen. To my surprise, Shu knows and remembers all my colleagues he met, either at a game of golf or at a scientific meeting. He has an amazing ability to care about others and be interested in what they are doing.

We all know how much Shu has contributed to the establishment of the Whitaker Institute of Biomedical Engineering. From writing proposals, sourcing of funds, and finally to the groundbreaking ceremony of the building a year ago. He has been tremendously successful in raising money to fulfill his vision of building the Institute of Biomedical Engineering into a world class institution.

My connection to bioengineering was really through Shu. About a year and half ago, during Professor Y.C. Fung's 80th birthday party, I had an opportunity to discuss with Shu about my family's desire to do something in honor of Dr. Fung. We came up with an idea of establishing an endowment for the Y.C. Fung Auditorium at the Powell-Focht Bioengineering Hall. Su and I would provide the matching fund and Shu will have to raise the remaining half. He was very thorough in following up and also timely in raising the remaining funds. It was such a great pleasure working with him. I don't think one can talk about Shu's accomplishments without also remembering how humble and self-effacing he is. I think he epitomizes the quotation: "There is no limit to the amount you can accomplish if you don't care who gets the credit."

I remember that one time I mentioned to Shu that I would love to play a round of golf with one of his brothers (the one who was Minister of Foreign Affairs) when I visited Taiwan. Then a few months later, I happened to be talking to Shu about some other matters and happened to mention that I planned to visit Taiwan. I did not want to impose on him, so I did not mention anything about golf during that conversation. I figured it probably didn't make sense to be pushy. Then a few minutes after hanging up the phone, Shu called me back

Shu & K.C.; Y.C. & Luna Fung; Su & Ernie Huang at the dedication of the Fung Auditorium, 2003.

and said he forgot to ask me whether I would like to play a round of golf with his brother. Shu is such a thoughtful and caring person with an amazing memory!

I also remember a conversation Shu and I once had. He mentioned to me that all three brothers have different personalities. In fact, according to their personalities, each appears to have wound up in the wrong position. For example, the elder brother (Minister of Finance) actually is more of a scholarly type. The younger brother (Minister of Foreign Affairs) is more of a finance and administration person and Shu is actually more skillful in diplomatic and human relationship areas. However, I believe that Shu possesses strong skills in all three areas. I believe he must have all these qualities in order to accomplish what he has so far. In other words, it takes scholarly, diplomatic skill and the great talent in finance and administration to run a highly successful institution.

Once, Shu told me that if he was to retire today, he would try to play golf everyday. But I am not sure if he will ever retire. Even at 70,

Shu is as active as ever, and I am sure he will continue to play a lot of golf as he is juggling his work at the institute together with all his other interests and responsibilities.

So, Shu, enjoy your 70th birthday. Natalie, Su and I pay our deep respect to all you have accomplished, and to all the lives your have touched in such a positive and inspiring way. We wish you a long future of health and happiness.

Section II

More Tributes for the 23 June 2001 Celebration

A Fun Poem

Chia Hui Shih[*]

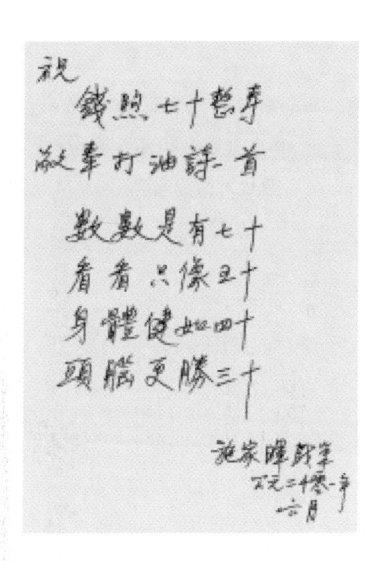

Congratulations to Shu and K.C. on their 70th Birthday.

    By counting, you are indeed seventy.
    But you look only like fifty.
    Your bodies are as healthy as those at forty.
    And your brains are even better than those at thirty.

By Chia-Hui Shih
June 2001

---

[*]Chia-Hui is the sister-in-law of K.C.'s brother George Kuang-Chi Hu.

A Poem

Suli Yuan[*]

鹤踌教

新世纪元年之初夏，逢业师钱煦公七十寿诞。古人云：人生七十古来稀。钱公古稀无胜少壮。幸得受言传身教近八载，获益匪浅。挚填此词，永贺亦示敬也。

弟子莘州来四面，
喜聚一堂，
贺钱公寿诞。
旧闻新趣一串串，
一生辉煌众人赞。

精神耕耘五十年，
创业树人，
研教皆贡献。
身正学高情无限，
尽显宗师之风范。

Suli Yuan 袁素莉 2001.6.23

---

[*]Staff Research Associate, Department of Bioengineering, UC San Diego.

Bird Stepping on Tree Branch*

At the beginning of the summer of the first year of the new century, we gather to celebrate Professor Shu Chien's 70th birthday. The old Chinese saying is that "it is rare for people to reach seventy since ancient times." But Dr. Chien is stronger than younger people as he turns 70. I have had the fortune to learn from him through his words and actions during the last eight years, which benefited me greatly. I wrote this poem to express my congratulations and respect:

>Disciples, relatives and friends came from all corners of the world.
>Gathering happily in the same hall
>To celebrate Professor Chien's birthday.
>Reminiscing the old days and talking about new interesting events,
>Everyone praises his glorious life.

>Sowing seeds and plowing the land for fifty years,
>He has educated many students like planting trees.
>With remarkable contributions to research and teaching.
>Model behavior, superb knowledge, and boundless feeling,
>He fully demonstrates the style of a great scholar.

Suli Yuan
23 June 2001

---

*The title is that of a particular type of poem, which has sentences of defined lengths and rhymes. It does not refer to the content of the poem.

A Birthday Card

Shlomoh Simchon, Ph.D.[*]

To Dr. Shu Chien: Happy Birthday!

Example is not the most important thing in influencing others; it is the *only* thing (Albert Schweitzer).

We can see farther than the ancients not because we have better vision or greater stature, but because we have been lifted up by giants (Bernard De Chartres).

---

[*]Research Associate Scientist, Department of Medicine, Columbia University, St. Luke's-Roosevelt Hospital, NYC.

With Shlomoh Simchon at the celebration symposium, 2001.

The above quotations apply to Dr. Chien as a Mentor, Leader and Friend.

Many returns from:

Shlomoh and Arlette Simchon

A Birthday Tribute to Shu

Sidney S. Sobin, M.D., Ph.D.[*]

Dear Shu,

I first met you, on paper!

When I was on the NIH Cardiovascular A Study Section in the 1960s, I reviewed your research grant applications. I recall being impressed with the original questions being asked and the ways of answering them. At that time you were involved with blood, both the erythrocyte and the fluid around it; I remember because our own laboratory, with CalTech, was also involved and competing in blood viscosity studies. Your pursuance of the red cell was novel because you teased information from all parts of the red cell — membrane, contents, flow properties and others. Later, I chaired the Cardiovascular Sciences Study Section and saw your application for studies on the endothelial cell, which you have continued in so many variations to this day: membrane, contents, reactions, effects of shear stress, and…. In many ways, currently, you are synonymous with vascular endothelium.

I first met you at Atlantic City during those years. I have forgotten who introduced us, probably Y.C., we shared food on a number of occasions not associated with the various society dinners. One of the pleasantries of Atlantic City in those days was establishing friendships, initially science based, then personal; our friendship here demonstrates this. When you came to La Jolla to take the chairmanship of the

---

[*]Adjunct Professor, Department of Bioengineering, UC San Diego.

With Venus and Sid Sobin, 1992.

Bioengineering program, you developed it into a full department in the Jacobs School of Engineering, and your vision and creativity changed the program into what is now the premier academic department of Bioengineering in the world.

Your vision for our department extended to the American Physiological Society, FASEB, and the American Society of Medical and Biological Engineering — whose presidencies you have successively held. Phenomenal! And the newly established National Institute for Biomedical Imaging and Bioengineering at NIH could not have been accomplished without you! All this you have accomplished before the 70th year! For your future many productive years, please guide your vision and creativity into channels as successful as the past. May age 70 be a signpost for your many future years. Welcome.

Very sincerely and fondly,

Sid

An Appreciation of Shu Chien's Work

Kitty Fronek, M.D., Ph.D.[*]
and Arnost Fronek, M.D., Ph.D.[†]

We followed Shu's academic achievements long before he came to UCSD, mainly from the literature. We admired how he moved easily from one research field to another. From hemorrhagic shock to rheology and to microcirculation. And now to molecular biology.

With Kitty and Arnost Fronek, 1999.

---

[*]Research Physiologist, Department of Bioengineering, UC San Diego (1968–1990).
[†]Professor Emeritus, Departments of Surgery and Bioengineering, UC San Diego.

We have to admit that we started to appreciate fully Shu's work and intellectual capacity only after he came to UCSD. He secured a fast national prominence of our bioengineering group, he achieved the establishment of a department of bioengineering as well as the important long range support from the Whitaker Foundation. Surprisingly, he had enough energy to help his colleagues in Taiwan to establish the Institute of Biomedical Science. It is a positive aspect of one's personality not to forget the roots.

We wish you Shu, many more productive years and we regret only that you didn't come to UCSD at least ten years earlier.

Kitty and Arnost

A Tribute

Sangeeta Bhatia, M.D., Ph.D.[*]

Dear Shu,

You are a very special role model and mentor for me.

 I admire so many unique aspects of your character — your loving family, your scientific rigor coupled with innovation, your intuitive sense for people and organizations, and your vision.

 It is a great privilege and pleasure to know you and work alongside you.

 Please accept my best wishes for the future.

Warm Regards

Sangeeta

---

[*]Assistant Professor, Department of Bioengineering, UC San Diego.

Tribute to Shu Chien

Ruth L. Kirschstein, M.D.[*]

On behalf of the National Institutes of Health, I salute Shu Chien for all his accomplishments and service to the biomedical research enterprise in general, and to bioengineering and physiology research in particular. I first met Dr. Chien through his activities in the American Physiological Society and then the Federation of American Societies of Experimental Biology (FASEB). He served as a Board Member of FASEB, as its President from 1992–1993, and as Chair of its Public Affairs

With Ruth Kirschstein, 2001.

---

[*]Acting Director, National Institutes of Health, Bethesda, Maryland.

Executive Committee in 1993–1994. It was in the latter capacity that he called me one day in 1993 to tell me I had been selected for the FASEB Public Service Award. Nothing could have honored me more than to have Shu, who had served NIH so selflessly, on study sections, review committees, councils and other advisory bodies, present me with this honor.

And, so, on his 70th birthday, on a personal note and for all of us at NIH, we salute Shu Chien, M.D., Ph.D., a distinguished scientist and public servant of science.

Profiles in Physiology

Ewald R. Weibel, M.D., D.Sc.[*]

Dedicated to Shu Chien for His 70th Birthday

For four years Shu Chien and I served the International Union of Physiological Sciences (IUPS), as Treasurer and President, respectively. In this service we shared an anomaly: we both were not truly "card carrying physiologists." Shu earned his bread under the label of bioengineer[a] and I under that of anatomist. Indeed, when I was elected President of IUPS a friend and properly "professional" physiologist, chairman of a University Department with that denomination, remarked: "How low has physiology fallen that we have to choose an anatomist as President?"

But is this so wrong or even anomalous? If physiology is interpreted as the "Study of the Logic of Life" — as it seems to read in Chinese script[1] — then physiology and those representing it must reflect all the complexity of Life. Daniel E. Koshland[2] has recently described this complexity as the "Seven Pillars of Life" which are: Program, Improvisation, Compartmentalization, Energy, Regeneration, Adaptability and Seclusion. For each of these pillars one can easily identify building blocks, which are physiological in nature. Compartmentalization and

---

[*]Professor Emeritus, Department of Anatomy, University of Bern, Bern, Switzerland. Past-President, International Union of Physiological Sciences (IUPS).
[a]Editor's note: Shu received his Ph.D. in Physiology and served as Professor of Physiology before serving as Professor of Bioengineering.

Seclusion are the very basis of all exchange processes that lead to polarity and tensions, the most prevalent occupation of physiologists. Energy — its supply, local production and use in the cells — is the driving force for all life processes and therefore central to all physiological reasoning: there is no function without energy. Regeneration — the provision for replenishing resources and for renewal by birth and death allowing life of an individual organism to be finite while preserving life as a system in all its diversity — includes all preoccupation of physiologists with processes of maintenance of life at the cellular and organismic levels. Adaptability is intimately linked to functional performance under varied constraints and leads to the wealth of biological diversity that can be so valuable in the research strategies that aim at understanding basic functional processes. Even the Program — the genetic blueprint of the organism — and Improvisation — its variation by mutation and selection — have become preoccupations of physiologists in the quest for understanding the origin of life processes from first principles. So physiology pervades all these pillars in one way or another.

But there is more: physiology is fundamental for integrating these pillars into a whole.[3] From the point of view of an integrative physiologist one may perhaps postulate an eighth pillar: Integration. But perhaps this should not stand parallel to the other pillars but be rather a "horizontal binder" at the base and on top of the row of pillars. Integration means to control the operation of the different elements, the essence of the pillars, in the overriding interest of the whole. Some of this integration is "hard-wired" in the structural design of the system. Thus a good physiologist will rarely limit his quest to one or the other pillar: he or she will build bridges between them. And there is inherent beauty revealed in the process of inquiring into the foundations of a well-integrated functional system.[4]

In the first place, however, physiology is a way of thinking about life and life processes, about how organisms work and how we can find this out by studying nature. Accordingly there will be many different profiles of physiologists, and the Logic of Life will be found out only if all these facets are taken into consideration and brought together. There will be:

- those who study various functions such as respiration and oxygen supply from lung to cells, or the heart beat in relation to ion channels, Physiology in the narrow sense;
- those who study humans in health and disease making physiology an important part of the Medical Sciences;
- those who look how different species are made and how they react to their environment, the basis of Biodiversity;
- those who study the link between genomics and function and open the new era of Physiological Genomics; and
- those who look at how the design and construction of cells, tissues and organs affects functional performance.

With the last profile we are very close to that of a Bioengineer — or of a Functional Morphologist or Anatomist. Both busy themselves with understanding the structural underpinning of functional processes and in doing so they will locate their endeavor to understand life into several of the Pillars of Life, from Compartmentalization to Seclusion, from Adaptability to the Program, and Energetics as well as Regeneration will be constant considerations.

Thus I am convinced that a bioengineer of Shu Chien's orientation and performance is rightly considered a Physiologist, and one who can make — and has made — very significant contributions to the advancement of the Physiological Sciences. There is a lot we can learn from his approach and his insights.

Serving an international scientific union such as IUPS, however, is a different matter as Shu Chien and I have soon realized. It was not enough to promote the noble goals of our science from a philosophical platform. Sometimes those in charge have to deal with very mundane problems, money matters and political issues, for example. I recall two particularly difficult political problems where a solution was only possible because Shu and I joined forces and worked towards a common goal: to achieve the best for the worldwide community of physiologists.

The first case came up during our first year of office. The International Physiology Congress in St. Petersburg in August 1997 had been a difficult exercise in many respects. But we would not have expected that in the end IUPS would be faced with the financial

With Ewald Weibel and Denis Noble, 2000.

burden of a quarter of a million dollars, even though the financial liability was not with IUPS but with the Russian Academy of Sciences. The Russian trick had been to promise — in the name of the IUPS Congress — substantial subsidies to a large number of invited speakers from all over the world, and then not to honor these promises when they ran out of money. A few months after the Congress, IUPS began to be bombarded with letters from frustrated physiologists who had, in vain, tried to pursue their claims. I immediately entered into negotiations with the Russian Academy of Sciences but it soon became clear that we could not expect anything from this partner as they were engaged in an internal quarrel about responsibilities and trust. IUPS would have to solve this problem.

So I was very glad when Shu Chien took over the office of Treasurer at the start of the year 1998. IUPS was in a serious crisis. What mattered was to find ways of satisfying at least part of the rightful claims of our constituency, but the till of the Union was virtually empty because we had put all our resources into the preparation of the Congress. Usually this would be refunded from a levy on the registration fees — but this too was not honored by our Russian partners. So Shu and I launched a campaign to obtain some special funds; this was successful and we could cover about 50% of what was owed the speakers, and in the end most people were satisfied. IUPS got out of this crisis.

The second instance occurred at the end of our term and it was of a very different nature. In 1997, there had been a fierce competition between four national physiological societies that wanted to organize the 2005 IUPS Congress. The US National Committee won that race by a margin with their proposal to hold the Congress in Washington D.C. in August. This was to be confirmed at the 2001 General Assembly of IUPS. A few weeks before that meeting IUPS was informed that the American Physiological Society had suddenly decided to change both venue and date in order to hold the IUPS Congress in conjunction with the large Experimental Biology meeting in San Diego in April 2005. It was not the change in venue and date that alienated Council of IUPS, it was rather the fear that the projected merger of the comparatively small IUPS Congress with the very much larger EB meeting would lead to a loss of international identity of the former for which we were responsible.

IUPS was again in a crisis, this time of a political nature. It is thanks to Shu Chien's diplomatic ingenuity and imagination that a solution could eventually be found that allowed for sufficient international identity of the IUPS Congress while still accepting the undisputed advantages of the joint meeting. This solution was mutually acceptable to both IUPS and its U.S. partners for the 2005 Congress. IUPS must indeed be grateful to Shu Chien for having played a most significant role as a skilful negotiator serving the best purposes of the physiological sciences.

So together with all the Physiologists of the world, I am looking forward with the keen anticipation of an engaged physiologist to the 2005 IUPS Congress that will be held under the skilful leadership of Shu Chien.

REFERENCES

1. Boyd, C.A.R. and Noble, D. *The Logic of Life: The Challenge of Integrative Physiology.* Oxford University Press, U.K., 1993.
2. Koshland Jr., D.E. The seven pillars of life. *Science* **295**: 2215–2216, 2002.

3. Weibel, E.R. The future of physiology. *News Physiol. Sci.* **12:** 294–295, 1997.
4. Weibel, E.R. *Symmorphosis: On Form and Function in Shaping Life.* Harvard University Press, Cambridge MA, 2000.

Bern, 8 April 2002

Shu Chien and the International Union of Physiological Sciences (IUPS)

Denis Noble, Ph.D.[*]

The St. Petersburg Hotel (formerly called the Leningrad) dominates the waterfront opposite the mooring place of the battleship Aurora that fired the famous cannon shot during the Russian Revolution. It reeks of the atmosphere of the old Soviet regime. Matronly minders on every dimly lit floor, waiting to be "sweetened" to allow a vodka and champagne party to occur, public rooms that never seemed to be used, restaurants that either didn't serve anything or presented a menu that was essentially a diktat. And much of the hotel was, probably still is, a building site: a memorial to lost dreams as the Soviet Union collapsed just a few years before (and with it any vestige of guarantee that the old USSR Academy of Sciences had given as underwriting of the Congress budget). Yes, there were also the elegantly furnished apartments that must have been the preserve of the high party officials in the old days. The Presidents of IUPS (Masao Ito as outgoing President and Ewald Weibel as incoming President) and I as the Secretary were allocated these oddly magnificent suites and we had marvellous views of the White Nights as the sun hardly seemed to set all night.

---

[*] Professor, Oxford University, Oxford, U.K., Past-Secretary General of IUPS.

This was 1997, and the occasion was the IUPS International Congress in Russia. We did drink a fair amount of Soviet Champagne (if you knew enough Russian to know where to buy it from it was incredibly cheap), but our spirits were not exactly high. As the chief officers of IUPS we already knew that the organisation was facing a major crisis, though we didn't yet know how really serious this was going to be financially. The truth came out later. The loss was more than the organisation could bear. And the professional congress organisers hadn't even paid the speakers their expenses! Into this terrible situation walked Shu Chien, just elected as the new Treasurer of IUPS.

The old and new Executive first met each other over a simple dinner in the only subterranean restaurant that worked. So, there on the other side of the table I met this beaming Chinese American who smiled and exuded the kind of enthusiam that we had almost had beaten out of us. We set to work immediately thinking about how to run a lower-cost but innovative and successful Congress. This had a double significance since Shu was to look after IUPS finances for the next turbulent four years, and the American delegates to the IUPS Assembly had just, narrowly, won the vote to host the Congress in Washington in 2005. He and I and our fellow officers in IUPS were clearly going to sink or swim together during the years that followed, so we needed to get to know our shipmates well.

First, we mounted the salvage operation, a task in which Ewald Weibel took a commanding and powerful lead with Shu and I as the support troops. Somehow the money was raised to reimburse every speaker at the Russian Congress who still needed expenses paid. Next, we met together in the city of Christchurch to see what the New Zealand organisers were planning for the 2001 Congress. Unfairly, I think, they had to take much of the brunt of our retrospective concerns about 1997. But they also impressed us with their innovation and endless enthusiasm. Shu, of course, had the responsibility to give their budget a critical eye and to ensure that IUPS did not face another financial crisis of a level that would, quite simply, have been the end of the organisation. We faced the prospect of bankruptcy.

With Denis Noble, Ewald Weibel, Sue Orsoni and Raymond Latorre, 2000.

That long trip in 1998 to the South Island of New Zealand was an important one, not only for the planning of the Congress there, but also for the coalescing of the relationships within IUPS Executive that were going to prove of vital importance three years later in the run-up to the 2001 Assembly, and the need to confirm the American bid for 2005. Shu came with his charming wife, and I remember getting the guitar out one evening to raise our spirits with some multilingual serenading that she greatly enjoyed.

In the event, the 2001 Congress was a great success (see the IUPS Newsletter on the website *www.iups.org*), though it did make a modest financial loss. But we never expected it to be the financial saviour of IUPS. What was crucial to the reputation of IUPS was that it should be shown that it could still hold such a Congress and make it a *scientific* success. On this, our New Zealand colleagues did a magnificent job. With the prospect of an American-hosted Congress to follow, we were surely out of the woods.

Or were we?

I shall never forget a telephone call I received from Shu Chien at my home in Oxford just as I was to leave for a family holiday in France before going on to the Christchurch Congress. What he told me was the equivalent of another bombshell: the American delegates were going to propose that we abandon the Washington proposal (which in any case had been won by only a narrow margin) and, apparently, fuse the IUPS Congress with the annual FASEB meeting in San Diego. Readers of this article need to know that the papers for the 2001 Assembly had already been circulated! A fundamental change at this late stage was unprecedented. The Assembly could conceivably reject the American confirmation, even leave IUPS with no Congress at all. I remember Shu and I discussing the matter in the brief time we had available and we ended up agreeing that this was going to be a tough ride. The rest of the world was going to be highly suspicious of the Americans' motives, and even think that, this time, the death knell of IUPS really had been rung.

While in a remote part of southwest France I tried to remain in touch via a primitive internet facility at the library of the local market town as e-mails, steadily getting more and more explosive and urgent, flew across the ocean. I remember a telephone call with the President that left us both feeling even more depressed than before. It looked as though we had jumped out of the hot frying pan of Russia only to land in the even hotter fire of the U.S.A.!

Yet — and here is where I come back to the bonding session in Christchurch three years earlier — I also knew that something I could be certain of was that Shu's motives, and for all I knew, those of the whole American delegation to the Assembly, were very well intentioned. As I reflected over an e-mail note that I drafted wearing my legal role as Secretary, I decided to include this carefully chosen phrase as well as doing what I had to do: to lay out the procedural and legal difficulties.

I believe that phrase was crucial in the events that followed. As soon as we arrived in Christchurch, Shu and I met for breakfast for a "cards on the table" and totally non-committal chat. This meeting confirmed my judgement. "Well intentioned" was precisely the right description, and Shu had not only marshalled the arguments to deal with my own doubts, he had taken fully on board the discussion we had before my holiday: that the Americans were going to have a particularly tough time, so deep was the suspicion that this amounted to an American takeover. The situation was complicated even further by the news that the U.S. National Committee had appointed Shu

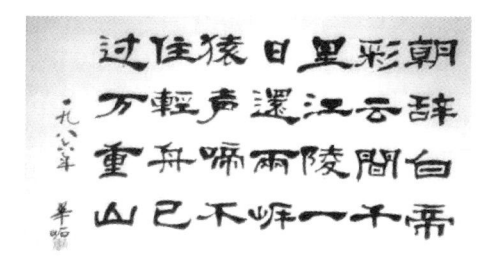

Poem by Li Po on "Sailing down to Chiang-Ling".

Chien to be the 2005 Congress Chairman! He was no longer a neutral IUPS officer. He had also to be the chief advocate for the new proposal.

Those few days as the Executive and Council met before the Assembly vote were the scenes of intense negotiations as the conditions for a San Diego Congress were hammered out. It is in such situations that friendships either deepen or wither. What came across to me from Shu was total enthusiasm and commitment. I have no doubt that his role in the diplomacy leading up to the successful vote at the Assembly played a major role. What eventually emerged is a very exciting proposal and one which I feel sure will succeed. I look forward to San Diego 2005 with great expectations. It could be the largest Congress since Glasgow 93, and it could reconnect physiology with the other physiological sciences.

But what a craggy journey to have to make to get there! I am reminded of that great poem by the greatest of Chinese poets, 李白 (Li Po), when he sailed down to 江陵 (Chiang-Ling) after being released from prison:

The last line translates as "my light boat skimmed past a thousand sombre crags" (Innes Herdan, 1973).[a]

I was Chairman of the 1993 Congress in the U.K. So as one Congress chairman to another, "good luck, Shu!" — I hope that 2005 will be the great event you and we all want it to be. Li Po sailed the thousand leagues down to Chiang Ling in a single day. Your journey

---

[a]Herdan, I. *300 T'ang Poets*. Far East Book Club Ltd., Taipei, 1993.

will be even longer — four years. Don't listen too much to the ceaseless chatter of the gibbons on the riverbanks. Enjoy the morning-glow of the clouds, but bring the fragile boat of IUPS to its safe haven.

Denis Noble
Oxford University, January 2002

The full translation for the peom by Li Po:

> Po-ti I left at dawn in the morning glow of the clouds;
> the thousand li[b] to Chiang-Ling we sailed in a single day.
> On either shore the gibbons' chatter sounded without pause,
> while my light boat skimmed past a thousand sombre crags.

---

[b]Chinese mile(s).

A Tribute to Professor Shu Chien on His 70th Birthday

Rakesh K. Jain, Ph.D.[*]

I met Professor Shu Chien 25 years ago at Columbia when I was Assistant Professor of Chemical Engineering and Shu was Professor of Physiology at the Medical School. Meeting Shu was very important in shaping my career. I had heard Professor Marcos Intaglietta's talk in 1974 about the use of dorsal windows to study tumor microcirculation and had read about the rabbit ear chamber in the literature. However, it was in Shu's lab that I saw my first live microcirculatory preparation, I became enamored with it and this started our research collaboration. With the help of Shu, his then post-doctoral fellow Geert Schmid-Schönbein, and my first graduate student David Zawicki, we studied the dynamics of vascularization during wound healing in the ear chamber. That was my first paper in the area of microcirculation.

I moved to Carnegie Mellon in 1978. With the unyielding support and encouragement of Shu, and the help of his co-workers Geert, Herb Lipowsky and Paul Sung, I was able to start my own research program in tumor microcirculation there. At the start, my lab was an extension of his lab. He was always available by phone when I needed advice. He made time at each year's Annual Microcirculation Society

---

[*]Andrew Werk Cook Professor of Tumor Biology, Director of Edwin L. Steele Laboratory, Harvard Medical School, and Massachusetts General Hospital.

260 *Tributes to Shu Chien*

With Rakesh Jain at the Conference on Vascular Endothelium in Health and Disease. In Front of IBMS, Taipei, 1987.

meeting to review my progress, listen to my plans and offer constructive criticism.

When I received an offer from Harvard and Massachusetts General Hospital, I again approached Shu for his advice. I was contemplating a move from an engineering school to a medical school, and Shu had just moved from a medical school to an engineering school. He shared his own experience and encouraged me to make the move. We have continued our tradition of keeping in touch by visiting as well as setting aside time at conferences to share our experiences.

If I had to sum up my sentiments about Shu, I would say he is one of the most generous, modest, optimistic and visionary persons I know. Every organization he has been a part of reached new heights under his leadership. Every student or colleague who has collaborated with Shu has benefited from his insight. Every area of biomedicine he has worked in is richer for his seminal contributions.

Shu, it has been a pleasure and honor knowing you for a quarter of a century! I thank you for the support, encouragement and friendship you have always shown me, and I join all your family members, friends, colleagues and students in celebrating this joyous occasion. Happy Birthday and many more!!!

A Sketch for the Lab People at Columbia

Emily Schmalzer, M.D.[*]

Note by Shu: Emily Schmalzer did this beautiful sketch for the Lab people at Columbia, with an attached explanation. When we moved to our present house in 1994, somehow this precious picture was misplaced. I told Emily about this on 22 June 2001 during the pre-birthday party

---

[*]Research Associate, Department of Physiology and Cell Biophysics, College of Physicians and Surgeons, Columbia University, NYC (1980–1988).

and apologized sincerely to her. I was extremely happy when she told me that she had kept a copy of the sketch, though not the who's who list. Emily then sent me the image by e-mail after returning to New Hampshire, with the list from memory. My warmest thanks to Emily for this lovely and memorable picture.

Message from Emily Schmalzer on 14 July 2001:

Dear Shu,

I don't think I kept the who's who list that I wrote out when I made the picture. I remember a few things, and can offer a few more:

The bird (upper left corner): Ann Baldwin, of course, since bird is British colloquial for attractive young woman, which she certainly was back then and still is. Also she always looked charmingly fragile to me, though of course she's not at all fragile. Grey jays (which this is supposed to be) are sort of like that, beautiful and fragile looking but actually quite assertive and tough.

The bigger eagle: Usami sensei, since he always kept a sharp eye on the lab.

The smaller eagle: Sylvia (Rofe) I think, she is very sharp too.

The elephant: Herb (Lipowsky) no doubt. Larger than life. I wanted to make him an Indian elephant but only had a photo reference for an African one. Indian elephants are more tractable (useful and reliable) than African ones. Good animal, good person.

The cheetah: I'd forgotten it wasn't a jaguar. Still, Walter (Reinhart).

The flamingo: Me. Danny (Batista) and Juan (Rodriguez) gave me that nickname presumably because I'm tall, pink and I flap.

The kangaroo: Amy (Sung), because she was a mother of young children when I knew her.

The African antelope: (forgot which kind) Reminds me of Steve House. I always thought his looks were very elegant. And of course he's very swift too.

The raccoon: Maybe Paul (Sung), because he is so clever with his hands.

The porcupine: Steve Gallik, maybe. Not smooth like the other Steve, but he was cute.

The lion: As I said, to me it's Kung-ming (Jan), although I never saw Kung-ming either roar or yawn. I think the lion that I used as a reference was yawning, actually. Still, Kung-ming spent a lot of time teaching me and I was very much in awe of his knowledge so he is kind of king of the beasts to me.

The giraffe: Well, George (Schuessler) was the tallest, so that must be him. Yellow and brown is sort of his coloring too.

The zebra: One of the animals has to be Dagmara (Igals), how could I leave her out? I would have preferred an animal that is more intelligent, like the raccoon, but at any rate her art is as abstract as the zebra's coloring.

The bighorn sheep: Probably Mike Kent. He was undoubtably only a blip on the screen of your life, but I am still very fond of him and we see him once or twice a year. I figure wild sheep are smarter than domesticated ones, so the bighorn may be much smarter than is advertised for sheep. At any rate, Mike is much smarter and more interesting and creative than his mannerisms would imply.

Missing:

Shlomoh (Simchon), like a badger, I think, because he's low to the ground and furry. He sort of moves that way. The only problem is that badgers are loners and unfriendly, which Shlomito is not.

Anne Chabanel, whom I would portray as a ferret. She's long and lean like they are, and likes physical activity. I've had them as pets and they are perky and playful, and Anne was fun to be with. One ferret I took care of was forever getting herself into scrapes just by dint of her good nature. Do you remember Anne's problem with the subway turnstile that she did not jump?

Mary Lee: A pretty little lap dog, I think. Very domesticated.

Bob Schmukler: Maybe a moose, though I'm not sure why. Aggressive but not malicious, possibly.

Ethel (Goodrich): Guard dog I guess.

Michelline (Faublas): I have run into her several times in the last decade. She had a job at NYU down in the village and was shopping for office supplies in an art store I had happened into. The job was working out well. A coolish personality, very able in her sphere, maybe a seal.

And there are so many other great people you had in the lab. The anesthesiologists, and the technicians. What a wonderful world it was!

But these are only my associations. Your relationships and feelings about people were different of course. It's your picture — you decide.

Take care, Emily

The Chien Laboratory at Columbia, 1978. (Left to right) 1st row: Syngcuk Kim, Shlomoh Simchon, Dagmara Igals, Shu, Ann Thurn, Kung-ming Jan. 2nd row: Chris Lin, Clare Ringold, Daniel Shaviello, Juan Rodriguez, Foun-chung Fan, Amy Sung, Digna Rodriguez. 3rd row: Robert King, Anne Kaperonis, Danny Batista, Ken Hampar, Paul Sung, Mary Lee, Shunichi Usami. 4th row: Geert Schmid-Schönbein, Herbert Lipowsky, Ethel Goodrich, George Schuessler, Lee Letcher, Benjamin Peng, Leslie Arminski, Richard Chen.

My Role Model

Gerhard M. Artmann, Ph.D.*

Dear Dr. Chien,

Over the past seven years we have collaborated scientifically and met on several occasions around the world. During this period, you have supported us in many ways. I want to thank you for that. Your help, your kindness was and is truly unusual. I remember many things

The cell biophysics team of the University of Applied Sciences, Aachen cordially congratulates Dr. Shu Chien for his 70th birthday. Gerhard Artmann is at the center.

---

*Professor of Applied Cell Biophysics and Bioengineering, University of Applied Sciences Aachen, Germany.

which I have learned from you by listening, by discussing and by just watching you. It has not been about science only. There was much more that I have seen and learned. This is why you have turned a role model for me and you will always be.

Sometimes I lean back and think of you: how calmly and smoothly and not at all hectically you take care of your job. This is still a role model for me. Sometimes I think: perhaps I need to spend only 50% of my energy but control myself better as you do, then I could do a better job.

I am happy that you have been honored as you deserve it and that friends have been around. Take care and best wishes to Mrs. Chien. I am looking forward to seeing you both in Aachen.

Gerhard

With Gerhard Artmann and Dean Siegfried Pawelke of University of Applied Sciences, Aachen, Germany at Shu's home in La Jolla, 1997.

Calligraphic Tributes

Kung Chen Loh[*] and H.S. Fang, M.D.

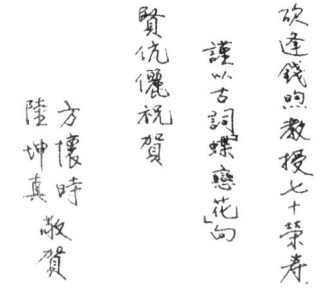

It is our pleasure to send this ancient poem *Butterfly Loves Flowers*[a] to express our sincere congratulations to Shu and K.C. on Professor Shu Chien's 70th Birthday.

H.S. Fang &
Kung Chen Loh

Mrs. Kung Chen Loh Fang and Dr. H.S. Fang, 1996.

---

[*]Kung Chen Loh is Mrs. H.S. Fang. Her father was President of National Taiwan University (Taita) from August 1946 to July 1948. Dr. H.S. Fang is Professor Emeritus of Physiology, Taita College of Medicine.

[a]The title is that of a particular type of poem which has sentences of defined lengths and rhymes. It does not refer to the content of the poem.

(Calligraphy in 2001)

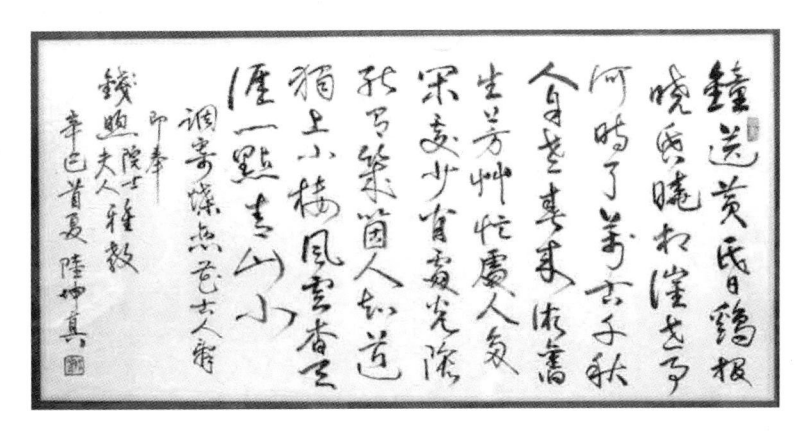

鐘送黃昏雞報曉，昏曉相催世事何時了？萬古千秋人自老，春來依舊生芳草。忙處人多閒處少，閒處光陰能有幾個人知道？獨上小樓風雲杳，天涯一點青山小。

調寄蝶戀花古人辭
即奉
錢煦 院士
夫人 雅教
辛巳首夏陸坤真

The bells send off dusk and cocks usher in dawn.
With dusk and dawn chasing each other, when will be the end of our daily chores?

Over centuries and millennia, people do age.
When spring comes, however, new grass will grow again as always.

There are more people in busy places, and less in leisurely places.
How many leisurely people care about whether time passes?

On top of the little tower, one sees the silent clouds and feels the wind,
The corner of the earth looks like a dot, and the green mountains are so small.

After ancient poem *Butterfly Loves Flowers*[a]

To: Academician and Mrs. Shu Chien
Beginning of Summer, 2001

From: Kung Chen Loh

(Calligraphy in 1987)

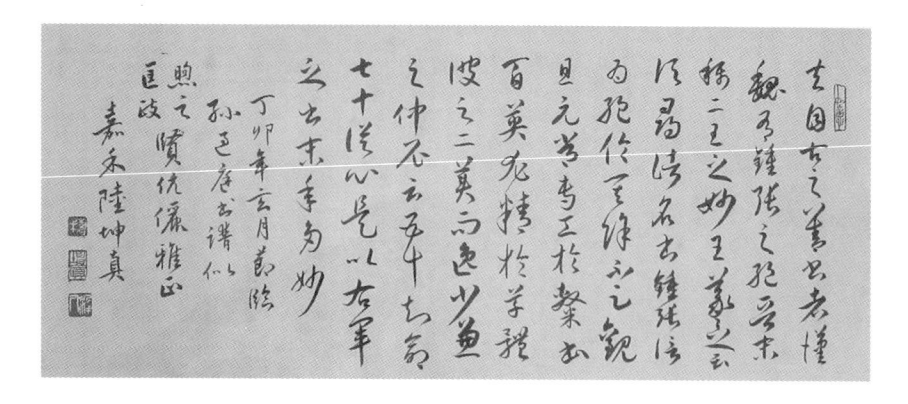

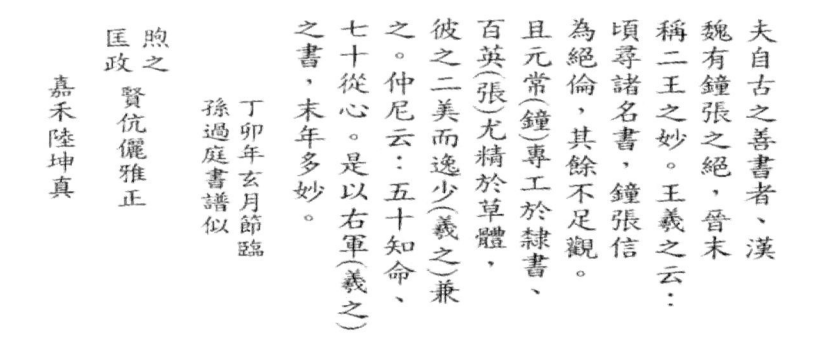

夫自古之善書者、漢
魏有鍾張之絕，晉末
稱二王之妙。王羲之云：
頃尋諸名書，鍾張信
為絕倫，其餘不足觀。
且元常(鍾)專工於隸書、
百英(張)尤精於草體，
彼之二美而逸少(羲之)兼
之。仲尼云：五十知命、
七十從心。是以右軍(羲之)
之書，末年多妙。

丁卯年玄月節臨
孫過庭書譜似

煦之
匡政　賢伉儷雅正

嘉禾陸坤真

Of all the calligraphy virtuosos since ancient times, Chung and Chang were the utmost in Han and Wei Dynasties. At the end of the Jin Dynasty, the best were the two Wangs. Shi-zhi Wang said: "I just searched for all the best writings. I believe Chung and Chang were peerless, and the others are not worth looking at."

While Chung specialized in the Li type of calligraphy, Chang was especially skilled in the script type. Shi-zhi Wang was able to combine both of these two beautiful styles. Confucius said: "One knows his destiny at 50 and can do whatever he wishes at 70." Thus, Shi-zhi's writing became even better in his later years.

— Partial transcript from the Calligraphy Manual of Guo-ting Sun in the Tang Dynasty (September 1987).

To: Dear Shu and Kuang-Chung

From: Kung Chen Loh of Chia Ho

Mrs. M.T. Peng, K.C., and Mrs. H.S. Fang, Taipei, 2000.

Section III

Additional Tributes Written by Speakers at the 23 June 2001 Celebration

A Tribute to Shu Chien

Paul La Celle, M.D.[*]

During my professional career, I have the unusual good fortune of meeting several U.S. Navy Admirals, including Fleet Admiral King and Admiral Schoeffeniels; five Nobel Laureates including Professors Paul, Smith, Brown, Cheh, Gilman; and a number of exceptional individuals in national and international leadership roles. When reflecting on these personalities and their contributions, I am repeatedly impressed by them as people and by the value of their contributions. Their personal characteristics also are uniformly evident: intelligence and thoughtfulness, visionary outlook in their fields, insight in distinguishing the important from secondary matters and unusual communication ability. Finally, all of these have exhibited the attitude of Galileo: others, no matter how humble can teach one something and deserve the respect as a person and for the potential of their ideas. Importantly, these individuals did not change in their personal characteristics as a result of their recognition, indeed several seem to have grown in awareness of younger colleagues and have been increasingly generous in contributing to their development.

My observations of Shu Chien over the years unequivocally distinguish him in the same ways that these individuals are distinguished from their peers.

As a scientist he has produced many original contributions and has clarified previous studies by many scientists through his careful,

---

[*]Professor of Pharmacology and Physiology, Senior Associate Dean for Graduate Education, University of Rochester, Rochester, New York.

uniquely designed experimental approaches. Certainly, the volume of his scientific contributions attests to his unusual insights and perceptiveness, which have permitted him to change in a seemingly effortless manner to new research areas within his general field. Very few individuals have been able to move into new complex areas with the clarity of direction and unique accomplishments, which have stimulated new fields to such an extent as Shu Chien has done.

His unusual leadership ability, noted clearly in his early days at Columbia, has become progressively evident in the developments in the Department of Bioengineering and the Whitaker Institute in San Diego. Concurrently, he has been a key national leader not only in physiology and the general biomedical sciences, but also in major developments such as the new National Institute for Biomedical Imaging and Bioengineering. In his leadership, he is visionary and his energies have been directed to the common good, not his own immediate interests.

An articulate man, Shu Chien has consistently brought thoughtful, insightful support to the work of scientists in training, to colleagues and to the agencies and institutions key to biomedical research. His

Thanking Paul La Celle after his speech.

communication skills have been complemented consistently by the ability to analyze situations and energetically pursue goals to completion. His enthusiasm and thoughtfulness have allowed him to influence others and secure their support in local and national goals.

All his achievements are illustrious and have contributed to us all; however, Shu Chien's personal qualities qualitatively parallel these achievements. He is a humble man aware of his abilities and achievements but also able to place himself in perspective. He is ever generous to others, in enthusiastic encouragement toward their goals and a willingness to recognize merit wherever he observes quality in a person or accomplishment.

Shu Chien has achieved at a very high level in the full range of activities in modern biomedical science and has been recognized for his exceptional contributions. A truly great man Shu has not been changed by his success: he remains Shu Chien. It is a privilege to know him.

Vascular Biology, Tissue Engineering, NIH and Shu Chien

Robert M. Nerem, Ph.D.[*]

Dear Shu,

It is a pleasure to be able to participate in this symposium on your 70th Birthday recognizing your many contributions. You are an individual who is not only a gentleman and a scholar, but also a statesman.

I do not think that I need to explain in anyway why I call you a gentleman. You are just one of those fine individuals who always is constructive in your approach and in your interactions with the research community and your students. I also do not feel a need to explain in any great detail why I call you a scholar. I am sure that this will be addressed by others, but it is clear that over your career you have made enormous contributions to our understanding of vascular physiology. These contributions have ranged across the biological scales including your more recent work on gene expression and its regulation. There can be hardly anyone working on the cardiovascular system who hasn't referenced work by you and your collaborators. You are a pioneer in the field, and you have not only taught your students, but

---

[*]Institute Professor, Parker H. Petit Distinguished Chair for Engineering in Medicine, and Director, Parker H. Peitit Institute for Bioengineering and Bioscience, Georgia Institute of Technology, Atlanta, Georgia.

At pre-birthday party in Shu's backyard, 2001. Y.C. Fung, Van Mow, Savio Woo and Bob Nerem (seated). Luna Fung, Barbara Mow, Pattie Woo, Shu, K.C. and Marilyn Nerem (standing).

you have taught all of us through the insight you have brought to your research.

In addition to being a gentleman and a scholar, you are also a statesman. You have had a number of leadership positions within the research community, but there is probably none more important than the leadership you have provided this past year in helping to shape the new National Institute for Biomedical Imaging and Bioengineering. Shu, you have been the steadying force that has provided a balance between the territorial interests of the radiology and imaging community and the more integrative approach of the bioengineering community. You have provided effective, diplomatic leadership, and have been the key person in working collaboratively with NIH in order to bring this new institute into being, while at the same time maintaining an important role for bioengineering across the entire NIH campus. This has not been easy, particularly in light of some of the "outrageous" demands of a few from the radiology community. Only someone like you could have effectively kept the dialogue with NIH going and in the right direction as well. For this, the entire bioengineering community owes you a tremendous vote of thanks. The younger members of this community in particular will benefit from your diplomacy, as their careers will be critically affected by how this new institute is implemented within the NIH system of institutes and centers.

It is thus a pleasure to be able to participate in this recognition of someone who I consider not only a friend, but an individual who also has an unusual combination of gifts, ones that set you apart from all others in being a gentleman, a scholar and a statesman for bioengineering.

Best wishes, not only for a happy birthday or even a happy birthday year, but also for all the good things yet to come. Thanks for everything!

With Bob and Marilyn Nerem at symposium honoring Bob in Atlanta, Georgia, 2003.

Tribute To Shu Chien

Harry Goldsmith, Ph.D.[*]

We have known each other for about 35 years. We have seen the slide rule give way to the primitive electronic calculator, the punch cards for the big IBM central computers at the University give way to the desktop personal computer, and the trusty GDM viscometer for measuring macroscopic blood rheological properties was left behind as microrheological and finally molecular rheological techniques were developed. Throughout all this time, there was Shu, steady as a rock, friendly as only he could be, always full of praise for other people's work, and incredibly hardworking in the service of his laboratory, his University, and the wider biomedical community.

I have very happy memories of our association. There were the trips to Columbia University for the Friday evening meetings of the Biomaterials Seminar when I spent time, earlier in the day, in the 168th Street P&S laboratories and met with the students, Shunichi Usami, Dick Skalak and Shu. A most memorable occasion was his visit to Montreal in late October 1970, and the experiments we did together with the famous Congo snake (Amphiuma) erythrocytes. Shu had organized the transport of the snake to Montreal from a company in Alabama. The animal was housed in a large tank in the McGill zoology department. Fearlessly, he advanced on the reptile with a syringe and large needle, and took blood from a vessel in the tail. Back in our laboratory, we took movies of the shear-induced motion of

---

[*]Professor, Department of Medicine, McGill University, Montréal, Quebec, Canada.

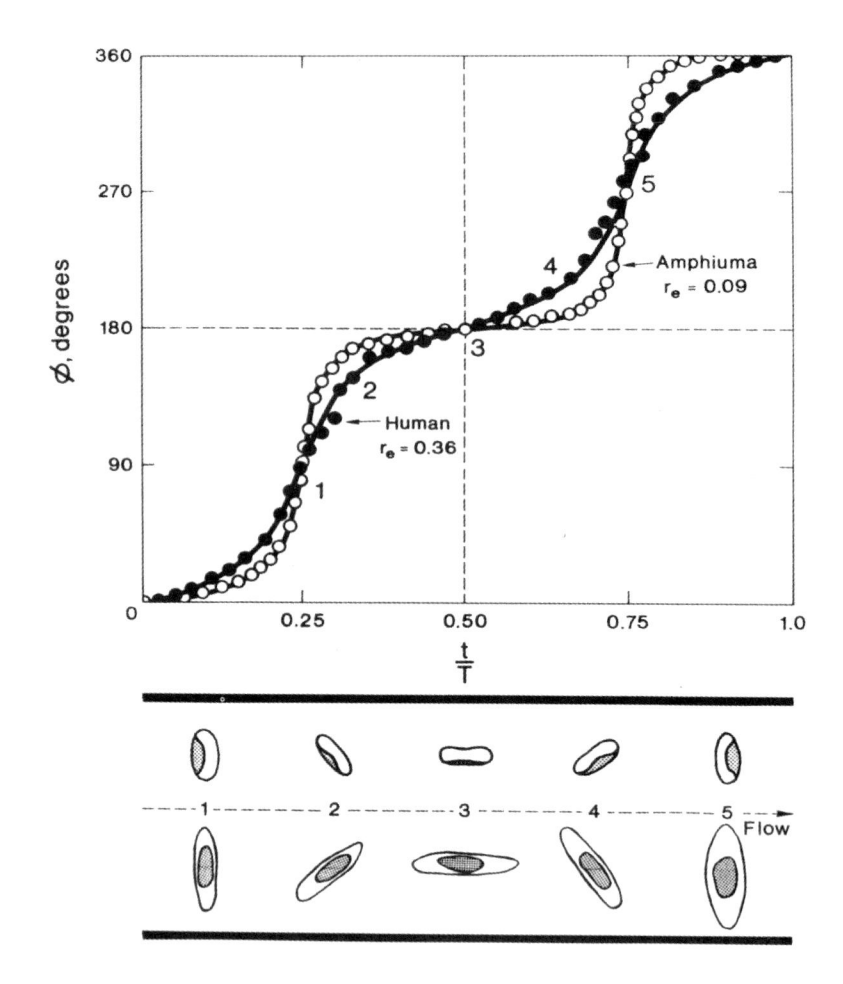

Do you remember a snake?
Dr. Chien, a Congo snake?
That a journey from Alabama did make
And her red cells that tumbled in flow
Through the tube, and bent like a bow
Around the nucleus, just so!

With Harry Goldsmith at the coffee break of the celebration symposium, 2001. At right is the special issue of Biorheology in honor of Shu.

these huge cells which are eight times the size of human erythrocytes. Using the travelling microtube, we tracked them in flow through tiny tubes, and watched them rotate and deform, as they bent around the rigid nucleus.

As a tribute to Shu, the man of science, Geert, Amy and I [Editor's note: virtually, Geert and Harry] compiled an issue of Biorheology to which several of your colleagues and former students contributed 15 articles. It was a real pleasure for all of us to put the issue together, since we hold you in great esteem. Yet, even more important are your qualities as a real "Mensch," a man of love, kindness, generosity and uprightness. I wish you a very happy 70th birthday, and may God grant you and K.C. many more joyful years together.

Shu: You Are the Greatest

Savio L.-Y. Woo, Ph.D., D.Sc.[*]

Writing a tribute that will do justice to a giant like you is not an easy task. So, I chose not to do it, but instead came to La Jolla to participate at the celebration and to congratulate you personally. After attending the special symposium honoring your 70th birthday, and the spectacular banquet party afterward, I was even less certain on whether there are adequate words that would justly describe your significant and lifelong contribution to the fields of biology, physiology, and in particular, bioengineering. Nevertheless, at the insistence of your staff, I will give a few words of my admiration of you.

As I sat at your symposium, I kept thinking about what I might have missed after I chose to leave UC, San Diego shortly after the arrival of you and Dick. In spite of the very gratifying experience and many worthy things we have done at both the Departments of Orthopaedic Surgery and Bioengineering at the University of Pittsburgh, I do wonder, once in a while what it would be like if I were working under your helm. Nevertheless, life has been kind and things have a way of working out to my benefit. Shu, you were kind to invite me to participate at the National Health and Research Institutes of Taiwan's activities almost ten years ago. There, I have had the opportunity to make a contribution to young people but also to personally witness your immense efforts on ROC's national health — an enormous and unparalleled achievement indeed. Your dedication, hard work and

---

[*]Ferguson Professor and Director, Musculoskeletal Research Center (MSRC), Department of Orthopaedic Surgery, University of Pittsburgh Medical Center.

With Savio and Pattie Woo at the celebration party.

福如東海
寿比南山   **from the MSRC!**

leadership have shown so brightly they paled the prestigious National Health Medal you received from the ROC's Department of Health. It is also my great fortune that we should have many other opportunities to work together at the National Academy of Engineering, Biomedical Engineering Society, American Institute of Medical and Biological Engineering, and others. Your extraordinary energy, deep perception, and unique ability to get things done and done right, are exemplary — and I've learned and benefited. You have helped effectively to get someone like me — who does not have very strong ties with Taiwan — to become a member of its most prestigious organization — the Academia Sinica.

Thank you so much! I will be forever indebted to you.

At the dawn of your 70th birthday, you can now choose to do whatever you wish, based on Chinese tradition. Would you consider spending a little time in orthopaedic research? With the advent of your significant interest in the game of golf, I'm sure you can help the sports medicine field to move forward in leaps and bounds.

Happy birthday Shu! Pattie and I wish you and Kuang-Chung happiness and health, always.

Professor Shu Chien: My Dearest Advisor

Si-Shen Feng, Ph.D.*

Professor Shu Chien was my Ph.D. advisor from 1984 to 1988 at the College of Physicians and Surgeons, Columbia University. He is a pioneer in bioengineering — which applies engineering concepts and principles to solve problems in biological and medical sciences with the goal to improve human health care. In the late 70s and early 80s he had become well established in blood rheology and had begun solving the blood flow problem on a cellular level. The topic of my thesis is *Tank-treading Motion of Red Blood Cells in a Simple Shear Flow*, which was to solve the internal and external flow problem of a red cell with moving boundary of the cell membrane under the condition of global and local area conservation. Under his sophisticated guidance, the problem was successfully solved by combining the classical Jeffery's solution of a rigid ellipsoid rotating in the simple shear flow field and a boundary collocation solution of a tank-reading cell membrane with area conserved in a quiescent fluid. The result and prediction of this theory have been proved by experimental measurement. As far as I know, the solution is still, after 13 years, the most complete and sophisticated solution of the red cell tank-reading motion and its effects on the blood viscosity. Another advisor of this work was Professor Richard Skalak, the Director of Bioengineering

---

*Associate Professor, Department of Chemical and Environmental Engineering, National University of Singapore.

At the pre-celebration dinner, 2001. Song Li, K.C., Shu, S.S. Feng (seated), S.Q. Liu, Laurence Canteri, Jin Gang, Xiong Wang, Jackson Wu and Daniel Ou-Yang (standing).

Institute at Columbia at that time. Professors Chien and Skalak demonstrated a perfect match in collaboration between medical and engineering scientists. Under their close collaboration, a lot of medical problems have been successfully solved; which include the active motion of leukocytes, the rheology of platelets, blood rheology, cell and tissue engineering, engineering guided cosmetic surgery, effects of mechanical forces on gene expression, etc.

My career development has been greatly benefited from Prof. Shu Chien. In my early years after my Ph.D., I continued the research on erythrocyte rheology to find the molecular basis of the red cell membrane deformability. I successfully applied the entry spring model to explain the role of the cytoskeletal protein network in cell deformability and developed a tethered adhesive model of intramembrane proteins to include the role of their distributional entropy in cell deformability. My students successfully applied AFM to visualize the inner and outer membrane structure of the red cell membrane. Their work has direct linkage with my Ph.D. thesis.

A few years ago, I realized that I should extend Prof. Chien's philosophy to solve more complicated and more practical problems in human health care. A new area of chemotherapeutic engineering caught my attention. I noticed that most anticancer drugs have limitations in clinical administration due to their poor pharmaceutical properties. Adjuvants are required in current clinical administration, which often cause serious side effects. We are thus developing advanced engineering devices such as polymeric nanospheres and lipid bilayer vesicles for cancer chemotherapy of better efficacy and less side effects and, with further modification, for oral delivery of anticancer drugs, which has not been available so far. Whatever I and my students have achieved and could achieve in the future, it can always be traced back to the education and training Prof. Shu Chien gave to me at Columbia. He is my dearest advisor, the man who changed my as well as my students' lives, and helped us build our careers in bioengineering. Professor Chien has been playing a leading role worldwide in bioengineering. We are inviting him for the Temasek Professorship in Singapore. Happy birthday, Prof. Chien!

The 70/30 Philosophy: A Tribute to a Teacher, Friend and Colleague

John Y.-J. Shyy, Ph.D.[*]

I have always thought that I am a very ordinary person. A special encounter or occasion usually can change the fate of an ordinary person. To me, it is Dr. Shu Chien and his "70/30" philosophy, which not only helped me establish my career but also guided Jenny, my wife, and I through our daily lives.

In the fall of 1990, when I was about to finish my postdoctoral training at the Ohio State University, and was just starting to consider my next step, Dr. Shu Chien's name was mentioned and was always linked with comments such as "what a nice guy." It was these compliments and Shu's academic reputation that led me to his lab at UCSD. During the next eight years, I myself experienced and learned Shu's 70/30 philosophy.

The spirit of this philosophy is to contribute more and take less. As long as a common goal needs to be achieved, Shu takes initiatives, and more impressively, he is willing to contribute 60% of the effort and receive 40% of the credit if it is necessary. In several situations that I know of, Shu's investment/reward ratio, in fact, is 70/30. It seems that Shu does not know how to manage business since he is a

---

[*]Associate Professor, Division of Biomedical Sciences, University of California, Riverside.

With Suli Yuan (p. 236), Gang Jing, Ben Chen, John Shyy and Pin-Pin Hsu, 2001.

lot of time in debt. Quite the contrary, he is always a winner. He wins respect and friendship that make him a most beloved person, and many other fantastic things. Furthermore, this 70/30 philosophy creates a win-win situation that benefits many of the people around him.

What would be the essential elements of the 70/30 philosophy? It requires 3SHU: sincerity, succinctness and sustenance; honesty, heartiness and honorability; understanding, unselfishness and ungrudging service. SHU has a big heart, and only people with big hearts can have room for SHU. This world would be more harmonic and more would be achieved if every ordinary person could practice 3SHU with his or her family members, friends, colleagues, or even a stranger on the street, just like the way Shu treats people. There used to be two Chinese terms to describe the distinct ways by which teachers instruct students: "Yen-Shih" is influencing people with virtue and principle, and "Gin-Shih" is instructing with knowledge and skill. Shu displays both "Yen-shih" and "Gin-Shih."

I would also like to present a 70/30 perspective. We celebrate your 70th birthday and in another 30 years, we will gather together to celebrate your centennial birthday.

Happy Birthday, Chien Shih (Teacher).

A News Report

*Daming Li**

The Chinese newspaper *The World Journal* published a report on 25 June 2001. The picture associated with the article (below) shows the presentation of the Biorheology Festschrift (p. 283) by Dr. Harry Goldsmith to Dr. Shu Chien. The Chinese title of the article says:

"15 Senior Scholars Praising his Contributions to Scientific Research; Shu Chien's 70th Birthday. Teachers, Students, Relatives and Friends Celebrated with an Auspicious and Festival Party."

Harry Goldsmith presented the special issue of Biorheology in honor of Shu [photo taken by Daming Lee and published in Chinese Daily News (June 25, 2001)].

---

*Staff Writer, Chinese Daily News, Monterey Park, CA.

世界日報

美西要聞　　MONDAY, JUNE 25, 2001

15位資深學者分別推崇學術研究建樹
錢煦70壽辰 師生親友祝壽隆重熱鬧

【記者李大明聖地牙哥報導】廿三日（上週六）是著名華裔科學家、聖地牙哥加州大學（UCSD）生物醫學工程研究院院長錢煦博士的七十壽辰，該校師生及其親朋好友以一整天時間，舉行一場大型學術報告會與一場隆重熱鬧的晚餐會，向這位學界泰斗表示崇高敬意。

當天上、下午的學術報告會在校園內的嘉倫禮堂（Garren Auditorium）舉行。由錢煦博士的同事宋國立教授等多人輪流主持。十五位資深學者分別從不同的角度介紹錢煦在生物工程、醫學等方面的建樹，以及為創建該校生物醫學工程系所作出的特殊貢獻。雖然演講內容多涉及高深的科學理論，但講者們仍不乏幽默與調侃的天才。有的稱錢煦對微血管功能的研究是該領域的「臥虎藏龍」；有的說自動化技術在醫學研究中的應用「可以提高錢煦的高爾夫球技」。華裔科學家馮元禎教授更熱情讚譽錢煦的學術成就「有若天際的彩虹」。

當晚，為錢煦祝壽的活動轉移到學校附近的拉荷亞酒店（Marriott Hotel La Jolla）。到會的一百多位華洋人士興致勃勃地觀賞了現場展示的錢煦生活照片、幻燈，及聆聽多位親友回顧錢煦的生平軼事。錢煦一家名人輩出，父親錢思亮是名滿中外的大學者，長兄錢純曾任國民府財政部長，現任監察院長。而錢煦一向謙虛勤勉，律己甚嚴，更積極投身公益事務，迭述他如何在長輩家初識錢煦夫人胡匡政醫師當晚發表感言，細編織四十餘載美滿婚姻。錢煦在外地就業的兩個女兒也專程趕回聖地牙哥，在晚會中發言，形容她們眼中的父親。最後，錢煦致答詞，對大家的一片熱忱表示衷心感謝。

當晚的來賓都獲贈一本裝幀精美、厚達兩百五十多頁的紀念冊，同時也在留言簿寫下了各自的賀詞。其中最令人難忘的當屬袁素莉女士的一首「鷓鴣枝」，概括錢煦的人生「一生輝煌眾人讚」形容當晚的氣氛，更以「舊聞新趣一件件，盡顯宗師風範」、「身正學高情無限」，熱情的祝壽晚會，直到午夜才結束。

Tributes to Dr. Shu Chien on the Occasion of His 70th Birthday Celebration

Robert J. Dellenback, Ph.D.[*]

Dear Shu,

Your 70th Birthday Celebration is a wonderful time to be able to reminisce on the past, and recall the very exciting times, and the many exciting colleagues with whom we shared these times. For me, the most exciting colleagues in the Physiology Department of Columbia University were you, Dr. Shu Chien and Dr. Magnus Gregersen. It was a wonderful experience for me to be able to work with you in your lab during those very early days, as it gave me the opportunity to learn from a master. Not only did I learn techniques and concepts, but I also learned from you the art of teaching. Your perfection in every aspect of the study of Physiology was immediately evident, and the enthusiasm which you showed for your work was an inspiration to all with the Department.

Dr. Gregersen, who actually brought us all together, was a most remarkable man and one whom I wish I could have known better. His enthusiasm, not only for Physiology, but for so many areas of human knowledge was indeed stimulating. I shall never forget Magnus. His insistence on the study of blood flow and viscosity came at just the right moment, and your acceptance of the project and your leadership

---

[*]Professor, Department of Physiology, Columbia University, NYC (1954–1971).

Bob and Dine Dellenback, Shu and K.C., Magnus and Georgiane Gregersen, and Shinichi and Tazuko Usami, 1959.

With Dine and Bob Dellenback in La Jolla, 2001.

within this field can be seen by the extent of your vast accomplishments. You are to be heartily congratulated for all of the work and leadership which you have provided, and most especially for the colleagues and students who have contributed so much because of your leadership. The entire field of Bioengineering would not be where it is without your leadership. The move from Columbia P&S to the University of California at San Diego was wise and rewarding to both you and the University.

Dine and I remember with affection, the relationship which was established with you and Kuang-Chung which gave us the ability to watch your outstanding children, May and Ann, grow to maturity. These hours together at your Birthday Celebration have been rewarding to us, to once again hear of the accomplishments of your entire family. We would like to express our loving thoughts and best wishes to a remarkable family. We send our best wishes for a most Happy Birthday to you, Shu, and fond greetings to all in your family.

Affectionately,

Dine and Bob

Tribute to My Friend Shu Chien

Chien Ho, Ph.D.*

We have been friends for almost 30 years. I have always valued your friendship and your wisdom. I am delighted to be able to be with you, K.C., members of your family, and friends, on the occasion of your 70th birthday. I would like to add a few remarks that previous speakers have not made. I would like to mention the genetic factor in your career. You have, no doubt, inherited excellent genes from your parents, and in some ways, you have made improvements in your inheritance!

Your father, the late Professor Shih-liang Chien, was a distinguished organic chemist and received his Ph.D. degree in chemistry from the University of Illinois at Urbana-Champaign. He was a professor of chemistry and then the President of the National Taiwan University from 1951–1970. He was appointed to the Presidency of Academia Sinica in 1970. Unfortunately, your mother passed away when I first visited Taiwan in the early 80s, so I did not know her. From what I gathered, she was a kind and graceful lady. Your father was a great scientist and an academic leader. Your father was always very gracious to his friends and guests. During dinners, he often gave toasts to his guests and enjoyed his drinks. Like your father, you are always very gracious to your friends, but at the same time also very careful with

---

*Alumni Professor of Biological Sciences, Department of Biological Sciences, Carnegie Mellon University, Pittsburgh, Pennsylvania.

With Chien Ho and Savio Woo, 2001.

how much you drink. I asked you why, and you told me that you have your mother's genes for alcohol tolerance and not your father's! Also, like your father, you are a great scientist and a very able administrator. It is common knowledge amongst those of your friends who knew both you and your father that you are better at public speaking than your father. It appears that you have been able to make some improvements in your "gene" for delivering speeches!

We have worked together on a number of committees during the past 20 years. You are a model committee member and chairperson, very generous with your time. Things always work well when you chair a committee, in spite of some very difficult issues that need to be dealt with. You seem to have a way of steering people away from confrontation to solution!

It has been a real pleasure to be your friend for close to 30 years. I have learned a great deal from you. Nancy and I will always treasure our friendship with you and K.C. Nancy joins me in sending our very best wishes to you for A VERY HAPPY 70TH BIRTHDAY AND MANY HAPPY YEARS AHEAD!

Shu Chien's Influence on Our Lives

Ann L. Baldwin, Ph.D.
*and Timothy W. Secomb, Ph.D.**

In 1979, Tim went to Columbia University to work with Richard Skalak for two years as a postdoc. He soon met Shu and his friendly group at P&S. Two memories from his meetings with Shu stand out particularly. One was Shu's uncanny ability to understand a new project inside out at the first moment that he heard about it. The other was his advice on how to present a graph in a paper so that its message would be crystal clear. This is exemplified by his famous graph of blood viscosity as a function of shear rate.

Not long after Tim left Columbia, Ann arrived to work with Shu as a postdoc. Like Tim, she was impressed by Shu's lightning-fast mind. Shu provided Ann with a perfect environment for research: fine equipment, skilled (and very entertaining) technicians, helpful, friendly co-workers, and his thoughtful comments and guidance. Ann also enjoyed the luxury of ordering anything she needed for her research by just making a phone call (after Ethel's approval, of course!).

In April 1983, Shu received the Landis Award of the Microcirculatory Society at its annual meeting, in Chicago. Afterwards, Shu and several members of his laboratory went out to continue the celebration, including Ann. Making the most of his Columbia

---

*Professors of Physiology, University of Arizona, Tucson, Arizona.

Tim, Julian and Clair Secomb, Ann Baldwin (seated). The Chiens, Edward and Sheri Leonard (standing).

Columbia colleagues at La Jolla, 23 June 2001. (Front row, left to right) Paul Sung, Kung-ming Jan, Shlomoh Simchon, Shu, K.C., Syngcuk Kim, Shunichi Usami, Tazuko Usami. (All other rows, left to right) Richard Chen, Timothy Secomb, Cheng Zhu, Dan Lemons, Gerard Norwich, Emily Schmalzer, Ann Baldwin, Herbert Lipowsky, Sheldon Weinbaum, Walter Reinhart, and Amy Sung.

connection, Tim joined the group, and this also provided an opportunity for Ann and Tim to become more closely acquainted. A year later, Ann and Tim were married and now their children have to put up with discussions of microcirculation over the dinner table.

Shu, we thank you for your beneficial influence on our lives, as on the lives of so many other people!

Happy 70th Birthday to Our Dad

May Chien Busch and Ann Chien Guidera

May and Ann reading aloud this poem they composed at the banquet on June 23, 2001.

Those who know our father
Are all suitably impressed
In every aspect of his life
He's a man whom God has blessed

From intellect and wisdom
To sportsmanship and love
He's shown himself a Renaissance man
One who's a cut above

You know him as scientist and leader
To these accolades we need not add
But let us shed some light on Shu
As May and Ann Chien's Dad

Dad's approach to discipline
Is to show and not to tell
He always got his point across
He'd lecture but never yell

He took us to the movies
(Another chance to nap?)
We bought the extra large popcorn
Dad put it in his lap

Before the feature started
We had eaten half the stock
Dad had us grab our napkins
Between our hands to make a lock

Cooking is a skill he has
He taught our mother to brew
Authentic Chinese chicken soup
He's a real live "Cordon Bleu"

And as busy as he's always been
He nonetheless took time
To edit our college applications
And help us with our upward climb

He excels at just about everything
But there is one item we recall
For fear he might just hurt himself
He wasn't allowed to touch at all

That item is the lawnmower
He should be able to handle that, but
With those scientific theories distracting his mind
More than grass might just be cut!

We're lucky to have Shu as our Dad
Taught us like we were PhD's
What other teenagers could recognize
A stack of RBC's?

But above all he's a modest man

So we're glad you've done the bragging for him
After all he is our Dad
And we forever will adore him!

Love,

May and Ann

To Shu and K.C. —
A Drum Song

Y.C. Fung (Singer)
and Luna Fung (Drummer)

Y.C. and Luna Fung singing and drumming the song they composed at the banquet on June 23, 2001.

To Shu and K.C. for the 70th birthday celebration,

Warm sun, cool breeze, San Diego is a lovely place.
On the edge of the Pacific, the coastal range dips its toes in the ocean.
Where in the world is winter so green, and summer so cool?
Where else can people work and play all year round?

A true scholar with perfect mastery of the art for human relationship, Shu is universally admired.
Meeting people as a spring breeze, K.C. is genuinely loved by all friends.
The two of you studied medicine together, and built your careers as one.
Your joint enterprise is envied by all.

You planted your roots in New York, where you impressed the world with your scholarly accomplishments.
The Wild West is, however, where you galloped to your satisfaction.
You reached great depth in your scientific investigations.
You obtained broad consensus in building up Bioengineering.
You established our Department and served the nation.
It is a credit people will never forget.

Shu and K.C.: Let your hearts bloom!
Your career and research are at their pinnacles!
Your friends all over the world are gathered here
On this most beautiful birthday, wishing you
Everlasting Health and Happiness.

Y.C. and Luna
23 June 2001

獻給錢煦和匡政 慶祝七十大壽

太陽暖，海風爽、　聖比亞哥是个好地方！
南加州的海邊上，　青山眺看太平洋。
冬天綠素夏天涼，　世界何處像這樣？
男女老少到此地，　莫不快樂而喜洋洋！

大事業，舉世欽，　我們敬賀錢先生。
對朋友們如和風，　各人不愛錢夫人。
既同學醫探奧秘，　又共創業東與西。
試看你們的成就，　世上誰不欽且佩！

東海峽，是根基。　学问成就驚世人。
西面來造新天地，　英才多匹任馳騁。
鑽研科學得神髓，　文藝演說尤擅長。
創新系又造学府，　功勞永記在人心上。

錢煦啊，匡政啊，請讓心花多開放。
學問事業造頂峰，口碑人人都崇仰。
全世界的朋友們，共為歡慶聚一堂。
在這美麗的生日，祝你們快樂又健康！

　　　　　　　2001年6月23日 馮元楨 喻嫻士

La Jolla Cove, at the edge of the Pacific Ocean.

Section IV

Tributes from the 8 August 2001 Celebration in Taipei

Section IV Tributes from 8 August 2001 Celebration in Taipei

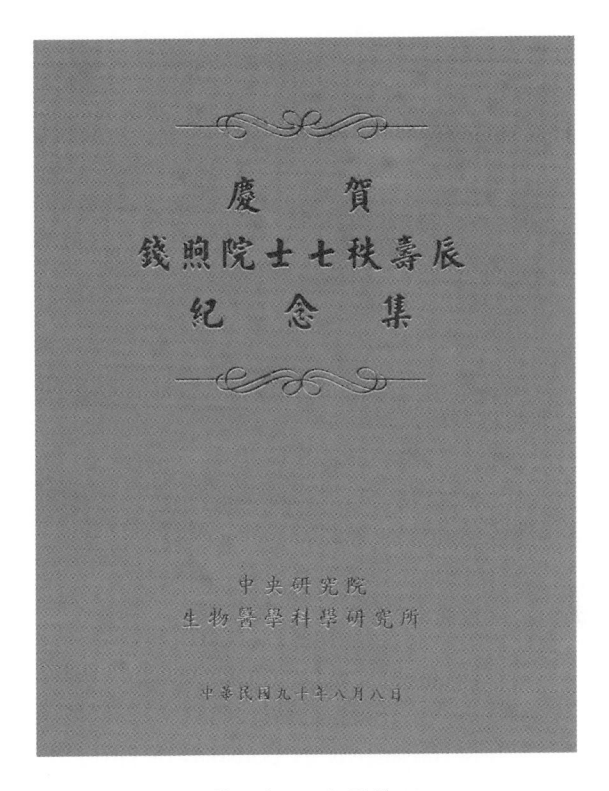

慶　賀
錢煦院士七秩壽辰
紀　念　集

中央研究院
生物醫學科學研究所

中華民國九十年八月八日

Collection of Tributes
in Celebration of Academician Shu Chien's 70th Birthday
Institute of Biomedical Sciences
Academia Sinica
8 August 2001

Chinese terminologies frequently used in the Chinese Tribute Book and speeches for the 8 Aug 2001 celebration:

[1]Lou-Shih, 老師 a respectful and honorable way of addressing a teacher.
[2]Shih-Mu, 師母 a respectful and honorable way of addressing the wife of a teacher.
[3]Eng-Shih, 恩師 benevolent teacher (with gratitude).
[4]Hsian-Sheng, 先生 a respectful and honorable way of addressing a teacher, a senior, or a learned person. It also can be used as Mister or Sir.
[5]Fu-Ren, 夫人 a respectful and honorable way of addressing Mrs., analogous to Madam or Lady.

To Shu Chien

Sunney I. Chan, Ph.D.[*]

To Shu Chien, M.D. and Ph.D.,

On the occasion of your seventieth
Happy Birthday and all best wishes
To the only Renaissance Man I know
Who could give a lecture on blood flow
With all those bloody RBCs
But also art and history of science
With all the penetrating insights
Few of us could even see
Whose Chinese is as classical as gold
Whose English is as polished as silver
All this he brought to Taipei from Peking
And then from Taipei to New York
Across America from Columbia to UCSD
Where he has taught biomedical engineering
But also mandarin to a Chan from Petaluma
To those of us in Academia Sinica
You are the wisest scholar we know
A counselor in prose and medicine
But also humanities and engineering
May your eternal wisdom continues
To flow from sea to sea

Sunney I. Chan
8 August 2001

---

[*]Vice President, Academia Sinica, Taipei, Taiwan.

Sunney Chan presenting his poem at the Taipei celebration, 8 August 2001 (sitting at the front row from left are Yuan Tseh Lee, Yuan-Tsung Chen, Shu, K.C., and Cheng-Wen Wu).

Academician Shu Chien — A Modest, Generous and Decisive Intellect

C.Y. Chai, M.D., Ph.D.*

The Chien family is extremely well known in academic fields and political circles in Taiwan. The talents and abilities of the three brothers Robert Chun, Shu and Fredrick Foo have long been widely acclaimed.

My contact with the Chien family can be traced back to my teacher, Professor S.C. Wang, who was Professor of Pharmacology at Columbia University College of Physicians and Surgeons. Professor Wang and President Shih-liang Chien (Shu's father) were classmates at Nankai University in Tiengin. As a result, I had back in 1956 the privilege of accompanying Professor Wang to visit the Chien family at their home on Foo Chow Street in Taipei.

In 1959, I was granted a fellowship from the China Medical Board of New York to work as a postdoctoral fellow in Prof. Wang's Laboratory at Columbia. At that time Chien Shu Hsian-Sheng[4] already had completed his Ph.D. study in physiology at Columbia and been appointed as an Assistant Professor. Therefore, although he and I graduated in the same year from medical schools in Taiwan (1953), I was a student in his class at Columbia. From 1963 to 1966, I was

---

*Former Dean, National Defense Medical Center (1975–1983). Fellow, Institute of Biomedical Sciences, Academia Sinica, Taipei, Taiwan.
[4]See p. 311.

fortunately granted an NIH International Postdoctoral Fellowship and matriculated as a Ph.D. student at Columbia. During that period, I took the physiology course taught by Chien Hsian-Sheng, thus formally becoming his student.

While at Columbia, because the National Defense Medical Center would not allow an extension of an overseas education beyond three years plus I had to work doubly hard due to my limited capacity, I spent almost all of my time working in the laboratory, leaving the lab only when I needed to attend classes. I rarely went out and did not get to visit the famous tourist attractions in New York. I remember when I was about to go back to Taiwan after my Ph.D. studies, Dr. and Mrs. Chien wanted to give me a farewell party. I still do not know where I got the courage to say to them that I regretted not having been to the world-famous Metropolitan Opera. As a result, they made a special effort to buy reserved tickets and spent the entire evening with me at the opera. We saw Madam Figaro by Mozart, which coincidentally is my favorite classical music composer. Therefore, I can still vividly remember the scenes of that evening.

As a professor at Columbia, Chien Hsian-Sheng was very much liked by his students. Like many medical courses at Columbia, physiology and pharmacology were jointly taught by several professors in the various departments. But Chien Hsian-Sheng was the one most admired by the students. He delivered his lucid and well-organized lectures with simple words and succinct concepts. There was little redundancy or irrelevance to the lecture topic. [Editor's note: occasionally he told a "relevant" joke and the whole class of a few hundred medical and dental students would just go wild: laughing and shouting, at the sametime pounding their desks and stamping their feet.] Some students taped his lectures and typed them out accordingly. The result was a series of complete lecture notes. Such an outstanding ability of verbal expression and organization is clearly demonstrated during his presentations at various scientific sessions and chairing of various types of meetings.

One thing that I most respect Chien Hsian-Sheng for is the way he can always grasp the key points when chairing meetings or scientific sessions. He can pick out the important points made by each lecturer

Academician Chien's 70th Birthday Celebration — Gratitude and Blessings

Eminy H.Y. Lee, Ph.D.[*]

I first saw Academician Shu Chien one day in 1984 when I had just returned to Taiwan to work at the Institute of Biomedical Sciences (IBMS). At that time, I only knew that Academician Chien was a member of the IBMS Planning Committee and that he was a world-renowned scientist. The temporary laboratory borrowed from the Institute of Physics was rather empty at the time, and we all worked hard to set it up. As I was eager to get our research started, I had not paid much attention to the visit of this "famous scientist." But subsequently I saw him several times a year. Mostly, he came with Academician C.Y. Chai, who was the Acting Director of the Preparatory Office (a few times Academician Paul N. Yu was also present), to visit our borrowed laboratories. The two often talked about the development of IBMS and Academician Chien would also ask us about the status of our labs and research. Though Academician Chien worked in the United States, he often appeared at the Institute of Physics. Then, I could tell that this senior scientist truly cared about us!

I only got the chance to know and interact with Academician Chien after he came back to IBMS as the Director of the Preparatory Office in 1987. As I was one of the pioneering scientists at IBMS and

---

[*]Fellow, Institute of Biomedical Sciences, Academia Sinica, Taipei, Taiwan.

was more familiar with matters and procedures at the Institute, I volunteered many tasks for IBMS besides my own research. These included assisting in the preparation of the Institute Bulletins and in the organization of several international symposia, summer student research training, and workshops in and outside of IBMS. As we began to work together, I felt right away that this scientist, who had left the country more than 30 years ago, had no distance from us in terms of communications and thoughts. He was able to adapt completely to the lifestyle in Taiwan without any complaints. Sometimes he seemed even more of a "native" than I, who had been out of the country for only five years. The lack of a cultural gap made it possible for us to work together with greater efficiency and mutual trust. The excellent mutual understanding between Academician Chien and his colleagues was extremely valuable for the successful development of the Institute.

Although Academician Chien did not work at the Institute for a long time (1.5 years), he worked wholeheartedly and was totally

Chinese and English versions of 1988 IBMS Bulletins produced by Shu in collaboration with Eminy Lee and other colleagues at IBMS.

---

[5]See p. 311.

devoted. I remember one stormy night, I had to go back to my lab to pick up some materials. I noticed that Academician Chien had just finished his work and was just getting ready to go home. Chien Fu-Ren[5] was there with him. I remember both of them shared an umbrella, which was pretty much useless in the storm. I thought at that time: "What was so important that he must stay this late in this kind of weather!" Such dedication to IBMS greatly boosted our morale in research work.

Because Academician Chien and I work in different fields, we did not have too many opportunities to discuss research. When he got together with research staff and students, most of whom were rather young, Academician Chien would talk about not only scientific subjects, but also the way of life. What impressed me the most was that he told us the Chinese way of being "loyal and forgiving." He said that one should be loyal to one's own country, loyal to one's group, and especially loyal to one self; we must set and demand very high standards for ourselves, but we should be forgiving toward others. These words were what we read in our textbooks in high school Chinese studies, and also what our parents told us when we grew up, but I had not heard words like these since I went abroad to study. That they would come out of the mouth of a scholar who had been out of the country for many, many years was not only incredible,

Eminy Lee presenting a T-shirt printed with a design of "IBMS" to Shu, 1988.

but also sounded much closer to heart and gave me a special warm feeling. I think of these words frequently during my days working at IBMS, and I find them increasingly more valuable as the years go by. Although Academician Chien worked as the Director for only one and a half years, to be with him and working with him was a lifetime experience. The first impression Academician Chien gives people is his modesty, forgiveness, gentility and grace. But most importantly he has provided a role model of a scholar and a gentleman.

I am very fortunate to have the opportunity to know Academician Chien as a consequence of my decision made 17 years ago, i.e., to return to Taiwan to work at IBMS. Academician Chien's words, deeds and his accomplishments in scholarly activities have provided very important inspirations for my research career.

The Chinese saying is that "kind people have longevity." On this wonderful occasion of Academician Chien's 70th birthday, I offer my sincere wishes to this kind and virtuous senior: May your longevity match the height of the Southern Mountain! May your fortune be as broad as the Eastern Sea!

Gratitude and Blessing

Lee Young Chau, Ph.D.[*]

Time really flies. It was 16 years ago when I first met Chien Hsian-Sheng[4] (I am still accustomed to calling him this way). I was doing my postdoctoral research at the Harvard Medical School, and the Institute of Biomedical Sciences (IBMS) was in a preparatory stage. I applied for a position at IBMS. Academician C.Y. Chai, who had just come to IBMS from his deanship at the National Defense Medical Center (often referred to as Dean Chai), forwarded my letter to Chien Hsian-Sheng. Chien Hsian-Sheng, who was at Columbia University and responsible for IBMS recruitment, contacted me to meet at the 1985 Federation Meeting. Although it was our first meeting and the talk was not long, I was deeply impressed by his gentle, graceful and scholarly demeanor. The next spring (1986) I returned to Taiwan, became the first to return to IBMS after being "interviewed" by Chien Hsian-Sheng in the United States.

During the first six months, I spent most of my time assisting negotiation and purchase of instruments. Although the construction of IBMS building was completed in the summer of 1986, most of the laboratories were empty, except those for Dean Chai, Dr. Eminy Lee and Dr. Lung-Sen Kao (who transferred from the Institute of Zoology), and there were only few people around. My own laboratory also was built from ground zero and got equipped gradually. Finally in the fall, after hiring the first lab assistant, we began conducting experiments.

---

[*]Fellow, Institute of Biomedical Sciences, Academia Sinica, Taipei, Taiwan.
[4]See p. 311.

In January 1987, Chien Hsian-Sheng came to IBMS as the Director. Soon after, several Principal Investigators (PIs) also arrived and more people were hired. Thereafter, several large international meetings were held at IBMS. The institute grew with a fast pace during that short period of one year and it became so vibrant. Chien Hsian-Sheng not only established the Cardiovascular (CV) Research Group, which was his field, but also helped initiate other research groups. I was not in the CV Group, and had little interaction with him research-wise. Nevertheless, whenever he saw me, he would ask me how I was doing and whether the research environment was satisfactory. Even after he left IBMS and went back to the U.S. in the 1988 summer, he would still ask my whereabouts when he visited the Institute.

About one and a half years after Chien Hsian-Sheng left IBMS, the Cell Biology/Cancer Group I originally belonged to was dissolved. Away in the U.S. he was very concerned about my situation. Through his colleague Dr. Usami, who was then at the CV Group, he asked whether I would be willing to join the CV Group. I had always thought that I was recruited by him, so I decided to join the CV Group without any hesitation. Thus I made a major change in my research subjects and direction. Till now I often think that Chien Hsian-Sheng had an important influence for me to stay at IBMS in conducting cardiovascular research, an area not familiar to me at the beginning. Without him, I probably would not have come to IBMS. Without him I probably would not have joined the CV Group and jumped into a new research territory. I am very grateful to Chien Hsian-Sheng for giving me opportunities, continuous encouragement, and care. I believe I am very fortunate to have met Chien Hsian-Sheng, a senior who is a teacher and also a friend, at the start of my independent research career.

It is hard to mention Chien Hsian-Sheng without mentioning Chien Fu-Ren.[5] In 1987–1988, they were my neighbors in the dormitory built for IBMS scholars. The two are a loving couple like body and shadow, so admirable and respectable. Like him, she is warm, kind, polite and easily approachable. While with them, I feel like being bathed by the spring wind, very relaxed and happy. Chien

---

[5]See p. 311.

Fu-Ren is very proud of Chien Hsian-Sheng. Her love and care for Chien Hsian-Sheng is often shown in their interactions. People say that behind a successful man there must be a great woman who devoted herself to his success. Chien Hsian-Sheng and Fu-Ren best exemplify this statement. Although having only spent a short time at IBMS, they have established a deep empathy with colleagues. Every time we see them, not only have their looks not changed much, but also the friendship has not faded, despite the time and separation. For IBMS to continuously develop to today's status over the last 14 years, the foundation established by Chien Hsian-Sheng with his full-hearted contribution is an important and indispensable factor. On this occasion of Chien Hsian-Sheng's 70th birthday, I would like to briefly outline the influence Chien Hsian-Sheng has exerted on the growth and development of myself and IBMS in the past 14 years, and to express my sincere gratitude. I also wish Chien Hsian-Sheng and Fu-Ren, while holding their hands together, reach another high peak of life, and be happy and healthy forever.

Mei-chi Jiang, K.C., Shu and Lee-Young Chau (seated). Danny Wang, Yu-Ling Chen, Shyng-Jong Len, Kung-min Jan, Cheng-Den Kuo and Cheng-Yau Su (standing), Taipei, 1988.

Feelings and Thoughts on the Occasion of Academician Shu Chien's 70th Birthday — Remembrance and Gratitude

Jang Jang, Ph.D.[*]

The first time I met Academician Shu Chien was during an interview in 1987, when he, as Director of the Preparatory Office of the Institute of Biomedical Sciences (IBMS), was on a trip to U.S. with the specific purpose of recruiting research personnel for the Institute. I was about to graduate from Yale University and felt rather uncertain about my future. Academician Chien arranged an interview for me at Columbia University in New York City. The moment I saw him I was immediately impressed by his gentle and graceful demeanor. Academician Chien told me that Academia Sinica would soon establish IBMS and would thus be recruiting a number of talented scientists. At that time, Academician Chien was initiating a Cell Biology Research Program, for which he had found a group of Principal Investigators (P.I.s) from the U.S. to take turns to go to IBMS (for two to three months each). He expressed the wish that I would go back to Taiwan as soon as possible as a long-term P.I. to take over the work started by these rotating P.I.s. I told him that I would have difficulty to return to Taiwan immediately and asked

---

[*]Fellow, Institute of Biomedical Sciences, Academia Sinica, Taipei, Taiwan.

With Tang K. Tang, Danny Wang, Cheng-Yoa Su and Jin-Jer Chen (p. 343) at the International Conference on Fibrinogen, Thrombosis, and Coagulation, Taipei, 1990.

whether I could do so after completing my postdoctoral training in the U.S. Academician Chien said that it was fine and asked me to submit my application immediately. I was surprised that he replied so quickly and definitively, for it would take at least a year for me to finish my postdoctoral program before I could return to work at IBMS. Academician Chien reserved the research position in IBMS for me and sent the appointment letter to me quite early. This would be very difficult today.

I came back to Taiwan on 4 June 1989, the very day of the Tian-an-men Square Event in Beijing. By then Academician Chien was already back in the States, and Academician Cheng-wen Wu had succeeded him as the Director of the Preparatory Office. On the first day of my meeting with Academician Wu, he told me, in good humor, that I was the first person employed by IBMS since his taking office without a formal interview.

I can still vividly recall that on the first day I stepped into my laboratory, I saw all the state-of-the-art instruments and a complete molecular biology lab, which were not far off from those of the top universities in the U.S. Before I came back to Taiwan, I had thought that I would not be able to begin full operations until a good half year later. I never imagined that I could actually start my research right on the second day after my arrival. I really wish to express my thanks to Academician Chien and previous rotating P.I.s (Dr. Kung-min Jan, Dr. Amy Sung, among others) for their excellent work in setting up the lab.

How time flies! I have been working at IBMS for 12 years now. Today, on the occasion of Academician Chien's 70th birthday, I would like to express my heartfelt gratitude for his contributions in establishing IBMS and his special help in providing me with the opportunity to work here. I hereby wish him health, happiness and success in his research. An old Chinese saying has it that "life begins at 70." I do hope Academician Chien will continue to contribute to the development of biomedical research in Taiwan, as he gets even stronger with age.

My Teacher and Friend

Danny Ling Wang, Ph.D.[*]

Hsian-Sheng[4] has long been famous in biomedical fields in the United States. I had the fortune to meet him because of the establishment of the Institute of Biomedical Sciences (IBMS) in Academia Sinica. I first met Hsian-Sheng at the IBMS recruitment meeting in Baltimore. What I felt at that time was that Hsian-Sheng was kind and courteous with people, being a modest scholar who is warm, gentle, learned and refined. During the preparatory meeting for the establishment of IBMS in early 1986 in New York and the seminar later that year in Taiwan, I further appreciated his depth and breadth in scientific knowledge. Later, I had the opportunity to visit his laboratory at Columbia University. Hsian-Sheng came downstairs to meet me, and accompanied me to my car at the end of our meeting. I was deeply impressed by his sincere and warm reception and hospitality during this interview. Subsequently, he phoned me several times at my laboratory in Temple University. His warmth and sincerity inspired me to make the decision to come back to serve my country.

Since I returned to work at IBMS in 1987, I have now become one of the "senior" members. When I first returned, I frequently met with Hsian-Sheng, Dr. Kung-ming Jan, and other colleagues to discuss our research work in the Cardiovascular Group. Hsian-Sheng is serious and meticulous in his scholarly activities. He would rigorously and

---

[*]Research Fellow and Professor, Institute of Biomedical Sciences, Academia Sinica, Taipei, Taiwan.
[4]See p. 311.

thoughtfully go over every presentation and scientific manuscript many times to improve the overall organization, sentence structure and word choices. I am deeply impressed by the fluency, simplicity and forcefulness of his writings. Hsian-Sheng's efficiency at work is even more admirable. I remember sending my manuscripts to him for review while he was in the U.S.; he would correct them and fax them back to me overnight. Such sincere devotion to science is extremely moving. Hsian-Sheng loves research, and he makes best use of time, down to minutes and seconds. Hence, during our Cardiovascular Group meetings, he disliked people coming in late. Whenever lunch boxes were served at the meeting, Hsian-Sheng was always the first one to finish eating. I was really concerned that he might have indigestion due to such rapid swallowing, almost without chewing.

Hsian-Sheng aims at excellence and elegance in his work, and he exemplifies that by what he does. I remember that, when we were preparing for the first Principal Investigators review at IBMS by

Seated: Shunichi Usami, Meei-Jyh Jiang and Shu. Standing: Tang Tang, Danny Ling Wang, Jeng-Jiann Chiu, Yaw-Nan Chang, and Jerry Hsyue-Jen Hsieh, IBMS, Taipei, 1990.

an external committee, he moved the chairs around himself, so that the arrangement in the room would be harmonious and effective. Whatever Hsian-Sheng does, he is always well prepared and chooses the most appropriate way. Furthermore, he clearly separates official and private matters. In the summer of 1987, the Cardiovascular Group had many visiting scientists from abroad, and IBMS held the First Summer Training Program for the medical students from the National Taiwan University (because of its excellent accomplishments, the National Science Foundation has continued to hold this program till today). Hsian-Sheng often paid from his own pocket to invite every one to have dinner at restaurants like the "Peiping Du-I-Chu." We would discuss various matters while eating, like a big family. We would then go back to IBMS to continue to work. I can still remember vividly this warm and joyful period of togetherness to this date.

While Hsian-Sheng demands high standards in research from his students, he always provides thoughtful advice and guidance. Even when he is very busy, he always participates in lab meetings. Hsian-Sheng would chose a book and ask everyone to read a chapter and report to the group. When Hsian-Sheng served as Director of IBMS, he was very kind to his staff and provided a role model through his actions. At that time, IBMS held many symposia, he and Fu-Ren often came to the administrative office to give encouragement and support for the staff. The administrative office usually was brightly lit at night, with everyone working happily without any complaints. What he brought to IBMS at that time was "momentum" and "vitality."

After Hsian-Cheng left his Directorship at IBMS, he still cared very much about the Institute. I also have kept close contact with him. I have benefited immensely in my scientific research, as well as the ways of working with people and on matters.

This kind and wise senior, who is my teacher and friend, is Academician Shu Chien, Director of the IBMS Preparatory Office at its inception. On this happy occasion of his 70th Birthday, I would like to send my special greetings and blessings. May your longevity match the height of the Southern Mountain! May your fortune be as broad as the Eastern Sea!

Forever a Boss in My Heart — Academician Shu Chien

Annie Su-Chin Lin[*]

I am greatly honored to be a part of the "Collection of Tributes to Academician Shu Chien" on his 70th Birthday. I now take a happy time-flight to return to more than ten years ago, when I was very green and did not know very much, an only very average person without much working experience. Thanks to the recommendation of Academician C.Y. Chai, I was transferred from the administrative office to become a secretary for Chien Hsian-Sheng[4] in January 1987. As a result of this working opportunity, I have been very fortunate to know this boss who treats people with modesty and sincerity and deals with matters by emphasizing both perfection and efficiency. The year and half that I worked for Chien Hsian-Sheng gave me the most important learning experience and became a turning point in my working life.

While Chien Hsian-Sheng's scientific contributions and his grace in dealing with people and matters are what everyone knows and likes to talk about, I would like to take this opportunity to share with everyone some of the small stories during the period I worked for Chien Hsian-Sheng, hoping that these light topics may help us to remember the past... .

Everybody knows that Chien Hsian-Sheng is very easy to get along with; he does not discriminate by people's class or level. Even when he

---

[*]Secretary to Director, Institute of Molecular Biology. Formerly Administrative Assistant, Institute of Biomedical Sciences, Academia Sinica, Taipei, Taiwan.
[4]See p. 311.

meets people in unexpected conditions, he can remember the other person's name, characteristics and the topics that they discussed. Do you know what is the secret for this? According to the observation of this secretary of his, besides his inborn talent and wisdom, Chien Hsian-Sheng's main secret weapon is sincerity. "Sincerity" means "true in his heart and honest in his mind." When Chien Hsian-Sheng talks with others, he always concentrates and listens carefully, is truly caring and never thinks about other matters. I believe the sincerity and respect with which Chien Hsian-Sheng treats people is the secret of his ability to remember everything that he encounters just once!

Chien Hsian-Sheng is a messenger of "truth, excellence and beauty" and a warrior with "speed, decisiveness and accuracy." When I helped Chien Hsian-Sheng to type letters, whether in Chinese or in English, or when I drafted documents for him, I was really very careful and proofread everything. The strange thing is that all of the missing or misspelled words could not escape from Chien Hsian-Sheng within one minute of his reading. The correction and improvement of the document was done in almost no time. Chien Hsian-Sheng's high efficiency really is admirable and surprising! I especially respect Chien Hsian-Sheng's "being strict to himself and forgiving to others." Although Chien Hsian-Sheng has always had very high standards and strict requirements in his work, he is kind and patient with his staff. When my work required revisions, he always kindly said, "You already typed fast and very well, but…. if you can slightly change a few words and rearrange the margins and spacing, the entire document will look even better." From my recollection, Chien Hsian-Sheng has never lost his temper on any of his staff; he always uses "encouragement" instead of "blame" in fostering the younger people. This warm "encouragement" is the motive force that has driven me to improve continuously.

Chien Hsian-Sheng keeps a low profile in work and does not seek to be famous. When Chien Hsian-Sheng worked as the Director of IBMS Preparatory Office, IBMS did not have an official car for the director. Chien Hsian-Sheng's duties at IBMS required him to travel to various medical schools, teaching hospitals, the National Science Council, the Department of Health and other government agencies for

The great team of administrative staff at IBMS, 1988. From left to right: Shirley Hu, Annie Lin, Wei-Shiang Tao, Amy Chu, Jamie Jan, Jenny Chen, Vivian Cheng, Nicki Chang, and Ivy Tu.

meetings, lectures, student guidance, etc. It did not matter whether it was under the hot sun in the summer or during stormy rain in the winter, regardless whether it was early morning, midday or evening, Chien Hsian-Sheng always carried his briefcase walking briskly to the front gate of the Academy to hail a taxi to attend to his busy schedules. When I was Chien Hsian-Sheng's secretary, I never heard him complaining about any matters, or try to get any personal kind of benefit. I have never seen Chien Hsian-Sheng asking for reimbursement of any of the taxi fares. From these small matters, one can see that "a modest gentleman, incorruptible when facing wealth, respectful in dealing with matters and noble in personality" is the exact description of Chien Hsian-Sheng.

There are many small stories about Chien Hsian-Sheng! If you are interested in these, please wait for his 80th birthday to get together and have long talks! For now, let us all get together to sing loudly the Happy Birthday song to wish Chien Hsian-Sheng and Fu-Ren health and happiness forever.

Special Birthday Wishes to Chien Hsian-Sheng[4] on His 70th Birthday from IBMS Administrative Staff

The following four tributes are written by four current members of the IBMS administrative staff, who have been working there since the time Dr. Shu Chien was Director of the Preparatory Office.

Nicki Shu-Chuan Chang
Personnel Staff

I recall my interview at IBMS in July 1987. That afternoon I had already settled down in the reception room, and in came two senior scientists: one was gentle and gracious and the other was kind and approachable. The two were very humble, asking each other to take the seat, and that lasted almost as long as one minute! It gave me a very warm feeling, thinking that scholars in the scientific community are indeed very different: They not only possess extensive knowledge, but are also very polite. Only after I have reported to my position in IBMS, I realized that they were the Director Chien Hsian-Sheng[4] and Academician Chai, respectively. That scene of polite yielding has stayed, till this day, in my heart.

---

[4]See p. 311.

Ivy Su-Ching Tu
Cashier of the Administrative Office

Chien Hsian-Sheng[4] came to the Institute of Biomedical Sciences (IBMS) in 1987. At that time I had already been working in the Institute. Chien Hsian-Sheng was very kind to the administrative staff and he was unassuming and approachable. He is a gentle and graceful senior.

I remember that one day I had to work overtime to organize the program project grant application. That night, at nearly 10 p.m., Chien Hsian-Sheng and Fu-Ren[5] finished playing ping-pong and saw that I was still working when they returned to the office. They urged me repeatedly to stop working, as they were concerned with my safety at night on my way home. They insisted on sending me home in a taxi. In the cab ride, as I talked with Chien Fu-Ren about family matters, I saw Chien Hsian-Sheng was already closing his eyes. It hit me then that he was actually very tired but he still wanted to send me home because he cared for his staff's safety. Today when I recall that night, I am still deeply touched by his concern for me. Chien Hsian-Sheng really treats his staff as his own children and cares about them!

When Chien Hsian-Sheng worked in IBMS, he led all colleagues as a teacher and a friend. This let me feel that IBMS at that time was like a big family. Everyone worked together with all one's might toward the same common goal. I always recall this unforgettable period with extreme fondness in my heart. I would like to express my appreciation and my best wishes to Chien Hsian-Sheng.

[4,5]See p. 311.

Shirley Tsae-Shene Hu
Accountant of the Administrative Office

For my being able to step into a career of public service, I must sincerely thank the person who selected and appointed me — Chien Hsian-Sheng.[4] I am grateful that he gave me the opportunity to contribute my heart and my effort toward the growth of IBMS during my long period of public service career.

When Chien Hsian-Sheng was the Director of the IBMS Preparatory Office, he treated the administrative staff as if they were his family members. He cared about us, asking how we were doing, and cared about everybody just like the head of a big family. He was not intimidating as a boss. Instead, he was thoughtful and patient as a leader, making us feel as if we were bathed by the spring wind. As a result, all the administrative staff worked together with a common goal to establish a solid administrative foundation for IBMS.

I sincerely hope that IBMS can always maintain this momentum, and I also wish Chien Hsian-Sheng the best of health and happiness.

---

[4]See p. 311.

Amy Hsiao Hsuan Chu
Staff

When I first saw Chien Hsian-Sheng,[4]

Wow! This man had an unusual demeanor: gentle, gracious, friendly and modest.

He walked into the administrative office at a pace that was neither too fast nor too slow.

All the staff called him Chien Hsian-Sheng.

I, who stood quietly on the side, also smiled and nodded to him.

My colleagues introduced me, the newcomer, to him.

His eyes were bright with spirit, firm and penetrating.

Chien Hsian-Sheng stepped forward and shook my hand, "Hello, Hsiao Hsuan!" he said, in a warm and gentle tone.

To date every time he returns to IBMS, he shakes hands with every colleague as in the past and asks how we are.

When he sees me, who has gained weight and changed shape, without pausing or searching, he still says outright: "Hello, Hsiao Hsuan!"

The feeling of how he treats people with respect and warmth, Chien Hsian-Sheng — He is simply different! He is one of his kind.

---

[4]See p. 311.

The Academician Shu Chien That I Know

Winston C.Y. Yu, Ph.D.[*]

I have known Dr. Shu Chien for about 16 years. During this period I went from having only heard of his name but never met him to working with him when he was Director of the IBMS Preparatory Office, and to occasionally playing golf together more recently. It is really difficult for me to write about Dr. Chien because my writing is not lucid and vivid enough to showcase, in half a page, a person with such a rich experience and charisma. Perhaps I can add two small stories that I know to what we have known about him.

Once during a whole staff meeting when Dr. Chien was serving as Director of the IBMS Preparatory Office, we talked about some administrative chores that I thought might have bothered him. But the reality was that he had never been bothered, he told me afterwards. Because, he said, as soon as a problem occurred he would look for a way to solve it. And he would do everything he could, going all out. As a result, once difficulties are encountered they could usually be swiftly and smoothly solved. This is such a pro-active and positive attitude in life!

Another is the seriousness with which Dr. Chien does his work and the perfection he demands for himself. Whatever he does, whether it is writing a letter, giving a presentation, or even singing a song, he always devotes himself to it. Those of us who heard him sing must

---

[*]Secretary General, National Health Research Institutes, Taipei, Taiwan.

With Winston and Marjorie Yu, and K.C., Taipei, 1990.

have seen his sincere expression and attitude! His current trip to Taiwan (August 2001) had to be cut short because he needed to return to the United States to rehearse a site visit for his grant. His seriousness in working, his modesty in treating people, and his character of total devotion are truly a role model for all of us.

Dr. Chien, I wish you a very happy 70th birthday, despite my disbelief that you are already 70!

To Academician Shu Chien

Jackson Chieh-Hsi Wu, Ph.D.[*]

How fortunate I was that I was able to participate in the Symposium in honor of Academician Shu Chien at San Diego on 23 June 2001! There were no words that could express my heartfelt happiness when I was also able to attend Academician Chien's birthday party held in the Institute of Biomedical Sciences (IBMS) in Taiwan one and a half months later on 8 August. It was most appropriate to select this Chinese Fathers' Day to celebrate the birthday of the founding father of IBMS and a world-renowned father in academia. Six years ago, when I was looking for my first postdoctoral position after graduating from Ohio State University, I happened to learn from a senior schoolmate that UCSD, which is 3000 miles from OSU, had an Academician Shu Chien from Taiwan, and that he was a celebrated professor in molecular medicine and bioengineering and extremely ardent in helping young scientists. As a result, I began to actively seek the opportunity to study under Academician Chien's guidance. Through arrangements made by the aforementioned schoolmate, I finally fulfilled my wish at the beginning of May 1995. After a five-hour flight from Columbus, Ohio to San Diego, I met Academician Chien for the first time. The warm reception Academician Chien accorded me at our first meeting was truly unforgettable. I never anticipated that a scholar of such distinction would be so kind and approachable. In our

---

[*] Associate Professor of Pharmacology; Director of the Office of Registration; and Chairman of the Department of Biological Science and Technology, China Medical College, Taichung, Taiwan.

Chatting with Jackson Wu, Taipei celebration, 8 August 2001.

conversation, Academician Chien showed the kindness and care to a young person that can only be expected from a very familiar and close elder. The deep impression he gave me and the warmest feeling I have ever had since I went to study in the U.S. made me decide right there and then that this was the person I was going remulate.

During my one and a half years in Academician Chien's lab, I worked very hard and learned the way of doing scientific research. Under Academician Chien's leadership, we cooperated with each other under a common research theme, without any barrier or boundary and without any selfishness or fractionation. Post-docs and graduate students would exchange ideas freely. Although there was no requirement on the timing of attendance, the lab was always filled with people and sounds, full of vigor and harmony. When I was in Academician Chien's lab, he was always very busy, including his participation at international symposia, attendance at Academia Sinica meetings in Taiwan, and guidance of department administration, etc. However, whenever I turned to him for help on research problems, he would explain to me with the utmost patience. All the students and research fellows looked forward to the weekly lab meetings, because these were always marvelous academic feasts with Academician Chien's presence. He would listen attentively to each report and come up with suggestions, some on major principles and others on experimental details. He would teach us the important elements in giving a presentation and in answering questions. These became the guidelines that I adhered to when I give academic speeches at all types of national and international meetings.

Five years have passed since I left Academician Chien's lab and started my teaching career in Taiwan. During these five years, I have continued to benefit from Academician Chien's instruction and guidance. Through our discussions, I have continued to learn from him how to pursue scholarly inquiries with commitment and develop new research concepts with creativity. On this wonderful occasion of Academician Chien's 70th birthday, I would like to wish him all the fortune as broad as the Eastern Sea and longevity matching the height of the Southern Mountain. I would also like to take this chance to express my heartfelt gratitude for having the opportunities in the past, present, and future to appoint Academician Chien, a man with tremendous wisdom, to be my role model for life.

On the 70th Birthday of Academician Shu Chien

Jin-Jer Chen, M.D.[*]

I first met Academician Chien in 1983 during the 8th Asian-Pacific Congress of Cardiology in Taipei. That congress was the largest international conference in Taiwan at that time. As the Deputy Secretary General, I was very busy working on various Congress affairs and had very little time to go to the scientific sessions. I learned that Academician Shu Chien, an internationally renowned scholar, a professor from Columbia University, and a graduate from my Alma Mater, would give a special lecture on Rheology and Microcirculation. As a physician working in clinical cardiovascular medicine, I thought that such high-level scientific research would be too far for me to reach, just like the rainbow in the sky. So, I thought, "Perhaps I should not bother to attend," but then, on second thought, I felt I should not lose the opportunity to meet this maestro. Even if I may be like a duck listening to thunders [Editor's note: could not comprehend the meaning]; at least I should sense the magnitude of the thunder shaking! Therefore, I sneaked away from my Congress duties to go to the lecture. Little did I know that this slight shift in my thoughts opened up a new window for my medical career, allowing me to appreciate the broader and deeper fields of basic medical sciences, in addition to clinical medicine.

---

[*]Associate Fellow, Institute of Biomedical Sciences, Academia Sinica. Associate Professor of Medicine, National Taiwan University, College of Medicine, Taipei, Taiwan.

As a result, I have been deeply immersed in basic medical research and enjoying it thoroughly.

Since the days of preparing for the establishment of the Institute of Biomedical Sciences, I have deeply felt Academician Chien's dedication to the encouragement and education of the younger generation. Every week he would always be at the lab meetings to direct and discuss the progress of research and help solve all the problems encountered. Whenever we need to give a presentation, he would come and participate in the rehearsal no matter how busy he was, improving on our techniques of presentation and slide preparation. Every time we went abroad to present at international meetings, Academician Chien was always there to give us encouragement. This greatly enhanced our confidence and allowed us to give our best presentations. I am sure that everyone in IBMS felt Academician Chien's selfless care and support, just like being bathed by the spring wind. I have always been grateful for this. In addition, Academician Chien's broadmindedness and gracious attitude in dealing with people and matters, his not taking credit, not benefiting himself and his uncompromising attitude in the search for truth makes him the ultimate role model for us, the younger generation, to learn.

It gives me great pleasure to congratulate Academician Chien on the occasion of his 70th birthday celebration. Dear Academician Chien, I sincerely wish you health, happiness and everlasting youth! I also would like to take this opportunity to express my deepest gratitude to my benevolent teacher who started me in basic medical research.

A Tribute to Academician Shu Chien on Your 70th Birthday

Jerry Hsyue-Jen Hsieh, Ph.D.[*]

Dear Chien Lou-Shih,

I was very happy to hear that there would be a party for your 70th birthday at IBMS. I recall that I got to meet you through Dr. John Shyy's introduction at the FASEB meeting in Anaheim, California in April 1992. Later, I went to UCSD to pursue postdoctoral research under your direction before my return to Taiwan to assume a faculty position at the National Taiwan University. I have to thank you for your guidance and support through all this time. Your graceful demeanors and the way you pursue scholarly work are models for us to learn.

My first impression of Lou-Shih was your easiness and approachability. When I first saw you, I had the joy of fulfilling my lifetime wish of finally meeting a maestro that I had long respected. Through our conversations thereafter, I was able to feel your sincerity and kindness even more. I remember when I visited you at UCSD and presented my doctoral work to the faculty, you arrived early to the conference room and started to arrange the tables, chairs and

---

[*]Associate Professor of Chemical Engineering, National Taiwan University, Taipei, Taiwan.

curtains in the room so that the audience that arrived later did not have to bother with these things. Although this is only a small matter in life, we can usually see the larger perspectives from these small occurrences. Your true personality of being easygoing, kind and caring showed naturally through this event.

I have also been deeply impressed by how much you care about efficiency. At the time I was at UCSD, you were the President of the Federation of American Societies for Experimental Biology (FASEB). Lots of fax documents arrived everyday. Your strategy was to answer them as soon as they arrived with a superb administrative efficiency, digesting quickly one thick stack after another. You also had the same high efficiency for scholarly research, taking actions to solve funding, equipment, space and other problems. The establishment of the Department of Bioengineering at UCSD and the high ranking in many surveys are all definitive accomplishments resulting from your leadership, capability and diligent endeavors. Along these lines, even the speed with which you finish your meal leaves all of us far behind astonished; this perhaps is also a part of your emphasis on efficiency!

Your scientific accomplishments have attained the position of "a great scholar of this generation," covering medical and engineering fields in research areas related to the cardiovascular system, and publishing several hundreds of scientific papers. When I carefully read some of my manuscripts that have been corrected by you, I respect and admire even more your accomplishments in the English language and your organizational ability. You can use precise and understandable words and concise and fluent writing to make the article more vivid and interesting, capturing the attention of the reader. I also recall how you plan research projects and interpret the results of experiments. You always can get down to the key points very quickly and can also see what others cannot see, speak on things others cannot speak. I have had the fortune of following you; it is like bathing in the spring wind. I have received a great deal of benefits.

On this occasion of your 70th birthday, I would like to use this article to express my appreciation, respect and congratulations. Best wishes for a Happy Birthday and health and happiness to your whole family.

Respectfully yours,

Your student,

Jerry Hsyue-Jen Hsieh

With Jerry and Mrs. Hsieh, and K.C. at the Elario's Restaurant on top of the Hotel La Jolla, overlooking the La Jolla Shores beach, 1992.

Tribute to Academician Shu Chien on His 70th Birthday

Shing-Jong Lin, M.D., Ph.D.*

Academician Chien is the person who has exerted the most important influence on my life. Words cannot express even a small fraction of my gratitude to Eng-Shih.[3] As his student, I always admire and look up to Eng-Shih's graceful demeanor.

In 1985, I had the fortune to be recommended by Dr. Benjamin Chiang of Taipei Veterans General Hospital to study under Eng-Shih. In August of that year I brought my family with me to Columbia University in New York to receive advanced training. Although we had never met, Eng-Shih and Shih-Mu[2] came to meet us at the Kennedy Airport themselves. They gave our daughter Tsern Yun a book about the Little Deer Bamby and a box of chocolate chips. Tsern Yun was elated and held on to these gifts and would not let go. Eng-Shih and Shih-Mu also arranged the place for us to stay and prepared food, so that we felt the warmth of home the moment we got to New York. For this we are deeply grateful. Shortly thereafter, our son Shaffer Huei was born. Shih-Mu bought him a beautiful coat and the baby went through the severe winter in New York comfortably.

---

*Professor of Medicine and Director of the Institute of Clinical Medicine, National Yang-Ming University, Taipei, Taiwan.
[2,3]See p. 311.

Greetings from Shing-Jong and Ya-Fen Lin at the Taipei celebration, 8 August 2001.

During the period of my study, under the constant care of Eng-Shih and Shih-Mu, our whole family had a warm, memorable and lovely time in New York, and our daughter and son also had a happy childhood.

I originally planned to receive training in Columbia University for only two years. However, under the encouragement of Eng-Shih, I began my graduate works at Columbia at the age of 31 and then studied for my Ph.D. degree. Eng-Shih told me that he expected me to complete my Ph.D. studies in four years; this gave me a good deal of pressure. Fortunately, with Eng-Shih's constant guidance and with the help of Professor Kung-ming Jan, I was able to overcome all the difficulties and graduated in three and a half years, thus fulfilling Eng-Shih's expectations. Thereafter, I went to University of California, San Diego in 1989 to participate in the establishment of Eng-Shih's new laboratories there. This allowed me to apply what I had learned at Columbia to help Eng-Shih during his move to UCSD.

Eng-Shih is rigorous in his research and dynamic in his teaching. He is kind and sincere in treating people, and he attends to details and considers all aspects in handling matters. He is respected and loved by his colleagues and students. He provides a role model for me to follow. Every word and every deed of Eng-Shih always gives me a lot of inspiration. After I returned to Taiwan to teach at the Institute of Clinical Medicine of the National Yang-Ming University and subsequently became Director of the Institute to advise Ph.D. students, I always use Eng-Shih as a role model to guide and demand from myself. At the same time, I use the principles I have learned from Eng-Shih to educate my students, with the hope that I can apply Eng-Shih's teaching to cultivate more young medical scientists for research and education in Taiwan. This is how I can pay back the education Eng-Shih gave me.

On this wonderful occasion of the 70th birthday of my Eng-Shih, our whole family respectfully wishes you longevity and health.

Respectfully,

Your student Shing-Jong, with Ya-Fen,
Tsern Yun and Shaffer Huei

My Extraordinary Benevolent Teacher

Lang-Ming Chi, Ph.D.[*]

When I heard that the IBMS of Academia Sinica would hold a 70th birthday party for Academician Chien, the first thought that came to my mind was "Is he really 70?" In my mind, Academician Chien is always, as some 13 years ago, a warm, scholarly, gracious and lovely teacher who is full of vitality.

When I first met Academician Chien in 1987 I was starting my Ph.D. thesis, not quite clear about how to pursue scientific research. One day my advisor Dr. Wen-guey Wu mentioned that Academia Sinica was setting up IBMS, and that a micropipette technique in the research team directed by Academician Chien could be applied to our research. Thus, through the arrangement of my advisor I went, with an uneasy feeling, to Academia Sinica. At that time I figured an unknown individual like me should just quietly come and go to complete my experiments, and a "VIP" like Academician Chien should have no time to care about me. Never had I thought that he actually talked kindly and attentively with me when we first met. This was when I first felt that this "VIP" was extraordinary. Subsequently, during each meeting, he would participate actively in discussions and give valuable suggestions. Throughout the period under his guidance, I never felt an overbearing attitude or high pressure that one would

---

[*]Assistant Professor, Department of Medical Technology, Yuan-Pei Institute of Science and Technology, Hsinchu, Taiwan.

Lang Ming Chi and her husband, 2002.

expect of a "VIP." What I always felt was his respect for people and his genuine concern. This is another reason that I feel he is extraordinary: how could he give people the freedom to develop while asking for success and efficiency at the same time?

In 1988, Academician Chien returned to the United States, and I graduated two years later. Since then, I have not seen him very often, but when we did meet, he always talked with a very kind and caring attitude. Our teacher-student relationship of less than one year never stopped like a sign of rest in the music staff notation. After Academician Chien has experienced the peak of life and been bestowed many halos, he still treats people, matters and things with an extraordinary heart, this is another one of his unique qualities that I have experienced.

Now that I have become a teacher myself, I am even more moved and touched by Academician Chien's extraordinary demeanor. Every time I recall my meeting with him back then as a student, I always feel how fortunate I was! He is always the great maestro in my mind and will forever be the most benevolent teacher in my life. I just regret that my writing is not good enough to express my feelings to Chien Lou-Shih.[1] I sincerely wish Chien Lou-Shih happiness and health on his 70th Birthday, and joy to his whole family.

---

[1] See p. 311.

Longing for Professor Chien

Cheng-Den Kuo, M.D., Ph.D.[*]

"Dr. Chien will be coming back!" I was surprised and overjoyed by the news when I first heard it. On second thought, I wondered why I had such a feeling of surprise and joy. Then I realized that 12 years had passed since my graduation from Ph.D. studies at the Institute of Clinical Medicine of Yang-Ming University and that I had not seen Dr. Chien for nearly ten years. Time really flies! It is hard to believe that ten years passed by at the blink of an eye. How many decades are there in one's lifetime?

Chien Lou-Shih[1] was strict with his students, but he was extremely helpful when they had to face the outside world. When we ran into difficulties and turned to him for help, he would always try his best to solve the problem for us and make us feel comfortable and appreciative. For example, once I felt the need to acquire some computer skills. Shortly afterwards, when the Institute was going to purchase two new computers, Chien Lou-Shih wrote in the purchase order that the vendor must let me disassemble one of the two and that they must then reassemble it before the finalization of the transaction. From this, it is obvious how much help I received in terms of guidance and direction of important research studies. Every Monday morning at Academia Sinica, we were required to report in English on the progress made during the last week. Chien Lou-Shih and several experts from abroad would give their criticisms directly there and then without any

---

[*]Professor, School of Medicine, National Yang-Ming University, Taipei, Taiwan.
[1]See p. 311.

With Cheng-Den Kuo at the Taipei celebration, 8 August 2001.

reservations. When it came to the revision of our draft manuscript, it did not matter how busy Chien Lou-Shih was, he would always manage to return the revised draft with numerous corrections within the shortest possible time. Of course, his aim was to ensure that our manuscripts get accepted by academic journals as soon as possible.

The final oral defense of my doctoral dissertation was held at the conference room of the Institute of Clinical Medicine of Yang-Ming University. Before the oral defense, I drove to Academia Sinica to pick up Chien Lou-Shih. When we arrived at the Institute in Yang-Ming I discovered that I had left at Academic Sinica all the slides and relevant materials I had prepared for the presentation, probably because I was too nervous the night before and had gone to bed too late. You can imagine the expression of disbelief on the faces of committee members, when the oral defense was about to begin. At that moment Chien Lou-Shih said calmly, "This is O.K." He said that I could use the chalk and blackboard to make my presentation, instead of slides, and that this could really show how much I understood the content of my talk. To persuade the other committee members that my mistake was not that serious, Chien Lou-Shih continued that a similar incidence had also happened to him, i.e. once he had forgotten to bring the slides to an important speech and had to make the presentation with only his mouth. Besides getting me out of this awkward situation, Chien Lou-Shih also helped when I had to answer committee members' questions. He truly played the role of an advisor perfectly.

Chien Lou-Shih is 70 now. When I realize that we have not seen each other for ten years, I cannot help feeling melancholy. I feel melancholy for the ruthless passage of time, and also because, as a result of my busy clinical service and research studies, I have not been able to be with Chien Lou-Shih during these years to learn more from him on how to be a better person. Everyone agrees that Chien Lou-Shih has been extremely successful as "a person."

As the Chinese saying goes, "Seventy is just the beginning of life." I hope Chien Lou-Shih will come back more frequently in the future, and will continue to lead us, i.e. his students and students of students, to march forward and further forward on the path of scientific exploration.

On the Occasion of Academician Shu Chien's 70th Birthday: A Tribute

Yu-Lian Chen, Ph.D.[*]

Dear Chien Lou-Shih,[1]

How lucky I was to study under an authority in biology of this generation, you Professor Chien, when I was working for my Ph.D. in Anatomy at the National University of Taiwan! I still remember Lou-Shih's patient instruction and strict requirement on my experiments. Busied with lectures and conferences, you could always find time to take part in our colloquia and give us short but incisive, to-the-point comments, making our learning deeper, more solid and more perfect.

As an international maestro in biology, you are like an evergreen. You are endowed with knowledge and virtue. Your devotion to promote young generation and your influence on many students, far and near, have deeply impressed me. Now that I am also a teacher myself, I have been trying very hard to follow what you have taught me, to cultivate diligently and quietly, to search for the truth, and to sacrifice and contribute to the noble profession of education.

---

[*]Associate Professor of Anatomy, National Yang-Ming University, Taipei, Taiwan.
[1]See p. 311.

During my year in the United States, Lou-Shih led me gradually and patiently in research. Shih-Mu[2] gave me mother-like care, making me warm and comfortable in my first trip abroad. That relieved greatly the solicitude of my parents, who still feel indebted today. After my marriage, you both continue to care about us and gave my children the privilege of visiting you with me in Taiwan. On behalf of my family, I would like to extend to both of you our highest respect and deepest gratitude. Your sincerity, modesty and devotion to science provide a role model for the young generation: goals of my endeavor over the years.

On the occasion of Eng-Shih's[3] 70th birthday, with gratitude and joy, I sincerely wish Eng-Shih and Shih-Mu health and happiness, forever and ever.

Respectfully,

Your student,

Yu-lian Chen

Gratitude from graduate and medical students, presented by Yu-lian Chen at Farewell Ceremony IBMS, 1988.

[2,3]See p. 311.

Thoughts on the Occasion of Academician Shu Chien's 70th Birthday

Phoebe Yueh-Bih Tang, M.D., Ph.D.[*]

The Institute of Biochemical Sciences (IBMS) asked me to write a short testimony for the occasion of Director Shu Chien's 70th birthday. Although it was a short notice of only four to five days, I decided immediately that I ought to write for this wonderful occasion.

I got to know Director Shu Chien in 1987 when I was a Ph.D. student at the Institute of Clinical Medical Sciences of National Taiwan University College of Medicine. At that time I was rather troubled by the difficulty of finding someone to guide me on some of the methodologies for my dissertation entitled "Studies on Microcirculatory Blood Flow of Venous Flap." It was a god-sent that Director Chien and Professor Kung-ming Jan came right at that moment from IBMS to give lectures to all doctoral students in our Institute. After lectures they asked each of us whether we had any questions regarding our research and I told them the crux of my problem. To my surprise, they had an immediate solution for me. They suggested that I learn from Dr. Shlomoh Simchon, who had just joined them from Columbia University for this summer, on the use of microspheres in studying blood flow of the venous flap. As a result, I drove every day that

---

[*]Professor of Surgery, National Taiwan University College of Medicine. Director of Plastic Surgery, Department of Surgery, National Taiwan University Hospital, Taipei, Taiwan.

With Phoebe Tang at the Taipei celebration, 8 August 2001.

summer to IBMS to conduct research, together with Chong-hsian Li and Ben-long Yu, two 4th year medical students at the National Taiwan University, carrying rabbits from the animal center of the National Taiwan University Hospital to Academia Sinica in my car. We made rapid progress and soon were able to give demonstrations in educational workshops at IBMS. These two excellent medical students have, like myself, also become surgeons after graduation. Chong-hsian Li continued to work under my supervision and is now an orthopedic surgeon at the En-chu-gong Hospital. Ben-long Yu has become a general surgeon at the Ho-Shin Sun Yat-sen Hospital. After a year of hard work, I had made considerable progress in animal experiments for my dissertation. Director Chien arranged for us to attend meetings of the Federation of American Societies of Experimental Biology (FASEB) and the Society of Chinese Bioscientists in America (SCBA) held in Las Vegas in the spring of 1988. Busy as he was, Director Chien did not forget to entertain his students; he took us to the most popular show at the Frontier Hotel at night, where we greatly enjoyed the eye-opening performance of the white tigers.

Director Chien was on my dissertation committee. Because I was busy with work plus children and family matters, and because the printing took quite a while, I finished the dissertation only a few days before the defense. As I apologized to Director Chien when I finally handed it to him, he comforted me, to my surprise, by saying: "All dissertations are completed in a rush." These words were extremely heartfelt to me.

Director Chien and Chien Fu-Ren[5] (Dr. Hu) are a heavenly couple admired by everyone. I will remember forever what Dr. Hu said, "A woman should do whatever she can to make herself happy; only then she can make her family happy." With her quality and culture, she is indeed a role model for all women to emulate. I have always taken her to be my goal in learning.

Director Chien and Professor Jan were to me "angels" descending from heaven. Without them, I most probably would not have completed my dissertation. At IBMS, I learned something new everyday and was exposed to the rich academic atmosphere. I still remember that Director Chien went through the trouble of bringing reagents from the U.S. to Taiwan when we needed them in a hurry. His devotion to promote research and willingness to support young scientists were extremely moving.

Now I am fortunate to be Director of Plastic Surgery at the National Taiwan University Hospital, and was recently promoted to a Full Professorship of Surgery at the National Taiwan University College of Medicine, thus have assumed important responsibilities of following footsteps of my predecessors and leading the next generation. Being taught by Director and Mrs. Chien, Professor Jan and Dr. Simchon, is like being bathed in mild breeze and gentle sun, and I have felt the urge of passing this spirit and style onto young physicians and surgeons of next generations to come. I am determined to transmit without any reservations the scholarship, knowledge, experience and skill to them, so that they in turn can benefit more and more people. I would like to present this resolution to Director Chien on his 70th birthday as a modest gift from me.

---

[5]See p. 311.

Twenty Seven Years Ago

Yuan-Tsung Chen, M.D., Ph.D.[*]

The Institute of Biomedical Sciences (IBMS) at Academia Sinica had the privilege of holding a magnificent evening gala on 8 August 2001 to celebrate Professor Shu Chien's 70th birthday. Hundreds of outstanding leaders in biomedical sciences and distinguished guests gathered for the celebration. It was indeed a grand and rare occasion in the academic circle in Taiwan. At that time, I had just taken office as the IBMS Director. As I was familiarizing myself with the missions and operations of IBMS, I became deeply impressed by the thoughtful planning and execution by Professor Chien who, as the first Director at IBMS, established the solid foundation for research and provided the multitude of opportunities for the further development of the Institute.

My relationship with Chien Lao-Shih[1] can be traced back to 1974. That year, after my graduation from the National Taiwan University College of Medicine, I wanted to pursue my graduate study in the field of human genetics. At that time modern genetics was in its germinating stage, and there were only a limited number of universities with established departments in the field of genetics, even in leading countries such as the United States. In the end, I made my choice to go to New York City to study for my Ph.D. degree in Human Genetics at Columbia University. Shortly after my arrival at the campus, I heard that Professor Shu Chien in the Department of Physiology, who came

---

[*]Director, Institute of Biomedical Sciences, Taipei, Taiwan.
[1]See p. 311.

Accepting the Chinese tribute book from Yuan-Tsung Chen, Taipei celebration, 8 August 2001.

from Taiwan, was an outstanding faculty. He had the tremendous honor of winning the "Best Teacher Award" five times elected by the medical and dental students. I purposely rearranged my schedules in order to sit in several of Professor Chien's classes. Indeed, he was able to use simple words to explain difficult principles, and his lecture was very well organized, allowing students to really learn a great deal. Because I attended his classes and learned from him, I have always called him "Chien Lao-Shih." In my subsequent acquaintance with Chien Lao-Shih, I discovered that he not only had great depth and breadth in his scholarly knowledge, but also was modest in his demeanor and unhurried in his conduct. He is truly the best role model of a scholar and gentleman. Similarly, Chien Shih-Mu[2] is a caring, kind, sincere and gracious lady who over the years has taken care of countless friends and students. The two are indeed a heavenly couple in our eyes.

---

[2]See p. 311.

At Chien Lao-Shih's celebration party on 8 August 2001, I respectfully presented, on behalf of IBMS, a colored glaze by Mr. Sha-Chun Wang entitled "May your longevity be comparable to that of the South Mountain" as a celebration gift. In addition, Deputy Director Sho-Tone Lee presented a poem that reads:

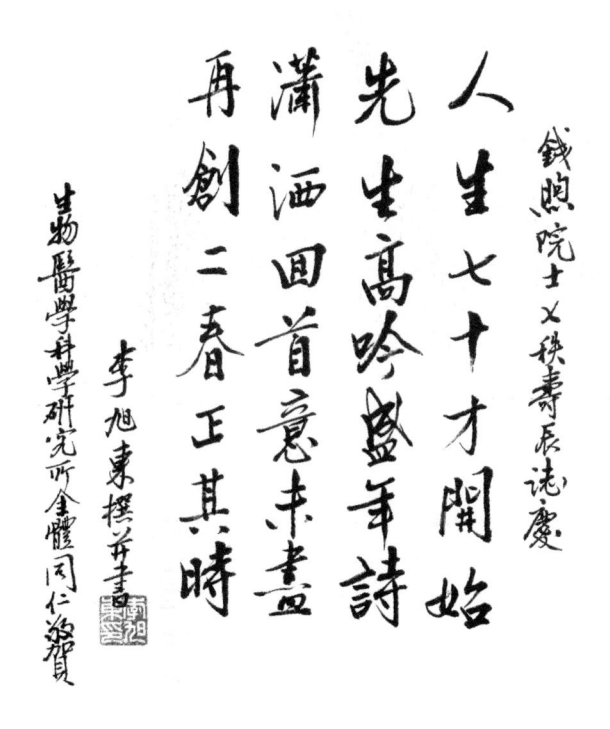

At seventy life is only beginning.
Hsian-Sheng recites aloud the poem of youth.
Looking back gracefully with thoughts still brewing.
This is just the right time to create the second spring.

Professor Chien probably will soon create the second spring, but our hearts have long been filled with warmth as we cherish the memory of the first spring of "blossoming flowers with ever-present fragrance" that he brought to IBMS.

Yuan-Tsung
31 May 2002

Tribute to a Role Model

Tommy Yung-Chi Cheng, Ph.D.[*]

It is my honor and privilege to congratulate Professor Shu Chien on his 70th birthday on behalf of the Society of Chinese Bioscientists in America (SCBA).

I first met Shu at the Grand Hotel in Taipei during the 1982 recruitment meeting for the Institute of Biomedical Sciences (IBMS) and the Institute of Molecular Biology (IMB). I was among the first group of Chinese bioscientists in America who went back to work at IBMS in January 1987. I remember that as I got off the airplane, Shu was going to fly back to the United States. He told me that the budget of IBMS was already in red when he arrived, but I should not worry, and that he would get the funds for the Institute. Shu really was able to do that. I appreciate very much his help in these difficult days and I admire his administrative ability in addition to his scientific achievements.

In 1985, C.C. Wang, Horace Loh and I had the idea that there are so many excellent Chinese biological scientists working in the United States that we should have an organization to enhance our communications, collaborations and contributions. We sent a letter to many Chinese scientists and received many positive responses, including that from Shu. At the first organizational meeting, there were a lot of discussions on the name of the Association. Finally the group adopted the name proposed by Shu, i.e. The Society of Chinese Bioscientists in

---

[*]Professor of Pharmacology, Yale University School of Medicine, New Haven, Connecticut.

Tommy Cheng congratulated Shu at the Taipei celebration, 8 August 2001.

America. I would like to represent SCBA to thank Shu for his many leadership contributions and particularly for his giving the name to our organization.

Shu has made outstanding contributions in Taiwan, and he has also done many wonderful things for Chinese everywhere, including the United States and Mainland China. On behalf of Chinese people, I would like to express our respect and admiration to you, Shu, on this occasion of your 70th birthday. You are indeed our role model.

Heartfelt Greeting and Blessing

Yuan-Tsun Liu, Ph.D.[*] *and Pauline Fu, Ph.D.*[†]

It is indeed a great pleasure and privilege for Pauline and me to participate in this celebration of the 70th birthday of Professor Shu Chien. I learned of Dr. Chien's 70th birthday from one of my students who told me that he read about the wonderful celebration in San Diego on 23 June 2001 in the Chinese newspaper in Los Angeles.

Thirty years ago, Pauline and I were only just graduate students at Columbia University when we were planning our wedding. Instead of a church wedding, we decided to have a traditional Chinese ceremony in Earl Hall at Columbia, and invite a distinguished person to officiate the wedding. Our friend Dr. Kung-ming Jan suggested that we invited Dr. Shu Chien, who was already a well-known scholar at the age of 40. Although we did not know Dr. Chien, we approached him, and mentioned that we were both alumni of the National Taiwan University. He seemed a little surprised but happily agreed. We had a wonderful wedding ceremony in July 1971 and that was probably the first time Dr. Chien ever officiated a wedding.

Dr. Chien's scholarly achievement is widely recognized and admired. He is a wonderful teacher. Anyone who has the opportunity of learning directly from him is most fortunate. As the Chinese saying goes, "It is like bathing in the spring wind." At this wonderful occasion of the

---

[*]President, Soochow University, Taipei, Taiwan.
[†]Chairperson, The Export-Import Bank of ROC, Taipei, Taiwan.

Shu officiating the wedding of Yuan-Tsun Liu (3rd from left) and Pauline Fu (2nd from left) at Earl Hall, Columbia University, New York City, July 1971.

Yuan-Tsun Liu and Pauline Fu (seated) at their 30th Anniversary, Taipei, July 2001. Standing are daughter-in-law Teh-ming, son Chunyi, and daughter Chunchi (left to right).

double 70th birthdays for Professor and Mrs. Chien, Pauline and I would like to present you with this card that says:

"As we remember the sweet and beautiful honor and fortune 30 years ago, we present you with our warmest and heartfelt greeting and blessing today." (see original card in Chinese)

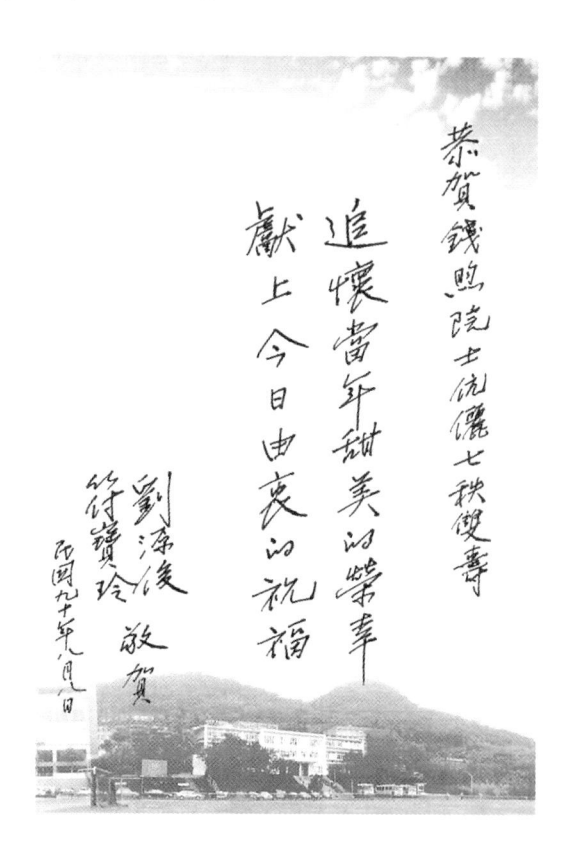

恭賀錢煦院士伉儷七秩雙壽

追懷當年甜美的榮華
獻上今日由衷的祝福

劉源俊
符贊玲 敬賀
民國九十年 月 日

P.S. by Dr. Pauline Fu:

I would like to add that, being an internationally renowned scholar and an excellent medical doctor, Chien Lao-Shih[1] and Shih-Mu[2] are however extraordinarily humble and compassionate. They are our role models and we are very proud of being blessed by them.

[1,2]See p. 311.

My Outstanding Student

H.S. Fang, M.D.[*]

I had the pleasure of having Shu in my Physiology class half a century ago. In the summer of 1954, after Shu graduated from National Taiwan University (Taita) Medical College and completed his military training, Dr. Tom Allen of Columbia University came to Taita as a visiting professor. Tom worked with our colleagues in the Physiology Department to study human body composition, especially the amount of body fat. In order to measure the body fat, the subject has to be immersed underwater, wearing only swimming outfits. According to Archimedes' principle, the difference in weights of an individual in air and in water provides a measurement of body density and hence the percent body fat. It was difficult to get subjects who were willing to be immersed underwater in their swimming suits or trunks. I called President Shih-liang Chien (Shu's father) to see if Shu would be a subject for this study. Shu's mother answered the phone and told me that Shu had a high fever resulting from the impurities in the glucose infusion given to him for the treatment of acute gastroenteritis. To my surprise, Shu showed up the next day for the experiment, saying that his fever had subsided and that he would very much like to be a subject for this interesting study. Twelve years later, another visiting professor came to Taita, and we worked together to repeat these measurements on the same subjects to understand the effect of age on body composition and blood volume.

---

[*]Professor Emeritus of Physiology, National Taiwan University College of Medicine, Taipei, Taiwan.

H.S. Fang gave his tribute at the Taipei celebration, 8 August 2001.

That visiting professor was Shu, who took a sabbatical leave from Columbia University in 1966–1967.

I have been deeply impressed by Shu's accomplishments in research and education. He has made outstanding achievements as a scholar and as a person. I would like to summarize Shu's extraordinary success with three points. The first one is his dedication: Whatever Shu does, he gives his total devotion and he is completely immersed. The second point is his ability to harmonize: Shu can bring everyone who works with him together as a unified team. The third point is his excellent interpersonal relationship. With these outstanding abilities, Shu has been able to make his contributions to the society anywhere in the world.

The success of any individual requires someone working closely together with him. For Shu, this person is Kuang-Chung. She has made wonderful contributions to Shu's success.

Dr. H.S. Fang with K.C. and Shu in their apartment in New York City, 1959.

Everyone has two kinds of age: a calendar age and a functional age. Although Shu's calendar age is 70, his functional age is at most 50. He still has a long way to go, and there are many important contributions awaiting him to make to the scientific community. Congratulations and best wishes for your health and happiness, dear Shu!

A Tribute from the Mainland

Zhu Chen, Ph.D.*

I am very pleased to have this opportunity to take part in this party to celebrate Dr. Shu Chien's 70th birthday.

As a relatively young person from Shanghai, I never had the fortune of meeting Dr. Chien until this SCBA meeting. However, I have been influenced by this world-renowned Dr. Chien since as early as 1978. I read many of his groundbreaking publications in the area of blood rheology and my wife, who studied hypercoagulability and blood rheology, has also benefited a lot in her research from Dr. Chien's publications. After obtaining recently a copy of the Tributes to Dr. Chien from his birthday celebration party in San Diego, I appreciate even more what he has done to contribute to science and society. More than a generation of scientists, though they possibly didn't physically meet Dr. Chien, has been greatly influenced by him.

Dr. Chien has made outstanding contributions to scientific exchange among the United States, Taiwan and Mainland China, especially to the education and training of young Chinese scientists. I have learned about his accomplishments from Academician Shan-Li Yang of our Academy and from Dr. Gang Jin, who has gone from Shanghai to Dr. Chien's lab for his Ph.D. education and will return to Shanghai to work on DNA microarray. Dr. and Mrs. Chien, like the spring breeze, have inspired many young people.

---

*Vice President, Chinese Academy of Sciences, Beijing, China.

It is my great honor to represent the Chinese people on the Mainland to express our most sincere congratulations to Dr. and Mrs. Chien on the wonderful occasion of their 70th birthday.

With Zhu Chen at the Taipei celebration, 8 August 2001.

Section V

More Tributes from Colleagues Worldwide

Expressing Our Feeling at Your 70th Birthday

Yun Peng Wu[*] and R.F. Yang[†]

We had the fortune of meeting Professor Chien in San Diego in the summer of 1979 when we were attending the World Congress of Microcirculation. Dr. Chen's scholarship and modesty truly express the virtues of our Chinese people.

With R.F. Yang, K.C. and Yun Peng Wu at Peking Palace Restaurant, San Diego, 2000.

---

[*]Former President of Chungqing University and Honorary Chairman of Chungqing City Technology Association.
[†]Professor of Bioengineering, Chungqing University, Chungqing, China.

Time flies, and more than 20 years have passed. During all this time, we continue to receive Dr. Chien's advice and the flow of his friendship.

On this occasion of Dr. Chien's 70th birthday, we would like to present this poem to express our feeling:

> Two decades passed at the blue sea and palm-lined park,
> Across the Pacific Ocean feelings are dear.
> Day after day you study from dawn and research till dark,
> Superior character and scholarly work are without peer.

> 椰林碧海二十年，
> 大洋兩岸情誼綿。
> 晨讀暮研如一日，
> 人品文章皆空前。

A Tribute to Professor Shu Chien

Fengyuan Zhuang[*]

Professor Shu Chien is internationally recognized for his contributions to biorheology, biomechanics and bioengineering. He has received great respect from colleagues all over the world. As a Chinese-born, internationally famous scientist, he has done the best to promote biorheology and biomechanics in China. To acknowledge outstanding and original research done by young Chinese scientists, a Chien & Fung Award was established in honor of Professors Shu Chien and Yuan-Cheng Fung. Professor Yuan-Cheng Fung was my adviser when I studied at the University of California San Diego (UCSD), and both Professors Chien and Fung have been my most important teachers and friends in my career.

On the 70th birthday of Professor Shu Chien, I am very happy to recall when I first met him. In 1980, I was studying Bioengineering in the Department of Applied Mechanics and Engineering Sciences at UCSD under the guidance of Professor Y.C. Fung. It so happened that a Chinese Biomedical Engineering delegation would visit several institutes in USA to gain experiences for Mainland China. Professor Fung suggested that I accompanied them. When the delegation visited Columbia University in New York, I met Professor Shu Chien for the

---

[*]Professor of Bioengineering and Dean, Institute of Bioscience and Bioengineering, Beijing University of Aeronautics and Astronautics, Beijing, China.

first time. He shared his experiences with us and showed us his lab enthusiastically. Since then, I have had several chances in international meetings to learn from his plenary lectures and to discuss the development of bioengineering in Mainland China.

In 1989, the 7th International Congress of Biorheology was held in Nancy, France. I listened to his plenary lecture entitled "Molecular Biorheology, Guiding Us Into a New Frontier." We also got together at Dr. Xiong Wang's home, sang Chinese folk songs and had a wonderful time. In 1995, I attended the 9th International Congress for Biorheology held near the Yellowstone National Park in USA. After that I revisited UCSD, especially the Institute for Biomedical Engineering and the Department of Bioengineering headed by Professor Shu Chien. This visit convinced me that the bioengineering research at UCSD is at the forefront, largely due to Professor Chien's efforts.

It was my great honor and pleasure to play host to Professor and Mrs. Chien during their visit to Beijing in September 1995. This was their first since leaving Mainland China in 1949. It coincided

With K.C., Fengyuan Zhuang and Hui Miao, who won the 1st Chien-Fung Young Investigator Award, at the 2nd International Conference on Medical Biorheology and the 2nd Chinese National Congress of Medical Biophysics, Shanghai, 1995.

with the 50th anniversary of the victory of anti-Japanese war and they attended the celebrations. To welcome him, a Workshop for Hemorheology and Microcirculation was held in the China-Japan Friendship Hospital. Most scientists in the fields near Tianjin and Beijing attended and presented their work. Professor Chien gave two lectures. In the one-day workshop, the exchange was intensive and enjoyed by all participants. Professor Chien appreciated the research work reported and all who attended had the chance to learn about the frontier research going on in the world.

The Chiens also visited several historic places, such as the Forbidden City, Summer Palace, Great Wall and Tiananmen Square, and were duly impressed. Afterward I accompanied them to Shanghai to attend the Joint Meeting for the 2nd International Conference on Medical Biorheology and the 2nd Chinese National Congress of Medical Biophysics. I was very happy that Dr. Hui Miao, my assistant at that time, received the first Chien-Fung award there.

Later I also had the chance to listen to Professor Chien's talk at the 6th World Congress for Microcirculation in Munich (1996), the 10th International Congress of Biorheology and the 3rd International Conference of Clinical Hemorheology in Hungary (1999), and Euromech Colloquium No. 420 Mechanobiology of Cells and Tissues in France. His lectures were all very instructive and inspiring to me.

Professor Chien cares very much about the development of Bioengineering in China. We often got together with other scientists from China and promoted exchange worldwide. He has given us many good suggestions. In the history of Bioengineering development in China and the world, the contributions of Professor Chien should never be forgotten.

On the 70th birthday of Professor Shu Chien, we wish Professor and Mrs. Chien great enjoyment of life, retain your vigor, and continue to make great contributions for the development of Bioengineering in China and the world.

Sincere Personal Congratulations from Gothenburg

P.I. Brånemark, Ph.D.[*]

June 23, 2001

Dear Shu,

At the celebration of your 70 year birthday there is a very special feeling, looking backwards and forwards.

We have put together pictorial recordings of some special moments that remind us of the particular excitement of our joint endeavours to bring observations at the cellular and molecular level to clinical applications in attempts to improve the quality of life of fellow beings.

To me, there is today some particular memories of our various visits and projects together with Dick. There is also a deep sense of respect for interdisciplinary generosity, with great expectations for future possibilities in reconstructive prosthetic procedures based on the recipe of Mother Nature.

We have tried to illustrate and visualize this philosophy in a piece of Swedish sculpture of glass, shaped by the senior sculptor Bertil Vallien, who symbolizes our efforts to reach the nucleus of truth in the forth dimension of time and space that Dick suggested to us.

I send you my brotherly greetings as symbolized by the sign that you yourself suggested in the eternal perspective.

Yours,

P.I. Brånemark

---

[*]Professor and Director, Brånemark Osseointegration Center, Gothenburg, Sweden.

We have found in our archives documents over more than three decades of personal and scientific interaction that remember us of moments of excitement in research. Particularly they remind us of your very generous personal support and encouragement, and the participation of your departments - both east and west as well as your international network - in our attempts to understand

blood as a mobile tissue.

Per-Ingvar and co-workers today and yesterday.

The Swedish-Gothenburgian branch
of
International Microvascular Research
would like to convey sincere personal
congratulations to you at your

70 year celebration

wishing you continuous harmony
and
health for all the years to come.

384 *Tributes to Shu Chien*

Shu at P-I's 70 year occasion.

A very special personal message and warm greetings, remembering the past and anticipating the future.

A good joke at an international dinner occasion.

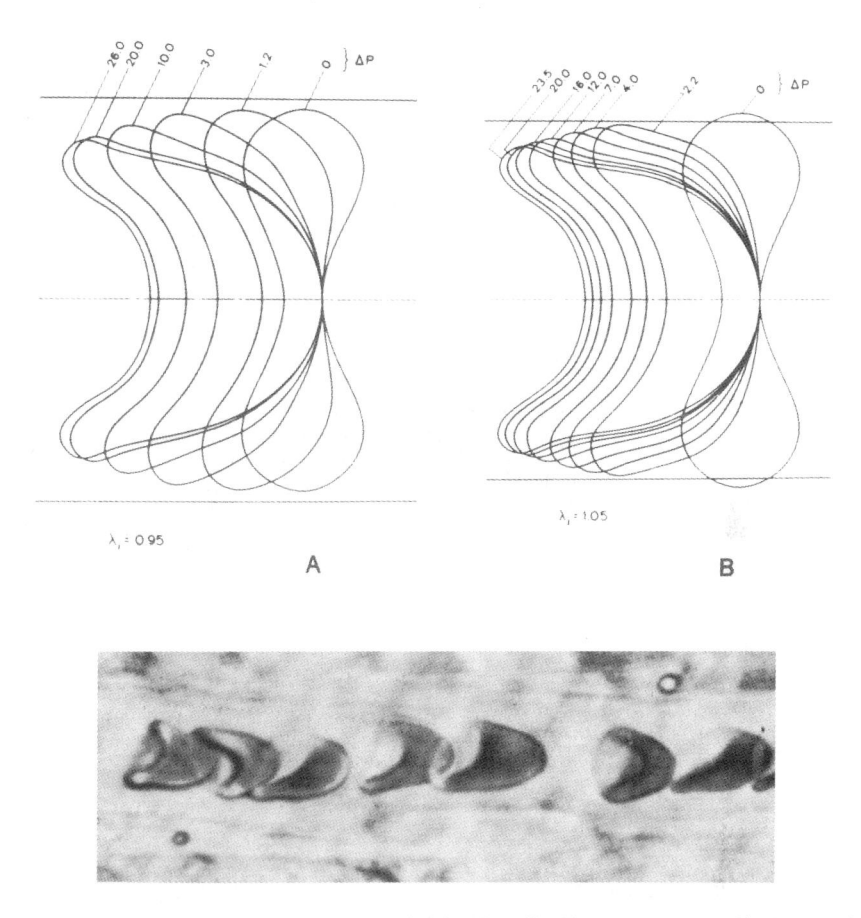

Top: Computed shapes of two red blood cells flowing in capillaries with initial cell/tube diameter ratios of 0.95 (A) and 1.05 (B). Each cell shows progressive deformation in response to increasing pressure drop (ΔP, in dyn/cm^2) across the cell. From R. Skalak and H. Tözeren, 1980, and included in R. Skalak and S. Chien, 1981 in Publication A145. Bottom: Deformation of red blood cells in a living human forearm skinfold flow window. Flow is from left to right. From R. Skalak and P.-I. Brånemark. Science 164: 717–720, 1969.

P.I. Brånemark, Dick Skalak, and Shu, 1991.

A graph composed by P.I. Each of the Chinese characters above "Shu," "Dick" and "P.I." is "person." The combination of the three characters is one way of expressing the word "a group of people" or "team."

Section VI
Shu Chien's Life

Appreciation, Gratitude, Reflections and Prospectives

Shu Chien

I. APPRECIATION AND GRATITUDE FOR THE CELEBRATIONS

A. The Birthday Celebration on 23 June 2001 in San Diego

I wish to express my most sincere appreciation and gratitude to everyone for the fantastic celebration of my 70th birthday on 23 June 2001 in San Diego. This is an event I treasure immensely and will always remember with tremendous pleasure and eternal gratitude.

I would like to thank most sincerely the committee, composed of Geert Schmid-Schönbein, Y.C. Bert Fung, Amy Sung, Paul Sung, Shunichi Usami, Julie Li and Jennifer Griffin (as well as Song Li and Elizabeth Hickman in the early stage), who made superb planning for this unforgettable event. When the committee started planning about a year ago, it was meant to be a surprise party. Geert and Shunichi were getting pictures from my wife K.C., and I wondered why she suddenly wanted to go over so many old pictures. A couple of months before the party, several friends seeing me at meetings and other occasions said to me that they looked forward to coming to La Jolla on 23 June, and others apologized that they would miss the event. By then, it became apparent what my friends on the committee were doing, and I surprised them by offering to help in finding pictures.

I am deeply grateful to everyone who worked on this celebration and greatly impressed by their energy and dedication. Geert, besides being the organizer of the program, was responsible for making the four precious posters of photo collections on my family, my homeland, and my life in science and bioengineering for exhibition at the 23 June banquet. His son Peter put in a lot of time helping to scan the photographs and compose them into a poster format as a computer file. Jerry Norwich and Tatyana Matusov also scanned many slides and pictures and Peter Chen took beautiful pictures of paintings and medals for the posters and for this tribute book. Frank Delano printed out the computer file as four marvelous 3-ft × 4-ft color posters. Alex Chuang, Executive Director of the San Diego Chinese Historical Museum, re-photographed many pictures he obtained from K.C. and others and added the photos he had taken over the years to prepare two wonderful posters with the help of his wife Agnes.

Amy Sung is responsible for the publication of this fantastic tribute book (L.A. Sung, Preface). The meticulous and thoughtful preparation included the invitation of friends and relatives to write articles, the reformatting and organization of these articles, the insertion and improvement of photos, the preparation of a table of contents and an index, and the arrangement for printing with a tight time schedule. An undergraduate student Thomas Kim provided superb assistance in computer work. In the final frenzy days, several members in Amy's lab, especially Carlos Vera and Lynn Truong, and Eric Sung, who was back from the Bay Area for a short time, also provided the timely help. When the book was ready for the first printing just before the 23 June party, the memory size was so large that it required Andrew McCulloch's printer and Elise Louie's assistance to print out the first complete color copy. Tribute articles continued to arrive after the book was sent to print, and speakers on 23 June who had not sent in a written tribute were invited to send in theirs. These are now included in this new edition of the book, and Amy also asked me to write this article. I appreciate greatly this opportunity to express my heartfelt gratitude to all those who have written articles in this tribute book and/or contributed to and participated in the celebration.

Harry Goldsmith, Geert and Amy [Editor's note: Virtually, Harry and Geert] edited the special issue of *Biorheology* as a Festschrift for me, with a very kind preface and 15 wonderful papers on "Perspectives in Molecular Biorheology" by friends and colleagues working actively in the frontier of the field. I appreciate immensely this special Festschrift, which will always be treasured.

I am deeply moved by the warm responses of friends and colleagues to the invitation to the celebration. Many of them came to La Jolla from far away, including Gothenburg, Sweden (Ulf Bagge, who also represented P.-I. Branemårk), Zurich, Switzerland (Walter Reinhart), Nancy, France (Xiong Wang and Laurence Canteri), London (David and Winnie Hu), Singapore (Si-shen Feng), Taichung, Taiwan (Jackson C.H. Wu), Montreal (Harry Goldsmith), Atlanta (Bob and Marilyn Nerem, Cheng Zhu), New York City (Ed and Sheri Leonard, Van and Barbara Mow, Shlomoh Simchon), Pittsburgh (Savio and Pattie Woo), University Park, PA (Herb Lipowsky), Bethlehem, PA (Daniel Ou-Yang), Houston (Larry McIntire), Memphis (Mike Yen), Seattle (Jim Bassingthwaighte), Bay area (Ben and Alice Chen, Song Li, Pin-Pin Hsu), Riverside (John and Jenny Shyy), and Irvine (Bill and Nancy Craig, Gang Jin, Ann Lee-Karlon). Our daughters May Chien Busch and Ann Chien Guidera came to La Jolla from London and New York, respectively. On 22 June, K.C. and I hosted a pre-birthday party at our house in honor of the out-of-town guests and members of the committee planning for the celebration. This provided a wonderful opportunity for us, as well as May and Ann, to meet with the friends who came from outside San Diego and to thank the committee members.

The Symposium on 23 June was a tremendously moving and memorable event. The well-organized program proceeded smoothly, thanks to the committee and to the valuable help by the Department of Bioengineering administration, including Chair David Gough, Acting MSO Loretta Smith and her mother Carmen Field, and staff members Linda Browning, Csilla Csori, Kay Faris, Lore Meanley, Katheryn Ritten and Linh Vu. The program was beautifully organized (pp. 549–558). Geert opened with a kind introduction of my career. This was followed by sixteen warm and generous speeches, starting with Dean Bob Conn and ending with Bert Fung. I was deeply touched, humbled, and grateful.

The banquet in the evening on 23 June at the La Jolla Marriott Hotel was marvelous. I had never seen anything like it, let alone experiencing it myself. The four posters prepared by Geert and associates and the two posters by Alex and Agnes Chuang captured all the highlights of my life, from the time I was a baby to the present. They were beautifully and thoughtfully done. I received a gorgeous digital camera as a birthday gift, and I immediately used it to take pictures. Harry Goldsmith presented me with the special Festschrift issue of Biorheology (p. 281). Daming Li from the Chinese newspaper *World Journal* wrote an excellent report on the birthday celebrations and printed the picture of the presentation of the Festschrift (p. 292). There were again wonderful speeches by friends and colleagues, bringing back so many precious memories. Then May and Ann read a lovely poem they composed (p. 302), and Bert sung a marvelous song that he wrote (p. 306), with Luna doing a beautiful tympanic accompaniment; they are so precious. After K.C. made a very moving speech, I sincerely thanked everyone from the bottom of my heart.

Throughout the evening, many friends were taking pictures, especially Shlomoh Simchon (p. 238), Alex Chuang and Fah-seong Liew. Shlomoh sent me over 100 wonderful digital images. Yingli Hu took the video pictures throughout the day, including the Symposium and the Banquet, and these will allow us to frequently review the wonderful happenings on that marvelous day.

B. The Birthday Celebration on 12 August 2001 in Taipei

One unforgettable event was followed by another. In the evening of 8 August 2001, another marvelous birthday party was given in my honor at the Institute of Biomedical Sciences (IBMS), Academia Sinica in Taipei by the National Health Research Institute (NHRI, President Cheng-wen Wu), IBMS (Director Yuan-Tsong Chen) and Society of Chinese Bioscientists in America (SCBA, President Ing-Kang Ho), which was holding its annual conference there. The party was attended by nearly 200 former teachers and students, colleagues, friends and relatives. Danny Wang and Eminy Lee of IBMS were the main organizers of the event, with the participation of many friends and colleagues, and Wen-Harn Pan was the Master of Ceremony. There were lovely songs and beautiful performance by the children of the IBMS employees with whom I had the pleasure of working in 1987–1988. Following a sumptuous buffet dinner, the program started in the IBMS auditorium. Eminy Lee showed lovely slides that summarized my work and life at IBMS, as well as some of my family pictures. After Cheng-wen Wu's wonderful opening speech, warm, kind and memorable remarks were given by President Yuan T. Lee and Vice-President Sonny Chan of Academia Sinica (p. 312), C.Y. Chai (Past Dean of National Defense Medical Center and one of the founders of IBMS, p. 314), Monto Ho (Past Chair of Microbiology of University of Pittsburgh, Head of Clinical Research at NHRI), Prof. H.S. Fang (my Professor of Physiology in National Taiwan University College of Medicine), Tommy Cheng (Past President of SCBA and Professor of Pharmacology at Yale, p. 364), Yuan-Tsun Liu (President of Soochow University; I had the honor of officiating his wedding with Pauline Fu in 1971, when both of them were students at Columbia, p. 366), my sister-in-law Julie Chien (Fred was in Africa as a Special Envoy for the President of R.O.C.), Yuan-Tsong Chen (newly arrived Director of IBMS, p. 361), Zhu Chen (Vice-President of Academy of Sciences, Beijing, P.R.C., p. 372), and Danny Wang (p. 328). After K.C. gave her words of appreciation and reminiscence, I expressed my warm thanks and sincere gratitude and summarized my life experience, especially in relation to Taiwan. I received beautiful flowers, marvelous gifts, lovely songs and dances by the children, and a precious tribute book in Chinese, which has been translated into English as Section IV of this tribute book. I feel extremely blessed and am most grateful to all my friends and relatives on both sides of the Pacific Ocean and everywhere. I started to write this article on 23 June 2001 which is my official birthday, but the date actually was based on the lunar rather than the western calendar (see next page). I completed the first draft on 12 August, which happens to correspond to June 23 for the lunar calendar in 2001.

C. The Tribute Books

As mentioned above, Amy Sung and many other friends got the first version of this Tribute Book (255 pages) finished before the 23 June 2001 celebration in San Diego. Because of the subsequent arrival of additional articles, an updated version (381 pages) was printed in September 2001, and Amy arranged to have this version categorized by the Library of Congress. Both of these versions are in English. At the 8 August 2001 celebration in Taiwan, I received from Yuan-Tsung Chen another Tribute Book in Chinese (58 pages) which was prepared by Eminy Lee and many other friends there. Amy and K.C. thought it would be nice to translate these articles into English and combine them with the second version of the Tribute Book prepared in San Diego, as well as several additional articles received. During a trip to Singapore in August 2001, I had the pleasure of meeting Joy Quek, an Editor of World Scientific Publishing Co., which produces high-quality books such as *Collected Works of Nobel Laureates* and the *Compendium of Bert Fung's Writings*. After seeing the second version of the Tribute Book, she kindly offered to publish the final version, which includes all the updated materials. Amy and K.C. have arranged to get all these materials together, and Jennifer Griffin has obtained permissions from all the authors. I am grateful to everyone involved in making it possible to publish this book, which is the best and most precious birthday gift I have ever received.

II. PERSONAL REFLECTIONS

In resonance to the wonderful tribute articles in this book on my life from childhood till I became 70, I have written my personal reflections to connect the precious pieces with references to them. I am grateful to have the opportunity to reflect on my life, and I am grateful for everything that I have experienced growing from a baby to 70 years of age.

A. 1931–1957

1. Childhood and Early Education

I was born in Beijing in 1931 to my father Shih-liang Chien and mother Wan-tu Chang as their second of three sons. The official birth date that appears on my passport and all other documents is 23 June but it was the date for the Chinese lunar calendar, which is usually more than one month later than the Western calendar. According to the Western calendar, my actual birth date was 6 August 1931. Because the dates corresponding to 23 June of the lunar calendar move around from year to year, my birthday according to the lunar calendar would usually be some other day, generally in August. For example, in 2001 it was 12 August. Therefore, I could have three birthdays per year, i.e. more than 200 birthdays over the 70 years. Since 23 June is my official birthday, it is wonderful to have the celebration of my 70th birthday in San Diego on 23 June 2001. The celebration in Taiwan on 8 August 2001 was right in between the other two birthdays (6 August and 12 August).

My paternal grandfather was a criminal court judge living in a town house in the French settlement off a street called Avenue Foch, which was named in honor of the famous French Marshall in the First World War (C. Hu, p. 3). The Chinese translation of Avenue Foch was Foo Shu Road. Foo means fortune, and Shu means warmth. So grandfather named me Shu. When K.C. and I went to Paris with our two daughters for a tour in 1971, K.C. took a picture of me in front of Marshall Foch's statue.

My father studied chemistry at the National Tsinghua University in Beijing. Two weeks after my birth, he came to the United States to pursue his graduate studies in chemistry at the University of Illinois at Urbana-Champagne. Because my elder brother Robert Chun (born in 1928) was staying in Shanghai with our grandparents, I was the only one at home with mother and received a lot of love and attention (T.W. Hu, p. 6). Mother was extremely loving, but also very strict in guiding us with clear principles and stern disciplines (see her in Grandpa Chang's family, p. 396). She was a lovely, vivacious lady, and she would bring life and spirit to any gathering, large or small. In contrast, father was a man of not too many words, but everyone liked him. My parents were a wonderful couple with complementary personalities. My brothers and I were indeed very fortunate to have them as parents to receive their wonderful genes (C. Ho, p. 297) and to be raised by them.

My father got his Ph.D. in less than three years and went back to Beijing in 1934 to become a full professor at Peking University

Grandpa Chang's family, Taipei, 1953. Grandpa (center, seated), Parents (left, seated), Brothers Fred and Robert, and Shu (1st, 3rd, and 5th from left, respectively, standing).

(Beida) when he was 26 years old. He taught there for three years. My younger Brother Fredrick Foo was born in 1935. When the war with Japan broke out in 1937, my father went to Kun-ming, where Beida was combined with Tsinghua and Nankai universities to become the Southwestern Union University. Mother and the three sons stayed with grandfather (picture on p. vii) in Shanghai. My father taught chemistry in Kun-ming, traveling back to Shanghai during summer and winter vacations. Although Shanghai was under Japanese occupation after the Chinese Government moved to Chungqing in 1937, the various international settlements were still free from control by the Japanese force and the puppet government it set up. Thus, my grandfather continued to carry out the judicial functions and report to the Chinese Government in Chungqing. In the summer of 1940, my grandfather was assassinated by gunmen hired by the puppet government because of his righteous handling of several spy cases. After this unfortunate incident, my father stayed in Shanghai with his family, worked in a pharmaceutical company, and did not go back to Kun-ming. After the war, he returned to Beijing in 1946 to become professor and chairman of the Department of Chemistry at Beida.

I grew up in Shanghai during the war. After one year of kindergarden, I was admitted to the second grade of Copper Elementary School, skipping the first grade. Then I attended Yu-Tsai Middle and High School until 1946. During the war, the economy was difficult and the schools were only able to offer the fundamental instructional courses, with little facilities for recreational education such as art and music. I liked math very much. Numbers, geometry and the quantitative aspects of how things work have always fascinated me. I can readily see matters in terms of quantity, space and time; these seem second nature to me and I like to work on them. My main recreation was sports that would not require much equipment, particularly playing ping-pong (with a wood stick resting on bricks as a net) and soccer (with a little rubber ball about 2-inches in diameter). It was a difficult time, but I enjoyed very much my childhood in Shanghai living with my parents and two brothers. My parents instilled in me the important elements of life, to be honest, diligent, humble and considerate and respectful of others. They rarely used strong words or gave lengthy lectures, but they taught us by examples (T.W. Hu, p. 6). I am grateful to my parents for who I am, not only from a genetic point of view, but also for their marvelous upbringing. Of course, this applies to my brothers Robert and Fred also. The three of us pursued different lines of careers, and I am extremely proud of what they have done.

2. Higher Education

After my father went back to Beijing in 1946, the remainder of the family followed. A few months after I transferred to the Yu-Ying High School in Beijing to continue my junior year, there was a joint entrance examination offered in 1947 by the major universities in China, including Beida, Tsinghua, and Nankai. In the two postwar years 1946 and 1947, junior-year students were granted special permission to take the examination one year before they finished high school. I decided to take the examination; I did not expect to pass it, but I thought it would be a good idea to get some experience that would help me the next year. When I registered for the examination,

I wanted to select mathematics as my first choice for a major. Then I realized that the entrance examination for mathematics majors included analytical geometry, which was a course I could not take until my senior year. I looked at all the other subject areas I was interested in that did not require analytical geometry, and found premedical studies to be the one. That is how I selected premed as my first choice in taking the examination. Even then, I was still not proficient in several subjects, especially physics. My elder brother Robert helped me a great deal and I did fairly well for that subject (R.C. Chien, p. 10). To my great surprise, I passed the entrance exam and was admitted to the premed program. Because I often felt the tremendous comfort a physician can give to a patient whenever I or other family members was sick, I always thought that it would be wonderful to be able to give people that kind of feeling. By skipping another year, I entered the Beida premedical program at the age of 16, and I was two years younger than most of my classmates (Han, p. 37).

In Beida, the premed and medical programs are continuous for a total of seven years. Under normal circumstances, I would have moved on to the medical school after two years, but the civil war evolved to a state that Beijing was under siege and shelling by the Communist Army. In late 1948, the Nationalist Government sent two military airplanes to evacuate the professors from universities in Beijing. In retrospect, it is remarkable that my parents were able to decide to leave on the day of notice. The airport in Beijing was under shelling, and the little propeller planes had to take off on a strip of cement in front of the Temple of Heaven. We flew to Nanjing, which was the capital at the time, and then went to Shanghai. My father accepted an invitation by President Sze-Nien Fu of National Taiwan University (Taita) to become Dean of Studies, which is equivalent to a provost, and our family went to Taiwan by boat.

At the beginning of 1949, I transferred to the College of Medicine of Taita as a second-year student. The medical program at the National Taiwan University at that time consisted of only six years, with one year of premedical study, four years of medical school and one year of internship. In this transfer, my medical education was shortened by another year, i.e. I skipped three years altogether in my education.

Some of my classmates had worked for several years during the war, and then came back to study. For this reason, not only did my classmates generally look upon me as a kid brother (Lee, p. 44; Liu, p. 48; Chin, p. 50), but also those in one class later (H.C. Chen, p. 52). The education at Taita medical school was excellent, but some of the professors at that time had difficulty lecturing in Mandarin Chinese and several courses were taught in Taiwanese dialect or in Japanese. I was the student who transcribed most of the professors' lecture notes onto stencils for mimeography and distribution to the class (Lee, p. 44), and this helped me a great deal in learning. At Taita, I enjoyed playing ping-pong (on the university team), basketball and soccer (on the college team), especially the latter two because they are team sports involving groups of schoolmates. Though our teams lost more often than won, I learned a great deal from these sports activities; about how to work as a group, how to accept losses and try to improve constantly. I would say that this was a very important aspect of my education.

Following a year of internship, we had one year of ROTC military training, which was mandatory for college graduates in Taiwan. During that year I planned for my future. I decided to pursue further education and/or training in the United States, either in basic biomedical sciences or in clinical medicine. It was a very difficult choice because I liked both. In research, one is able to find interesting problems, to be innovative and create new knowledge that will eventually help people in terms of health problems. On the other hand, clinical medicine would give me the opportunity to treat patients, and I like very much working with people. I applied for medical internship at a few hospitals and also graduate studies with research assistantship or fellowship at a few universities. I got offers from just about every place I applied, and I finally decided to go into graduate study for a Ph.D. degree in physiology. I had excellent teachers in physiology at Taita: Professors H.S. Fang (pp. 369–371), M.T. Peng and T.F. Huang. I was particularly fascinated by the functions of the brain and the cardiovascular system, so I applied for physiology studies in these areas. Dr. Magnus Gregersen, who was my eventual advisor at Columbia University, visited Taiwan when I was in medical school. He helped me to decide to attend

Columbia. I was fortunate to win one of two Li Foundation fellowships given annually to Taita graduates to study abroad. The Li Foundation was set up by Mr. K.T. Li, who was a very successful tung oil industrialist and philanthropist. Looking back at my education, I can see that everything happened by chance. I could have gone into mathematics or clinical medicine or neurophysiology, and would then have a different career and life, but at bifurcations one has to make a choice, and life has its way of directing us to a certain path. Magnus Gergersen was on the board of an organization that was formed during World War II called the American Bureau for Medical Aid for China (ABMAC), now known as the American Bureau for Medical Advancement in China. This connection led to my serving on the ABMAC Board (Pierson, p. 62).

Shortly after I arrived at Columbia, Magnus Gregersen assigned me to assist in laboratory teaching in the physiology course for dental students. That was my first exposure to teaching, and I enjoyed it very much. I took several courses in physiology at Columbia, but I also took other courses. For example, I took a course on "Electrical Circuit Analysis" taught by a famous Professor of Electrical Engineering, Herbert Taub, in the City College of New York. This reflected my strong interest in mathematics and paved the way for my becoming a bioengineer later. My Ph.D. thesis was on "The Role of the Sympathetic Nervous System in Hemorrhage." When a person loses blood, his/her sympathetic nervous system is activated. Adrenaline flows and other regulatory mechanisms operate to compensate for the loss of blood to maintain the blood circulation in the face of this stress. The choice of this topic reflected my interest in neurophysiology in addition to cardiovascular physiology and represented an attempt to combine the two. The design of the study involves the removal of all the sympathetic nerves surgically (total sympathectomy). I compared the cardiovascular responses to various levels of hemorrhage between this group of animals to normal controls that had an intact sympathetic system. I learned that operation from Dr. S.C. Wang (M.K. Wang, p. 56), who was a Professor of Physiology (and later Pharmacology) at Columbia. I later learned that he had been my father's classmate in Nankai Elementary School in Tientsin, China. What a coincidence!

During my graduate study, I worked on the effects of x-irradiation on cardiovascular functions in addition to my Ph.D. research. I had the pleasure and privilege of working with my Taita physiology teacher M.T. Peng, who was taking a sabbatical leave at Columbia, and co-authored two of my earliest papers with him and Magnus Gregersen. My other collaborators in these early years were Ruth Rawson, Gerhard Muelheims, Zaharis Dische, Charles Pallavicini and Louis Cizek.

B. 1957–1988

1. Columbia University

After completion of my Ph.D. studies, I remained in the Department of Physiology as an Instructor for one year and then became an Assistant Professor. The results of my thesis study made me realize that, even when the sympathetic nervous system has been removed, there are still other factors that contribute to the compensatory responses to hemorrhage, e.g. humoral, endocrine and physical factors. The resistance to blood flow is attributed to the viscosity of blood, as well as the size and number of resistance vessels. When Magnus Gregersen attended a meeting in Europe in 1959, he was greatly impressed by a paper given by the late Dr. Lars Erik (Charlie) Gelin, who was a Professor of Surgery in the University of Gothenburg (Göteborg) in Sweden. Gelin showed that the flow properties of blood (blood rheology) could affect the cardiovascular responses to shock. Thus in the early 60s, we switched the emphasis of our program to investigations on blood rheology, though we had little previous experience on this subject. With the participation of Shunichi Usami (p. 70) from Kyoto, Japan and Bob Dellenback (p. 294), who started his instructorship at Columbia together with me, we began to determine blood viscosity and its role in health and disease. Magnus was able to obtain generous support from Mrs. Alan Scaife of the Mellon family to set up the rather expensive rheology facilities. We conducted investigations that were interesting not only from a scientific viewpoint, but also in terms of the types of animals studied, e.g. the Congo eel, which has the largest red blood cells (RBCs) (H. Goldsmith, p. 281), and elephants, which have a very large body mass but not especially large RBCs. We worked with many scientists on rheological investigations, including Richard Abbott, Tony Benis, Martin Blank, Brigit Blomback, Cy Bryant, Steven Burakoff, Paul H. Chen, Giles Cokelet, Al Copley, Maryann Farrell-Epstein, Mason Guest, Tom Impelluso, Bob King, John Lundberg, Sarah Luse, L.B. Nanninga, Allen Rakow, David Schachter, Bob Schmukler, Geoffrey Seaman, Duncan Sinclair, Lily Soo, Lou Soslowsky, Tim Springer, Harry Taylor and Husnu Tözeren. Magnus passed away in 1969. Bob has retired. Shunichi and I are still working together at UCSD; our fruitful and enjoyable collaboration has now exceeded 40 years.

In 1966, I was going to take sabbatical leave to go to the Medical Research Council in Mill Hill, England, to learn the molecular structure and function of fibrinogen, which is a plasma protein that can cause RBC aggregation to increase blood viscosity, in addition to its role in coagulation. Because my mother fell ill at that time, I changed my plans and we went to Taiwan instead. I worked in the Department of Physiology at National Taiwan University College of Medicine and the National Defense Medical Center. Besides teaching the bright medical students (Kung-ming Jan, p. 72, was one of them), I worked on several research projects. One of the projects involved follow-up measurements of blood volume and body fat in human subjects 16 years after the initial study performed in 1950 by Tom Allen (when he took sabbatical leave from the Department of Physiology at Columbia to go to Taiwan). I was a subject in that study and it was fun to do this follow-up study as an investigator working with my former professors Fang, Peng and Huang. The results showed interesting effects of age on blood volume, body composition, body fat distribution, and their relations. While in Taiwan, I also completed my review paper on "Role of the Sympathetic Nervous System" for *Physiological Reviews*, which is a journal with the highest impact factor among all physiology periodicals. The six-month sabbatical allowed K.C. and me to be together with my parents and our siblings, and May and Ann to learn Chinese by attending the schools there. My mother had recovered by the time we left Taiwan to return to the United States.

Photomicrographs of red blood cells (RBCs) from nine animal species (bottom scale is 10 μm). Left: Top view; Right: Edge view of mammalian RBCs in rouleaux and non-mammalian RBCs that have nuclei and present a spindle shape. From S. Chien et al., 1971, in Publication A62.

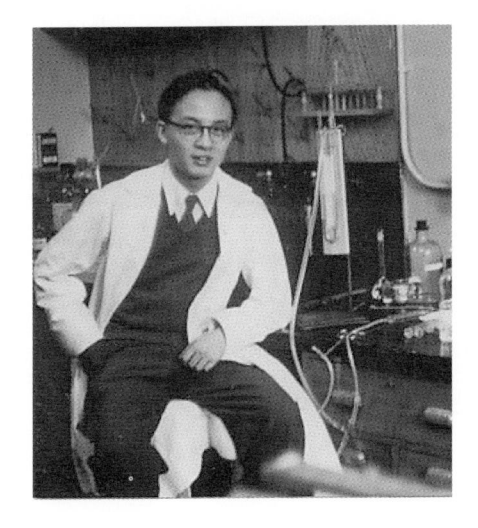

In the Hemorheology Laboratory, P&S 17th floor, in the mid or late 60s.

One of the greatest successes of our research group in Columbia at that time was the publication in 1967 of a series of three back-to-back papers in Science on the roles of red blood cell deformability and aggregation in governing blood viscosity. These papers are important because they formulated the fundamental determinants of blood viscosity, put our lab (a group of novices in rheology) on the map in this area of investigation, and triggered my 30 years of collaboration with Richard Skalak (A. Skalak, p. 108; T. Skalak, p. 166). Dick, who was a Professor of Civil Engineering and Engineering Mechanics in the Engineering School of Columbia University, was on sabbatical leave in Gothenburg, Sweden in 1968–1969, working there with P.-I. [Per Ingvar] Brånemark (p. 382). P.-I. was studying the flow dynamics of blood cells in the microcirculation (small vessels) using a titanium window implant, which was later developed into a highly successful titanium dental prosthesis called the Brånemark implant. Although Dick and I were both on the faculty at Columbia, we had not yet met. Dick read the three articles in *Science* when he was in Gothenburg and wrote to Magnus Gregersen, expressing his great interest in them,

In the 11th floor P&S office, 1983.

especially the paper on the flow of red blood cells (RBCs) through narrow filter pores. In 1968, I went to Gothenburg to attend the European Conference of Microcirculation. That was my first trip to Europe, and the first city I visited was Gothenburg. It was there that I met Dick thousands of miles away from Columbia, our common home institution. That initiated a most fruitful and precious collaboration for 30 years, during which we jointly published 59 full-length papers.

By working with Dick and through reading, I learned a lot about blood rheology and engineering simulation and modeling. That allowed me to put my math interests into practice. I found out only after we both had moved to La Jolla that Dick married Anna Allison (p. 108) in the Madison Avenue Presbyterian Church in New York City four years before our wedding in the same church. There are a countless number of cities in the world, and there are thousands of churches in the city of New York. Again, what a coincidence! I was blessed to have Dick as a friend and colleague for nearly 30 years. I am very happy that I organized his 70th birthday celebration in 1993 (Brånemark, p. 383), four years before his passing. I wish he could have been here for the celebration of my 70th birthday.

Because the majority of the researchers in my laboratory at Columbia were trained in biomedical sciences, rather than engineering, I instituted a series of talks on viscoelasticity at our lab meetings in the late 70s and early 80s. We had Dick or one of his students (especially Aydin Tözeren), Kung-ming Jan, and myself giving talks on this subject at lab meetings. We had a very effective continuing education program on biorheology.

In the 70s, we initiated four new directions of research that emanated from blood rheology. The first was that we moved from studies on the blood as a suspension to investigations on individual cells using both experiments and modeling. We isolated individual RBCs and white blood cells (WBCs), experimentally determined their rheological behavior and deduced their viscoelastic properties by theoretical modeling. In 1973 Dick, two of his students (Aydin Tözeren and Richard Zarda), and myself published a paper on the mechanical properties of the RBC membrane. In order to determine the viscoelastic properties of RBC membrane, Paul Sung (p. 98) set up the micropipette aspiration technique, which had been used by Paul La Celle (p. 275) and his colleagues for studying RBC deformability. This approach was later extended to the study of WBC rheology with Geert (G. Schmid-Schönbein, p. 90).

The second new direction was research on the mechanism of RBC aggregation. Dick, Kung-ming and I worked on the mechanism by which RBCs adhere to each other in a reversible way. By using a combination of experimental and theoretical approaches, we were able to analyze the energy balance in RBC aggregation. The net aggregation energy represents the algebraic sum of the macromolecular bridging energy balanced against the disaggregating energies due to electrostatic repulsion and mechanical shearing. The net energy is stored in the membrane as a change in strain energy and causes changes in cell shape.

By combining these two new areas of research with the experimental investigations by David Schachter on RBC membrane lipids and Jack Bertles on sickle cells (see below), as well as theoretical modeling by Dick, I organized a Program Project Grant (PPG) proposal to the NIH on "Red Blood Cell Rheology and Membrane Mechanics" and served as its Principal Investigator (P.I.). The PPG was a group effort that involved five projects tied together by a common theme and several core facilities, including ultrastructure (Mary Lee), computation (George Schuessler), instrumentation (Usami) and administration (S. Chien). Following very careful preparations in proposal writing and rehearsals for the site visit as a team, the application was funded in 1974 on the first trial. This was the first and only PPG in the Department of Physiology at Columbia and it continued until Dick and I left in 1988. In 1975, I succeeded in getting an NIH Training Grant on Cardiovascular Physiology and Biophysics for interdisciplinary predoctoral and postdoctoral training that involved many departments and centers at Columbia, also on the first attempt. These two grants provided important support not only for my lab, but also for many collaborators toward the common goal of enhancing interdisciplinary research and training in cardiovascular physiology and biophysics. Many investigators at Columbia and other institutions requested copies of these applications for use as models for their proposals.

The third new area of investigation was microcirculation. In order to assess the relevance of cell deformability and aggregation to flow dynamics *in vivo*, Shunichi and I decided that we needed to study the living microcirculation. He spent three months at UCSD with Ben Zweifach in 1972 to learn microcirculation research. In 1976, we were fortunate to be able to recruit two recent Ph.D. graduates from UCSD: Herbert Lipowsky (p. 83), who studied under Ben, and Geert, who trained with Bert Fung. Bert (pp. 137–145) and Ben (B. Zweifach, p. 146), both superb scientists, founded the Bioengineering program at UCSD in 1966. Bert came from an aeronautical engineering background and he is widely regarded as father of biomechanics. Ben came from a biology background and was widely regarded as father of microcirculation. UCSD was the center for microcirculation research, with Ben, Bert, Marcos Intaglietta, Arnost and Kitty Fronek (p. 242), Sid Sobin (p. 240), David Gough (p. 149) and Peter Chen (p. 164). Both Herb and Geert had excellent training and their coming to my lab in 1976 made it possible for our microcirculation research to take off quickly. As a result of their joining, we were able to obtain supplemental funding to our PPG and added microcirculatory research as a part of the Program activities. In 1980, we broke into this new field by having our first four publications in *Microvascular Research*, with Geert and Herb being the first authors on two each. Other scientists who collaborated on microcirculation research included John Firrell, Steve House (p. 96), Rakesh Jain (p. 259), Colin McKay and Roland Pittman.

The fourth new direction of study was macromolecular transport across vascular endothelium in relation to atherosclerosis, which was initiated in collaboration with Shelly Weinbaum and Bob Pfeffer of the City College of New York (CCNY). I first met Shelly and Bob when Bert Fung was invited to give a series of lectures at CCNY in 1969 (Weinbaum, p. 111). Later, Shelly and Bob each spent a sabbatical year with Colin Caro in the Imperial College of London. After their return, Shelly and Bob proposed to work with me on atherosclerosis and endothelial cells (Caro, p. 118). With Shelly, Bob, Colin and Kung-ming as collaborators and George Palade as a consultant, I submitted a research proposal entitled "Studies on Endothelium in Atherogenesis" to the NIH. This grant application was funded in 1976 on the first submission, and it has been renewed recently for four more years. In addition to the colleagues mentioned above, the other scientists who worked on this project included Cynthia Arbeeny, Leslie Arminski, Ann Baldwin (p. 299), Peter Ganatos, Steve Garrick, Dean Handley, Yaqi Huang, Kwang-Hee Lim, Chung-Yuan Liu, David Rumschitzky, Steve Gallik, Gung-Bi Wen, Yongui Yin and Fan Yuan. A great deal of the work being carried out in my lab at UCSD today stems from this program on endothelial biology. As we began to venture into the new area of endothelial biology in relation to flow mechanics, I started another reading club in the lab using the excellent book by Colin Caro and his colleagues on *Blood Flow in Arteries*, which combines fluid mechanics with physiology. This gave all of us an excellent foundation in an important area of bioengineering.

The ability of our getting into these new areas is largely attributed to my wonderful colleagues with biomedical and engineering training. Besides those mentioned above, we also had from Dick Skalak's group Cheng Dong, Si-shen Feng (p. 287), Tim Secomb (p. 299), Cheng Zhu (p. 105), and others. Other close colleagues in Columbia Engineering School were Ed Leonard (p. 66) and Jordan Spencer. We needed the cooperation and advice of all these colleagues to form the team. We had basic bioscientists (including molecular biologists, electron microscopists and physiologists) and clinical investigators, which included cardiologists (e.g., Kung-ming Jan), hematologists (Anne Kaperonis, Emily Schmalzer and Tien-Hsi Young), anesthesiologists [Jeff Askanazi, Richard Chen (p. 81) and Foun-chung Fan], neonatologists (Stuart Danoff and Tom Starc), neurologists (Jim McMurtry and Don Quest), neurosurgeons (Karen Muraszko, Bob Solomon and Bill Young), radiologists (Philip Alderson and David Seldin), and dentists (Lennart Edwall, Syngcuk Kim and Hank Trowbridge). We also had wonderful visiting scientists [e.g. Ulf Bagge, Magnus Braide, Ann Chabanel (p. 94), Chen Chang, Toshioki Itoh, Vojislav Magazinovic, Branko Peric and Walter Reinhart]. It really was a marvelous multidisciplinary team. In addition, I was fortunate in having a terrific group of technical staff [Juan Rodriguez, Danny Batista, Dagmara

Igals, Michael Kent, Jerry Norwich, Honor O'Sullivan and Mark Wang (p. 33)] and administrative staff (Ethel Goodrich, Michelline Faublas and Sandra Taylor).

The Ph.D. students who received training in my lab included Kung-ming Jan, Ron Carlin, Ann Thurn, Shlomoh Simchon, Syngcuk Kim, Shyng-Jong Lin, Margaret Vayo, Dan Lemons, Fan Yuan, Shi-shen Feng, and others for whom I served as a co-advisor with Dick or Shelly. I also had the pleasure of having several medical students (e.g. Larry Krakoff, Stu Billig, David Burton, Paul Gustavson, Alan Burke and Jay Heldman) and dental students (e.g. Syngcuk Kim and Ken Treitel) working in my lab.

As we developed the research at cellular and microcirculatory levels, I realized that we needed to integrate the knowledge obtained from different levels of the biological hierarchy by using the network control theory to understand how the physiological systems work. I took a course on "Control Theory" given by Bob Bernstein of the Department of Electrical Engineering of Columbia; this again reflected my interests in quantitative approach to physiological research. I found my earlier course on electrical circuit analysis and our discussion group on the modeling of viscoelasticity to be very useful in this control theory course. Thereafter, I applied the knowledge I learned from this electrical engineering class to initiate a graduate course on "The Application of Control Theory in Physiological Systems" in the Department of Physiology at Columbia. This elective course was the only quantitative one given in the department and was popular among graduate students and postdoctoral fellows. It brought together students with different research emphases, whether they studied neurophysiology, cardiovascular physiology or renal physiology.

Concurrent with the recognition of the need to integrate our physiology knowledge at the system level, I also realized in the late 70s and early 80s that the rapid advancement of molecular biology provides extremely powerful tools for us to probe physiological problems into greater depth. At that time, molecular biology was not yet recognized in physiology and bioengineering. I began to learn molecular biology by taking some courses and did a lot of reading on my own. In the mid-1980s, I instituted a molecular and cell biology reading club by using the excellent book by Bruce Alberts *et al.* on *Molecular Biology of the Cell*; thus starting another round of continuing education. As we were learning molecular biology, we also began molecular biology research in our lab. This was facilitated by having Amy Sung there; she had received her Ph.D. from the Department of Genetics and Development at Columbia under Dr. Elvin Kabat, with me as a co-advisor (L.A. Sung, p. 76). In mid-1980s, she started her research on the molecular biology of RBC membrane proteins in relation to cell mechanics.

After my election to Council of the American Physiological Society (APS) in 1985, I took it as a mission to promulgate the incorporation of molecular biology into physiology (Frank, p. 189). In 1986, with the help of Jay Gargus, I organized a symposium and a hands-on workshop on "Molecular Biology and Physiology" at the APS Meeting. At the symposium, which was attended by about 600 people, I gave an introductory talk to present the basic concepts to those who had not previously been exposed to molecular biology, and five other experts then spoke on the application of molecular biology to physiology. A book was published as a result of the symposium (Publication B5). I felt very gratified that several leaders in physiology have since told me that they have successfully adopted molecular approaches to physiological research through that symposium/workshop. In 1988, I organized a second symposium and workshop for APS that was focused on the cardiovascular system. That symposium was also published as a book (Publication B6). I continued to foster the application of molecular biological approach to physiology when I was APS president (1990–1991). Later I also spoke at the International Congress of Biorheology, the International Congress of Microcirculation, and the Biomedical Engineering Society Meeting on the potential application of molecular biology to these fields.

While I was fostering the probing of biological problems down to the molecular level, I did not lose sight of the importance of the application of molecular and cellular levels of knowledge to the understanding of functions at tissue, organ, system and whole body levels, as indicated by my learning and teaching of control theory as applied to physiological systems. I have always fostered an integrative approach, whether it is bioengineering or physiology. The final purpose of reduction down to molecular and genetic levels is to provide the foundation for synthesis and integration into the whole. We need to use both experiments and theories to integrate our knowledge, bridging biomedical and engineering sciences. The diagram I designed to represent this integrative concept (Upper diagram on p. 154) is still used by the APS for the pamphlets and web page for its publications and by the UCSD Department of Bioengineering for the cover of its departmental brochure.

It is my belief that the quantitative engineering approach is applicable not only to body systems, but also at other levels of the biological hierarchy. Thus, we need to understand the dynamics of the interplay between one set of molecules or genes and the next. We need to use quantitative approach and bioinformatics to determine how these molecules and genes cross-talk and interact with each other to accomplish the coordinated action of control and integration in optimizing the function of the system.

While most of our research activities are basic in nature, we always look for applications to clinical medicine, i.e. disease states. Here my medical training has been very helpful. For example, we applied our expertise on blood cell rheology to the study of sickle cell disease, which involves a reduced RBC deformability due to the crosslinking of sickle cell hemoglobin during deoxygenation. These studies were performed with our hematology colleagues Jack Bertles, A.K. Brown and Jeanne Smith. We also assessed the roles of blood rheology in clinical conditions such as myocardial infarction (Tom Bigger and Kung-ming Jan), cardiac tamponade (Bob Potter), hypertension (Micky Alderman, David Case, Richard Devereux, Harry and Irene Gavras, John Laragh, Lee Letcher, Bill Manger, Jay Meltzer, Tom Pickering and Jean Sealey), atherosclerosis (DeWitt Goodman, Alan Seplowitz, Frank Smith and Larry Witte), carotid occlusion (J.W. Correll), malaria (Louis Miller), muscular dystrophy (H. Somer and Alfred Spiro), hereditary elliptocytosis (Jiri Palek, J.T. Prchal and S.C. Liu), and surgery (Frank Gump, John Kinney, Fred McAllister, Peter Scholz and

Richard Simmons). We performed these clinical investigations in parallel to our fundamental research.

Our research program in the Department of Physiology at Columbia enjoyed the continued support by the Department Chairs John Taggart and Sam Silverstein (p. 121, 425), who succeeded Magnus Gregersen, the administrative staff, especially Kathleen O'Callaghan, and the technical staff.

During my years at Columbia, my learning and development in research were accompanied by those in teaching. I am very appreciative of the positive responses of medical, dental and graduate students to my teaching. I was extremely gratified when Bob Lefkowitz, the leading authority on adrenergic receptors, told me when we were both inducted into the Institute of Medicine in 1994 that my lectures and demonstration on the autonomic nervous system generated his initial interest in the field. The reward in teaching is enormous.

When I first started giving lectures, I wrote out every word and went through many rounds of practice, including the planning of the space allocation and temporal sequence of the use of the blackboard. When I used slides for lectures or research presentations, I would design the plan of attack for covering the content of the slide and track the time required for each slide, and this was very helpful in delivering the lecture (Chai, p. 314). Later, when a tape recorder became available, I taped my practice talks and listened to them for five to six times per lecture, including the timing of each subtopic. By making the last part of the lecture a little flexible in content, I was always able to end my lecture exactly on time. Following many years of rather intensive preparations, I have learned to estimate the length of a slide just by looking at it, and I can eventually give lectures with minimum preparations.

When I was young, I was rather shy and did not like to speak in public. The confidence I gained from lecturing allowed me to feel at ease not only in the classroom, but also in any situation. Now, I can get up and speak extemporaneously without any preparation in social occasions. This is something I never thought I could do even when I was 30 years of age. I learned all of this at Columbia. I owe a great deal to Columbia and all my friends and colleagues there. When the Department of Physiology gave me a farewell party (p. 425) before I left for San Diego, I said "Once a New Yorker, always a New Yorker. Once a Columbian, always a Columbian." This popped up when I thanked Van and Barbara Mow (p. 122) for the beautiful Tiffanys crystal apple they gave me at the birthday symposium on 23 June 2001.

In the following diagram, I summarized the various cities I have been in my life throughout my years of education and professional careers. It is obvious that the time at Columbia University in the City of New York is the longest. In the following sections, I will talk about my life and experience at the Institutes of Biomedical Sciences in Taiwan and University of California, San Diego. I am grateful that I had the pleasure of being at these places and with all the wonderful people there. I am most thankful for this blessing.

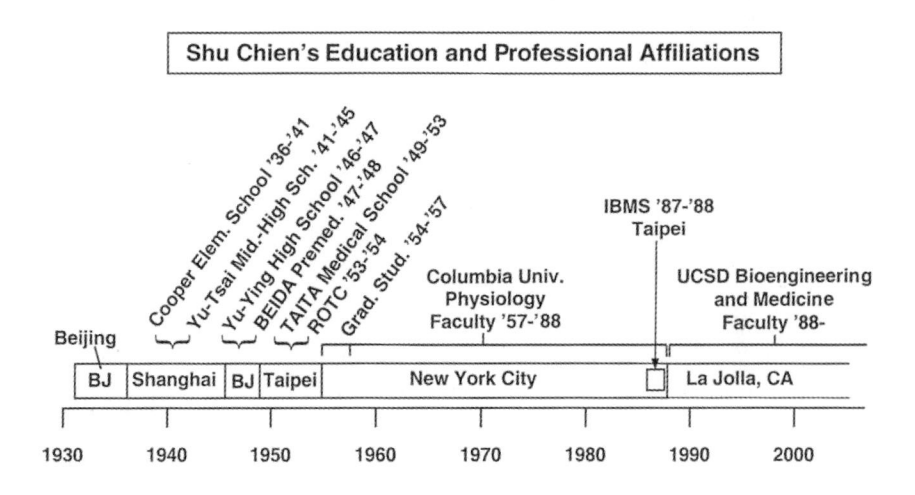

Shu Chien's Education and Professional Affiliations

Cooper Elem. School '36-'41
Yu-Tsai Mid.-High Sch. '41-'45
Yu-Ying High School '46-'47
BEIDA Premedical '47-'48
TAITA Medical School '49-'53
ROTC '53-'54
Grad. Stud. '54-'57

Columbia Univ. Physiology Faculty '57-'88

IBMS '87-'88 Taipei

UCSD Bioengineering and Medicine Faculty '88-

Beijing | BJ | Shanghai | BJ | Taipei | New York City | | La Jolla, CA

1930 1940 1950 1960 1970 1980 1990 2000

2. Institute of Biomedical Sciences, Academia Sinica

In the 1980s, I began to devote a good deal of time and energy to contribute to the advancement of biomedical sciences in Taiwan, where I received my medical education. I was elected as a member in

Academia Sinica in 1976. My father was President of the Academy from 1970 until his passing in 1983. Several members of the Academy had tried to nominate me for membership in 1972 and 1974 during the biannual elections, but they were persuaded by father not to do so. I admire immensely my father's fairness, sense of propriety and sensitivity to conflict of interest, real or perceived. I appreciate greatly his stopping these attempts for nomination. I wrote an article in his memory after his passing, entitling it "Father, you are the perfect person!" (see P.K.H. Cheng, p. 41) In 1976, the nominators insisted over his objection, and father and I became the first father-son members in Academia Sinica.

During the membership meeting of Academia Sinica in 1980, Dr. Paul Yu, who was President of the American Heart Association and Chief of Cardiology at the University of Rochester, proposed with some 30 members, including myself, the establishment of the Institute of Biomedical Sciences (IBMS). This proposed institute would not only pursue frontier biomedical research in Academia Sinica, but also take the leadership to promote cooperation among various universities and medical schools. An advisory committee, chaired by Paul, was formed. Five of these initial members were in the United States, including Paul Yu, Monto Ho (p. 133), S.H. Ngai (H.H. Wang, p. 59), Paul Ts'o (p. 131), and myself. Because Paul Yu was in the U.S., most of the decisions were made by the five overseas members and subject to the approval of the members in Taiwan, who were very supportive.

The advisory committee made careful planning and recruited scientists, mainly from the U.S.; I was in charge of the recruitment. During an advisory committee meeting in New York in November 1986, we decided that the institute would start functioning in January 1987. We had all assumed that Paul Yu would be the director. At the meeting, however, Paul said that he was not able to leave Rochester in 1987. Everyone on the committee pointed to me and said, "You are the one that will have to go to Taiwan to start the Institute." As a result, I had to ask for sabbatical leave from Columbia with only a two-month lead time. I had to arrange my lab, my home and everything. I did arrive at Taipei in the evening on 8 January 1987, and went straight to IBMS. IBMS had a six-story building, which was constructed under the dedicated supervision of C.Y. Chai and was almost totally empty at that time.

We had a few investigators that were already there (including C.Y. Chai, Lee-Young Chau, Lung-Sen Kao and Eminy H.Y. Lee). Recruiting people from the U.S. to do long-term work in Taiwan was very difficult. Senior people would have a lot of commitments, younger people would have young children and hence concerns about schooling, and often the spouse could not or would not go back to Taiwan. There is also the difficulty of matching their salaries in the U.S. We did finally recruit quite a few permanent members, and we also recruited some scientists who would work there for a year or several months. With the help of Chien Ho at Carnegie-Mellon University (p. 297), we assembled a team of cell biologists (Lan Bo Chen, Wen Tien Chen, George Cheung, Wallace Ip, Diana Lin, Jim Lin, Shin Lin, Henry Sun, Eugenia Wang, John Wang, Kuan Wang, Yu-Li Wang and Reen Wu) to go to IBMS for three months each, with two at a time and overlapping like a relay team. The group met in New York several times to plan the team's research. The topic chosen was cell biology of esophageal epithelium. This was based on (1) the expertise of the participants, (2) the area of interest to Taiwan (esophageal cancer is a prevalent disease), and (3) the lack of duplication with the scientists' research in their home labs in the U.S. (so that something unique can be established in IBMS). This team research plan worked out very well, and eventually two papers were published from the results of these studies. In the process, many young investigators were trained, new knowledge in cell biology was introduced into Taiwan, and it led to the successful recruitment of an excellent cell biologist Tang K. Tang (p. 325) as a long-term scientist in IBMS. It was a very meaningful endeavor.

Shortly after I arrived in Taiwan, I went to every medical school to introduce this new institute, telling faculty and students everywhere of the opportunities available for research and training at IBMS. We did not want to attract faculty away from the local institutions, because that would not help to enhance biomedical research in Taiwan. Rather, we asked their students and fellows to come to do research and receive training. Many medical students were trained at IBMS, especially in the summer time. Kung-ming Jan, who went from Columbia to IBMS for the year of 1987, was responsible for student training, giving them a series of excellent talks about ways of research, prior to their starting in the lab. The students were very enthusiastic about the wonderful learning opportunity, and they provided a lot of excitement to the Institute. Many of them presented their research findings at national meetings in the U.S. Raymond R.K. Chang, C.H. Chen, Hwei-Jong Chen, Jen-Chang Chi, Norman T.Y. Chien, Pao-Tien Chuang, Jih-Shuin Jeng and Chih-Hsiu Wu published full-length research articles in first-rate scientific journals. Many of them are now faculty members in the U.S. or Taiwan. The training was very fruitful and rewarding.

During my 18 months in Taiwan, I was fortunate to have several graduate students working with me on cardiovascular research in IBMS. The Ph.D. or D. Med. Sci. students who did their dissertation with me as an advisor or co-advisor (with Benjamin B.N. Chiang, Huai San Lin and Wen-Guei Wu) included Yuh-Lian Chen (p. 356), Lang-Ming Chi (p. 351), Cheng-Den Kuo (p. 353), Hung-Wen Pong, Yuh-Bi Tang (p. 358) and An-I Tsai. The research scientists in the Cardiovascular Laboratory of IBMS during 1987–1988 included Jin-Jer Chen (p. 343), Hui Chong Chiu, An-Li Huang, Kung-ming Jan (p. 72), Mei Chi Jiang, Chung Yuan Liu, Cheng-Yao Su, Tang Tang (p. 325), Ham Ming Tseng (p. 206) and Danny L. Wang (p. 328), as well as the Columbia visiting group mentioned below. The leaders in the other research groups in 1987 were C.Y. Chai (Neuroscience), Tommy Yung-Chi Cheng (Cancer and Virus, p. 364), Patrick Wong (Eicosanoids) and Chien-Jen Chen (Epidemiology and Public Health).

IBMS held a series of workshops in 1987–1988, particularly with the team of prominent cell biologists from the United States who could go there for only three months. We asked each one of them to give a seminar and/or workshop on their expertise. These workshops attracted people from all over the island. I remember clearly that one of the first workshops was given by Jim Lin (University of Iowa) and Yu-li Wang (now at the University of Massachusetts) on "Characterization and Utilization of Monoclonal Antibodies, Microinjection and

Videomicroscopy;" these were frontier areas in biology and medicine that aroused a great deal of interest throughout Taiwan. The IBMS auditorium, which can hold 250 people, was not large enough, and a closed-circuit TV room had to be set up to accommodate another 200 people. Other equally successful workshops were given on "Biophysical Approaches to the Study of Life Sciences" (Paul Sung and George Schuessler), "Microcirculatory Hemodynamics" (Herb Lipowsky and Steve House), "Microspheres for Regional Blood Flow Measurements" (Shlomoh Simchon), "Immunological Methods for the Study of Differentiation of Epithelial Cells" (Henry T.T. Sun), "Colloidal Gold Electron Microscopy" (Dean Handley), "Microvascular Casting Methodology in Scanning Electron Microscopy" (Kazuto Takahashi et al.), "Statistical Analyses of Case-Control Studies" (Kung Yee Liang et al.), "Medicinal Chemistry and Synthesis of Eicosanoid Antagonists" (R. Noyori et al.) and "Analytical Methods for Eicosanoids" (Granstrom et al.).

In the summer of 1987, nearly everyone from my lab at Columbia went to IBMS to help me build up the institute, whether they were Chinese in origin or not. In fact, most of them were not, e.g. Dean Handley, Steven House (p. 96), Herb Lipowsky (p. 83), Silvia Rofe, Geert Schmid-Schönbein (p. 90), George Schuessler, Shlomoh Simchon (p. 238), Aydin Tözeren and Shunichi Usami. They went there because of our friendship and their desire to help build this new institute in a land far away. During that summer, we jointly gave a course on "Microcirculation and Cell Biophysics." Eight students from four schools registered for credit, but the course was regularly attended by about 70 people, including some of my previous teachers. Several scientists subsequently initiated research projects on areas related to the course. In August 1987, taking advantage of the Microcirculation World Congress in Tokyo, I invited some 40 internationally renowned scientists in the field to come to Taiwan to participate in a satellite symposium on "Vascular Endothelium in Health and Disease." That satellite symposium, which was included as part of the Microcirculation course, stirred up a lot of interest and activity.

In April 1988, an "International Conference on Prostaglandin and Leukotriene Research," organized by Patrick Wong, was held at IBMS.

The conference was attended by about six hundred scientists, including two Nobel laureates (Bengt Samuelson and John Vane) and all the leading authorities in the field. It served to introduce the current developments in this area to Taiwan and promote the understanding of research developments in Taiwan by the visiting participants.

In addition to building up the research activities in IBMS and increasing its international communication, I also had the responsibility of enhancing cooperation among the medical schools and hospitals in Taiwan. To this end, I was able to obtain a significant budget line item from Academia Sinica to support the Clinical Research Centers at three major medical centers in Taipei, viz. the National Taiwan University Hospital, the Taipei Veterans General Hospital and the Tri-Service General Hospital. I initiated joint conferences among the hospitals and encouraged collaborative projects among different departments in each hospital and, more importantly, between the hospitals.

In 1987, with the help of Paul O.P. Ts'o, Jacqueline Whang-Peng and others, IBMS worked with these three medical centers to initiate a Medical Oncology Training Program, with Paul Carbone of the University of Wisconsin leading the effort. A number of outstanding clinical professors of oncology were invited to come to Taiwan from the U.S. to conduct this two-year program, which provided excellent training to six physicians (two from each hospital) on the current concepts and methods for diagnosis, treatment and prevention. In addition to introducing state-of-the-art oncology practice to Taiwan, this program also led to close interactions among the three hospitals through these six wonderful trainees. One year into the program, the success was clearly evident, and other major hospitals outside Taipei also wanted to participate. In 1988, the program was extended to include the National Cheng Kung University Hospital, Chung-ho General Hospital of Kaohsiung Medical College and Chang Gung Memorial Hospital. The trainees from this Oncology Training Program have made major contributions to the advancement of medical oncology in Taiwan. More recently, a similar and equally successful training program on Infectious Disease has been implemented by Monto Ho (p. 133) under the aegis of the National Health Research Institute (NHRI).

With all the activities in research, education and administration, and interactions with other institutions and government agencies to start a new institute, my schedule was extremely full and it was difficult to find time to sleep. What kept me going was that I was able to sleep in the taxicab while riding back and forth between the Academy and the center of Taipei City. IBMS did not have a car when I was the director; but I was able to receive the approval for the purchase of two institute vehicles, which were later delivered in fall 1988. During my 18 months at IBMS, I took the cab almost daily (Annie Lin, p. 331) for meetings with people in the universities, medical schools, hospitals and government agencies, for matters related to IBMS, and also to attend dinners, which are needed for establishing and maintaining connections with people related to IBMS. Herb Lipowsky is correct in saying that "What more could one ask for, besides a little sleep?" (p. 83), and I believe the picture he showed was taken during one of my trips rather than during his seminar.

Because I had a very active lab in Columbia whilst being very busy in IBMS, I needed to keep in close touch with both sides. In addition to the seven round trips I made during these 18 months, I kept in contact by using the fax, which had just become available in the U.S.

All these activities in IBMS needed funding support. When I arrived at IBMS in the evening on 8 January 1987, I discovered that the National Science Foundation (NSC) had not acted on the funding for the Cell Biology group because such a relay concept was unprecedented. That night I wrote a four-page letter in Chinese and presented it to NSC Chairman Li-An Chen and Vice Chairman Chao-Shiuan Liu the next morning. After reading my letter and listening to my explanations, they approved the request for funding this group. During the first few months at IBMS, I met frequently with scientists in the institute to prepare a PPG application to NSC. Our proposal was approved for funding together with that from our sister institute, the Institute of Molecular Biology (IMB), submitted by the then director James Wang (Professor of Chemistry at Harvard University). These were probably the first PPGs in Taiwan. The research programs in IBMS were able to progress smoothly because of the availability of funding from

NSC and from Academia Sinica. The late President Ta-You Wu of the Academy was very supportive of the growth and development of IBMS and IMB, not only in terms of intramural research funding, but also the IBMS extramural activities such as the Clinical Research Centers and the Oncology Training Program. In addition to government funding, I am very grateful for the generous support from private sources, particularly Mr. Yung Ching Wang, Mrs. Tien-Wen Niu, Dr. Ming-Shen Yang, Dr. Teh-Chou Chen and Mr. Samuel C.Y. Fang. These were particularly appreciated because such philanthropic contributions in Taiwan and other Asian regions are much less common than in the U.S.

In order to make the missions and activities of IBMS known to the academic institutions in Taiwan, as well as public and private sectors, I updated the brochure of the Institute as soon as I arrived there. The first version was produced by photocopying and consisted of ten pages of Chinese text. We continuously revised the pamphlet, and two editions later we published in November 1987 a bulletin that was typeset printed; it had a color picture on the cover, 26 pages of text, and an English, as well as a Chinese, version. By the time I left in June 1988, the 90-page brochure had nearly 40 action pictures (some of them in color) that depicted vividly the research activities and resources in IBMS. I had the help of many scientists and staff, especially Eminy Lee (p. 318) and Ching-Tung Chen, in preparing these brochures.

We built up excellent core facilities unprecedented in Taiwan, including computer center (Flory F.J. Wang), machine shop (Chou Pin Liu), electronics shop (Wen-Ching Wu), photo arts room (Ching-Tung Chen), animal facilities (C.C. Kao), library (Li-Ling Chen), maintenance office (Kuo-Chang Chen), general services (Jui Ho Lin), stock room (V.P.F. Sun) and administration office (Jenny L.C. Chen). Every unit was composed of extremely capable, diligent and devoted staff members. As the director, I was in contact most frequently with the administrative staff and appreciate immensely their excellent and dedicated work. In addition to Jenny Chen, and secretary Annie S.C. Lin (p. 331), the administrative staff who are still at IBMS or NHRI include Nicki S.C. Chang (p. 334), Tiffany H.C. Chen, Amy H.H. Chu

(p. 337), Shirley T.H. Hu (p. 336), Jamie M.L. Jan and Ivy S.C. Tu (p. 335).

While I was at IBMS, I also initiated the NMR program. Back in 1985, Chien Ho proposed to hold a symposium in Taiwan on "Nuclear Magnetic Resonance in Biology and Medicine" to introduce the recent advances in this frontier field. It was a very successful meeting and led to the publication of a book on this topic (Publication B1). I had little prior knowledge on NMR, and I studied this subject during the months prior to the meeting, which was successfully held in 1986. I was honored to have co-authored a chapter in this book with George Radda, Director of the Medical Research Council of Great Britain. In 1988, when I was IBMS Director, NSC called a meeting in Nan-Yuan, Taiwan to discuss the development of structure biology, which was a frontier area in life sciences. My knowledge learned from that symposium and book editing gave me the preparation needed to participate effectively in the discussion. The conclusion from the meeting was that an NMR facility will be established in IBMS and an X-ray crystallography facility will be established in IMB. There were some concerns about recruitment of the professional personnel for the NMR facility. My response was that, if we can have state-of-the-art facilities and provide excellent research opportunities, we should be able to get the kind of persons we need. With the help of Chien Ho, IBMS indeed was able to recruit. Tai-huang Huang and several other outstanding scientists to set up an excellent NMR structure biology program. In this and other events, I do not shy away from taking calculated risks.

The 18 months I spent at IBMS were extremely memorable. The dedication and diligence of the research scientists, students and staff were truly remarkable. Excitement was in the air and the morale was very high. People were working day and night. Anyone who joined would get charged up and become a part of a great team. People, whom I have known for a long time as hardworking and dedicated scientists, for example Kung-ming Jan, achieved at even higher, incredible levels. I remember that several scientists from other countries at a similar level of development as Taiwan were so moved during their touring of the institute that they had tears in their eyes, because they were wondering why they could not do this in their own countries. That was a special time, and IBMS is always a special place for me.

When I left Taiwan to return to the U.S. in July 1988, Cheng-wen (Ken) Wu (p. 126), who was a member of Academia Sinica and the IBMS Advisory Committee, took sabbatical leave from his position as Chair Professor of Pharmacology at The State University of New York at Stony Brook to succeed me as IBMS Director. His wife Felicia, who was also an outstanding cancer researcher, joined IBMS as a Chair Scientist (unfortunately she passed away in 1999 because of breast cancer).

3. Marriage and Family

Nineteen fifty-seven was a very special year in my life. I received my Ph.D. degree in May, but more importantly K.C. and I were married on 7 April. K.C. also went to Taita medical school. She was two years behind me and was in the first class with a seven-year medical education instead of 6 years. We knew each other in medical school and participated in many group activities such as outings, ping-pong, track meets, etc. As noted by K.C. (p. 16), Mrs. Hu Shih made special efforts to get us together by praising each of us to the mother of the other. Our dating started in 1953 during my ROTC training. From 1954 to 1956, while I was studying physiology at Columbia and K.C. was completing her last two years of medical education, we wrote to each other regularly. I was very happy that K.C. was able to come to New York as an intern in Mother Cabrini Memorial Hospital in November 1956. We were engaged in January 1957 in Mrs. Hu's apartment. Our wedding took place in Madison Avenue Presbyterian Church, with a lovely reception arranged by Mrs. Hu in Chinese Consulate General K.W. Yu's Park Avenue apartment. I am extremely fortunate to be married to K.C. In our four and a half decades of marriage, she has given me unconditional love, care and support. I have never had a thread of doubt about her total devotion and love. She has played a major role in whatever I have done. If I have to pick one most important event in my life, besides my birth, it has to be my marriage to K.C. I am most grateful for this.

Shu in front of the house in Englewood, NJ, where the family lived from 1963–1971. The picture was taken in 1955.

K.C. and Shu in front of the house in Englewood, NJ, where the family lived from 1971 to 1988. The picture was taken in 1976.

K.C. continued to receive her clinical training after our marriage, serving as a pediatrics resident at the Metropolitan Hospital in New York City. On 25 May 1959, we were extremely happy that May was born. K.C. stopped working after completing her residency training when May was about one year old. Another year later, we had the tremendous joy of having Ann born on 9 May 1961. After having the two lovely girls, we decided to move to Englewood, NJ in 1963, where we stayed for 25 years. Both May and Ann attended the public schools there and had outstanding accomplishments. May went to Harvard to study economics and later her MBA. She met Len Busch in Harvard and they got married in 1985 (p. 23). Ann went to Yale to study biology and studied medicine at Columbia University College of Physicians and Surgeons (P&S), where she met Steve Guidera and they got married in 1989 (p. 27). I had the pleasure of having both Ann and Steve in my class when I taught circulatory physiology at Columbia. When I was a premedical student in Peking University, I was taught by my father in Qualitative Analytical Chemistry (Han, p. 37). There are probably very few people who have been taught by their parents and have also taught their children in classes of some 100 students. K.C. and I feel that we are extremely blessed by having May and Ann, both being such wonderful children, and by having Len and Steve as our son-in-laws, both being such marvelous husbands. We are truly grateful!

The Chiens, 1979.

C. 1988–2001

1. University of California, San Diego (UCSD)

The Move to UCSD

In the mid-1980s, UCSD wanted to recruit a senior faculty in bioengineering because of Bert Fung's pending retirement in 1989 and Ben Zweifach's retirement a few years before that. I was surprised when Ben and Bert called in 1985 to ask me whether I would consider moving to La Jolla. I was deeply involved at Columbia with my large research group and a lot of other activities. An important factor was that K.C. was not prepared to leave New York because of our children and our friends there. Furthermore, she had a full-time job in New York as a pediatrician; to practice in California she would have to take the medical license exam again because the State of California does not have reciprocity with the States of New York or

The farewell party for Shu at Columbia, 1988. Sam Silverstein (left, p. 121) presented proclamation and Columbia P&S chair to Shu. Looking on are K.C., Shunichi Usami, and Eric Kandel, and Don Tapley. Standing in the back of the room beyond Shu is Steve Guidera.

New Jersey, where K.C. is licensed to practice. Therefore, I was not ready for the move. UCSD, however, kept inviting us to visit, and after several visits K.C. said that perhaps we should give it a try. In 1987–1988, I was very busy in Taiwan and could not focus on considering the move at that time. I remember one morning in early 1988 Ben phoned me in Taiwan (6 a.m. there) and asked me whether I would come to La Jolla. Half asleep, I answered "yes." So, what I had done to Herb Lipowsky (p. 83) was done to me by his Ph.D. advisor, but I am glad that I had the right mind to accept the position even though only half awake.

I returned to Columbia from Taiwan in July 1988, and left in September 1988 to come to UCSD. Probably because of my service of over 30 years and because I carried out all my teaching responsibilities during my sabbatical, Columbia was very accommodating in giving me the terminal sabbatical. Furthermore, I was given a two-year leave of absence to test out whether I would like to stay in San Diego. An important element which concerned me about leaving Columbia was the effects on my collaboration with long-time colleagues — particularly Dick Skalak. In response to my concern, UCSD asked Dick to also come to San Diego. UCSD got a good deal with two people at the same time. In addition, Amy and Paul Sung, Jerry Norwich (p. 205), and later Shunichi Usami also came to La Jolla. After having been at La Jolla for a few months, K.C. and I decided to stay. We were warmly welcomed in San Diego. In addition to colleagues in Bioengineering, we were very pleased to meet long-time friends. For example, my medical school classmate Chih Chao Chin (p. 50) and her husband Chien Sze Hsu, K.C.'s high school and college classmate Betty Hwa and her husband Kuo Chang Wang, K.C.'s brother George's sister-in-law Chia-Hui Shih (p. 235), and my Columbia colleague Helen Ranney, who was Chair of Medicine at UCSD. UCSD administration, including Chancellor Richard Atkinson, Vice-Chancellor Harold Ticho, Dean Richard Attiyeh of Graduate Studies and Research, Dean Lea Rudee (p. 135) of Engineering, Acting Dean Wayne Akeson of Medicine, and Chair Dan Olfe of Department of AMES, and my colleagues in Bioengineering, all gave me wonderful support and assistance. Several colleagues, especially in the School of Medicine, including Jim Covell,

Rick Lieber (p. 172), John Ross, Peter Wagner and John West, all helped me started in research and teaching at UCSD. The San Diego Chinese American Science and Engineering Association gave me a warm welcome party and asked me to deliver a speech. I chose the title of "A Tale of Three Cities" to describe my interesting experience during the past year, being a Columbia faculty and IBMS Director, while setting up my new lab in UCSD.

The incorporation of molecular biology into our research program was fully realized shortly after we moved to UCSD. In 1990, Amy Sung (p. 76) and I published in the *Proceedings of the National Academy of Sciences* our first paper in the area of molecular biology: "The Molecular Cloning of Human Protein 4.2: A Major Component of the Red Blood Cell Membrane." (p. 79) This work was done with the collaborations of Ann Rybicki, Ron Nagel, Long-Shan Chang and their colleagues. Russ Doolittle and Geoff Rosenfeld at UCSD provided important advice and help in many ways. This paper was followed by the paper by Amy and me, with the collaboration of Velia Fowler, on the molecular cloning and characterization of human RBC tropomodulin. These studies set the stage for our current research on the molecular

The Central Library of the University of California, San Diego. It has been used as the logo for UCSD, and renamed Geisel Library in honor of Audrey and Theodor Geisel (Dr. Seuss) in 1995.

basis of the modulation of signal transduction and gene expression by mechanical forces (Section II.C.1.g of this article).

Now, 14 years later, it is clear that the move to UCSD was marvelous. It is very gratifying that the bioengineering program at UCSD has developed to the state it is in today. I am very grateful to UCSD for giving me this opportunity to work with such wonderful colleagues, staff and students. I am grateful to the University of California for appointing me as a University Professor. I appreciate immensely the presentation of this appointment on 28 March 2002, form President Richard Atkinson, Regent Peter Preuss, Chancellor Bob Dynes, Senior Vice-Chancellor Marsha Chandler, Dean Bob Conn and Chair Dave Gough, and the kind and warm speeches by Peter, Marsha, Dave and Executive Vice Dean Frieder Seible at the reception.

The Program Project Grant

Based on the excellent foundation established by our founders of UCSD Bioengineering, I organized a research team, as I did at Columbia and IBMS, prior to my arrival at UCSD. We submitted a PPG application on "Biomechanics of Blood Cells, Vessels and the Microcirculation" to the NIH less than one month after my arrival in San Diego. Mainly because of the haste in its preparation, this application was not funded the first time. The critiques were fair and helpful, reinforcing my belief of the value of the peer-review system. Because the resubmission of a revised application, even if funded, would result in a loss of ten months, there was great financial stringency in the funding of personnel in my newly established laboratory. I decided to use all the funding at my disposal (including private gifts) to maintain the strength of the lab. It was a calculated risk and I believed that we would be able to satisfactorily revise the application. We worked very hard in revising the application, and we did succeed in getting our PPG funded on the next round in 1990. Here my experience of coming back from defeats in sports was very helpful. The project leaders were Bert, Geert, Amy, Paul and myself at UCSD and Larry Sklar at the Scripps Research Institute. We had core facilities on administration, ultrastructure (Geert), computation (Andrew McCulloch, p. 160), instrumentation (Shunichi) and cell culture (Bob Hoffman, later replaced by Bob Sah, p. 155). The Bioengineering program at UCSD had always been very strong, but this PPG was the first time a team research effort succeeded in receiving extramural funding after peer review. That was a big breakthrough, because it positioned us for all the later developments. We succeeded because of the strong scientific foundation at UCSD, the cooperative efforts of all participants, and the experience at Columbia, where our PPG had been continuously funded for 15 years, and at IBMS.

The Institute for Biomedical Engineering

After receiving the PPG and subsequently renewing the NIH training grant, which had been started by Bert in the early 70s, we began to address the identity of our Bioengineering program. From the beginning of our program in 1966, we had been a part of the Department of Aeronautical (later changed to Applied) Mechanics and Engineering Science (AMES). The AMES department was helpful in fostering activities related to bioengineering. However, as one of six groups in a large department, we always had to wait for our turn for resources, such as faculty positions and operating funds. Furthermore, to the outside world we were not a clearly identifiable entity. Therefore, the situation was not ideal. Most of my colleagues, however, were used to the situation and felt comfortable with the way it worked. Dick and I finally were able to convince our colleagues of the importance of having a departmental structure. It was wonderful that in the end everyone agreed on this important issue, and some colleagues who had considerable reservations initially became the strongest proponents.

Knowing how difficult it was to form a department, we worked first to establish the Institute for Biomedical Engineering (IBME) as an Organization Research Unit (ORU), which is a structure in the University of California system for fostering interdisciplinary research. ORUs cannot appoint faculty and do not have students, and its function is to coordinate and foster interdisciplinary research activities. More than 30 faculty members from schools of medicine, biology and engineering, and from the Scripps Institution of Oceanography, agreed to join in this effort. We held a retreat to discuss the research theme to be emphasized and agreed on tissue engineering science. Tissue engineering is a term coined by Bert and used by him in an earlier application to the National Science Foundation for an Engineering Research Center. We felt that this was a common theme that could tie all of us together, working on different kinds of tissues using common bioengineering approaches.

One of the first things I did when starting the IBME was to prepare a brochure that summarized all the institute members and their research activities. I used the experience I gained from preparing several editions of the IBMS brochure in Taiwan to great advantage. One of the special features of the IBME brochure was the preparation of an index based on research topics, in addition to a name index. This brochure received many positive comments, and I suspected that the newly created IBME and this brochure had some bearing on the sudden jump in national ranking of UCSD Bioengineering. In the U.S. News and World Report annual surveys, we went from being unranked prior to 1993 to No. 5 that year and have been No. 2 or 3 during recent years. The very comprehensive survey done every decade by the National Research Council of the National Academies of Sciences ranked UCSD Bioengineering in 1995 No. 1 in graduate education and No. 2 in faculty quality.

One of the missions of IBME is to foster academia-industry collaboration. Therefore, I established an Industrial Advisory Board, which has about 20 members. The Board has worked effectively with the institute to foster research collaboration, hold symposia and workshops, facilitate student internship placement, provide advice and consultation on curriculum of educational programs particularly relevant to industry, and sponsor various bioengineering graduate student activities such as their "Breakfast with Industry and Annual Symposium," etc. John Penhune of SAIC, who has been an outstanding Chair of the Board since 1995, will retire from the position in July 2002, and the Board has elected Bill Craig to succeed John. Bill is Vice-President for Research and Product Development of ISTA Pharmaceuticals and has made wonderful contributions to the Board and the institute. We look forward to continued development of our academia-industry cooperation under Bill's leadership.

The Whitaker Foundation Development Award

Shortly after the official formation of the IBME in November 1991, the Whitaker Foundation announced the request for proposals for the Development Award ($5 million for six years) to develop biomedical engineering. We were in the right place at the right time. We had just gotten people together to form the institute and to define a common theme, hence we were able to submit a well-prepared pre-proposal.

Among the 63 pre-proposals submitted, the Whitaker Foundation selected 14 for invitation to submit full proposals, and we were one of them. Immediately following the submission of the pre-proposal, I started the writing of the full proposal without waiting for the invitation, knowing that the timing would be tight. I gathered the needed information and materials from everyone involved and put the entire proposal together, making sure that all parts fit in a cohesive manner. Based on the review of the full proposals, we were chosen by the Foundation to be one of the five institutions to have a site visit. Finally, only three institutions received the Development Award that year, i.e. Georgia Institute of Technology, University of Utah and UCSD. The Development Award received in 1993 allowed us to recruit four new faculties and set up core facilities, and it was very valuable in helping us to form the department in 1994. The four new faculties would receive three years of salary support from the Whitaker Foundation, and UCSD would pick up their salaries thereafter. This is a wonderful win-win-win situation for the Foundation, the University and the Department. With the Whitaker Foundation Development Award, we were able to recruit four outstanding faculties, i.e. John Frangos, Bernhard Palsson (p. 152), Amy Sung (p. 76) and Sangeeta Bhatia (p. 244).

The Department of Bioengineering

The university administration was very supportive of the formation of the department, including Chancellor Atkinson, Vice-Chancellor Caserio, the Deans of Engineering Lea Rudee (till December 1993; p. 118) and Bob Conn (starting from January 1994), Vice-Chancellor Jerry Burrows and Dean George Palade of the School of Medicine,

Dean Attiyeh of Graduate Studies and Research, and AMES Chair David Miller. Dean Conn is a plasma physicist, but he very quickly identified bioengineering as a field he wanted to foster. The department was already in the process of being formed when Bob came, but he strongly facilitated that. It was, however, difficult to get the proposal for the formation of the department approved by the AMES department and the Academic Senate. In many group meetings and individual discussions with my AMES colleagues, I pointed out that a stronger Bioengineering program as a result of the formation of the department will enhance the entire Engineering School and benefit each of its departments, including AMES. AMES faculty finally voted to allow Bioengineering to leave and form a new department. The proposal then met some difficulties in the Academic Senate (because our proposal came up at a time of budget stringency at the University of California). Due to budget shortage, UCSD had just closed its School of Architecture two years after its opening, and no new department had been formed for at least five years. I presented our case to various committees on behalf of Bioengineering and agreed to use mainly existing university resources allocated to Bioengineering through AMES and extramural funding to start the department. The Academic Senate finally approved the proposal and forwarded it to the UC Office of the President.

On 20 August 1994, I received the official notice from the Office of the President that the Department of Bioengineering was established at UCSD and that I was appointed to be the Chair starting 1 July 1994. I did not know that I had been the Chair for 50 days before I was notified. It took a long time for the department to be established, but when it happened it came very quickly. Following the formation of the department, it was essential to recruit a Managing and Service Officer (MSO), who is the Chief of Staff in the UC System. We were very fortunate to have recruited Mary Hart as the founding MSO. We worked closely during the five years I was the Department Chair. Earlier this year, Mary accepted a wonderful offer by the San Diego Supercomputer Center as its Associate Director, and Loretta Smith is doing an excellent job as the MSO.

The governing of the Department of Bioengineering has always been by consensus, with the active participation of all faculty members.

Bioengineering Office staff at party for Shu's completion of his first term as Chair of the Department, 1999.

When we established the Department, we agreed to have a three-year term for the Chair, with the possibility of a two-year extension, just like the other departments in the Jacobs School of Engineering. In 1997, I received unanimous support to extend my term to 1999. I was very happy that David Gough (p. 149) agreed to accept the Chairship in July 1999 when my five-year service as the founding chair was complete. Dave has done a wonderful job as Department Chair.

The Whitaker Foundation Leadership Award and the Powell-Focht Bioengineering Hall

In the spring of 1997, the Whitaker Foundation announced the availability of a new Leadership Award program aimed at fostering Biomedical Engineering by supporting infrastructure at a level much higher than that offered by the Development Award. We submitted an application shortly after the announcement. (Ours was the only one submitted that round.) Because our application was focused too much on the building rather than the programmatic plan, it was returned for revision and re-submission. The revised application submitted in Spring 1998 received positive review, and a site visit by the Whitaker Foundation was held on 23-24 September 1998. The site visit team consisted of scientific consultants (Allen Cowley, Jr., Dominique Durand, Don Gibbons, Howard Morgan and Buddy Ratner), Foundation's Governing Committee (including Chair Burtt Holmes, Past Chair Ruth Whitaker Holmes, Portia Whitaker Shumaker and James Kelley), and Foundation's Executives (President Miles Gibbons, Jr., Executive Vice-President Peter Katona and Vice-President Jack Linehan). We made thorough preparations and extensive rehearsals for the visit. The presentations involved the Bioengineering faculty and graduate students, faculty from the School of Medicine (Ken Chien), the Supercomputer Center (Philip Bourne) and the Salk Institute (Tom Pollard), and industrial colleagues (John Penhune, Tony Ratcliffe, Ann Lee-Karlon, and others), as well as leaders from UCSD administration (Chancellor Bob Dynes, Senior Vice-Chancellor Marsha Chandler, Vice-Chancellor John Alksne and Dean Palade of the School of Medicine, School of Engineering Dean Bob Conn, Technology Transfer Office Director Alan Paau, and Abi Barrow for CONNECT). We did extremely well as a group during the site visit, which was the best I have ever seen, and this sense was shared by others. Bob Conn said that he had never felt the kind of intellectual power that filled the room that day. The event was totally unbelievable; everything just clicked. It was obvious that the Whitaker Foundation's site visitors were very impressed and got involved in the exciting and electrifying atmosphere.

The total amount requested in our written application was $13.8 million, including $0.8 million for faculty recruitment. We requested this amount because we did not want to risk losing it by asking for too much. Furthermore, there was a need for 1 to 1 matching, and it was difficult to get that much matching fund. Dean Conn was able to obtain a gift of $3 million from the Powell Foundation. Later, the Powell Foundation decided to give us an additional $5 million because of our high ranking in various surveys (see p. 137). Thus, without being asked, the Chairman of the Board Herbert Kunzel called Bob Conn and told him that the amount of the gift was raised to $8 million. (Herb Kunzel unfortunately passed away in 1999 and was succeeded by Joel Holiday, who has continued the wonderful support by the Powell Foundation.) The remainder of the matching required was provided by the university. The $13.8 million we requested from the Whitaker Foundation, together with the matching funds, would allow us to construct a new building to accommodate about three-quarters of the department, with one-quarter remaining in the EBU1 building.

Toward the end of the site visit on 24 September 1998, the site visitors asked, "Why do you leave some of the people in the old building? Why don't you include everyone in the new building?" I replied that we would have liked to propose a larger building, but we were concerned that the amount to be requested would be too large. The site visit committee then asked me to submit a supplemental proposal based on an estimate for constructing a building that would include the entire Department of Bioengineering. While the site visit committee encouraged us to all move into the same building, they also wanted to know what would be done with the space in EBU1 that we would vacate, and raised the question whether the Foundation would actually be supporting departments other than Bioengineering. I explained that the space vacated will be assigned to the Department of Electrical and Computer Engineering (ECE), which already has faculty working on bioengineering-related projects and will recruit additional faculty working on such projects. Thus, the space we vacate will be used in research contributing to advances in bioengineering (Now this proves to be indeed the case). I summarized the situation by saying that the moving of the Bioengineering Department into the new building will put "Bioengineering under one roof," while the performance of bioengineering-related research by ECE faculty in EBU1 will make "Bioengineering without walls." This explanation was well received by the site visit committee. I am very honored that Peter Katona (p. 184), now President of the Foundation, subsequently quoted these words on other occasions.

When we requested an additional $4.2 million to bring the total funding up to $18 million, the university would have to come up with another $4.2 million. This was approved by Chancellor Dynes, Vice-Chancellor for Development Jim Langley and Dean Conn. I wrote a letter to the Whitaker Foundation, giving the reasons for the additional request. A month later, the Governing Committee of the Foundation approved the total Leadership Award in the amount of $18 million. I had never had the experience of having the funding amount increased by a granting agency above the initial request. The letter requesting supplemental funding led to an increase in Whitaker funding by $4.2 million. The amount of funding generated by that letter (of course, it was due to everyone's effort at the site visit and the encouragement by the site visit committee) is truly fantastic.

We are very grateful to the Whitaker Foundation, not only from the point of view of our department and our university, but also as members of the bioengineering community. The Whitaker Foundation, formed as a result of the tremendous vision by the late U.A. Whitaker, is the main reason for the remarkable advances in biomedical engineering nationwide in the last two decades. The community is grateful to the wisdom of the governing committee, led by the Chairs Ruth and Burt Holmes, and the effectiveness of the executive leaders and staff in fostering biomedical engineering. The Whitaker Foundation is magnanimous in the support of bioengineering. Thus, despite its large share of support for the new bioengineering building at UCSD, the Whitaker Foundation agreed to name the building the Powell-Focht Bioengineering Hall, in recognition of the Powell Foundation gift and the late Judge Focht who played a major role in the Powell Foundation. To recognize the tremendous contributions by the Whitaker Foundation, UCSD proposed to name its IBME the "Whitaker Institute of Biomedical Engineering (WIBE)," and we are very pleased that the Whitaker Foundation approved this naming.

During the period of the Leadership Award, we have recruited additional outstanding faculty to the Department, i.e. Gary Huber (p. 159), Shankar Subramaniam (p. 158) and Michael Heller (joint appointment with ECE Department). Additional faculty members are being recruited, including a joint appointment with the Cardiology Division of the School of Medicine that is near completion.

With the support and help by UCSD administration, especially Boone Hellmann and Jay Smith, and JSOE, processes for the construction of the Powell-Focht Bioengineering Hall began in early 1999. Anshen + Allen Los Angeles (AALA) was selected as the architect, with Research Facility Design (RFD) as Lab Consultant. Jay Hughey of Anshen + Allen and Tom Mistretta of RFD interacted closely with the faculty in the design of the building. A groundbreaking ceremony was held in August 2000 and the actual groundbreaking started in January 2001 and the construction is proceeding smoothly. A dedication ceremony has been scheduled for 12 August 2002. We began the move-in in late November 2002. We all look forward with great enthusiasm to working in this state-of-the-art new building.

   In early 2002, we heard the wonderful news that the von Liebig Foundation has contributed $10 million toward the implementation of the technology transfer program in the Powell-Focht Building. In January 2000, through the introduction of John Watson (p. 177), Joe Bear and Jack Savage of the Development Office in the Jacobs School of Engineering (JSOE) arranged meetings with Linda Hamilton, Burt Kanter and Jean Goggins of the Foundation at UCSD. David Gough, Christophe Schilling (then a graduate student), and I made presentations on our plan for technology transfer, and it was well received. Subsequently, Joe Bear and Bob Conn visited the von Liebig Foundation in Naples, FL to continue the discussion. These contacts eventually led to the successful completion of this important gift for fostering technology transfer. For this and other JSOE development activities, I wish to thank Bob Conn and the members of the JSOE Development Office (Joe Bear, Kelly Briggs, Tatis Cervantes, Steven Flamm, Denine Hagen, Paul Laperruque, Jeff Nagel and MaryAnn Stewart, in addition to Joe and Jack) for their efforts in bioengineering development activities.

Research Activities in the Vascular Engineering Laboratory

As mentioned in Section II.C.1.a of this article, after moving to UCSD from Columbia, the research emphasis of my lab gradually shifted to focus on the mechanisms of mechanotransduction in endothelial cells and smooth muscle cells of blood vessels. That is, how do mechanical forces, such as the shear stress due to flow and the stretch due to pressure, act on these cells to modulate their signal transduction and gene expression. These studies have considerable significance in bioengineering because the molecular and genetic bases of cellular responses to mechanical forces is a fundamental problem at the interface of biology and engineering. They also have important clinical implications because the understanding of these processes may shed new light on the mechanisms by which atherosclerotic lesions have preferential locations in the vascular tree and may generate new therapeutic approaches.

This change in the area of emphasis was greatly facilitated by the joining of John Shyy (p. 290) from Ohio State University in 1990. He took a major responsibility in pursuing these studies, contributing many ideas and technologies. He also served as my key scientific associate during this period. When John accepted the faculty position at UC Riverside in 1999, I am very fortunate to have Julie Yi-shuan Li (p. 203) taking over these responsibilities and doing a superb job.

During my years at UCSD, the following graduate students received their Ph.D. degrees with me as an advisor or co-advisor: Cathy Galbraith, Janet Hansen, Pin-Pin Hsu, Shila Jalali, Gang Jin (p. 208), Song Li (p. 201), Kurt Ming-Chao Lin, Mohammad Sotoudeh, Yingxiao Wang (p. 218) and Yihua Zhao. The Masters students include Hsuan-Hsu Chen, Matt Petrieme and Michael Kim. The postdoctoral trainees and visiting scientists that have worked in my lab are Gerhard Artmann (p. 266), Matt Bartosiewicz, Indermeet Bhullar, Peter Butler (p. 216), Benjamin P.-C. Chen, Dennis K.D. Chen, Michael C.K. Chang, Hans Gregersen, Ute Henze, Jerry H.J. Hsieh (p. 345), Toshioki Itoh, Venus Labrador, Sylvaine Muller, Michael Pear, Kenji and Terue Sakakibara, Bo Skierczynski, Curt Thompson, Tsui-Chun Tsou, Franci Weyts, Lori Wickham, Xiong Wang (p. 223), Jackson C.H. Wu (p. 340) and Steven Wu. All of them contributed importantly to the research activities in the lab. I am also fortunate to have excellent administrative service and support for my lab by Jennifer Griffin, Tatyana Matusov and Elizabeth Hickman, as well as departmental support in finance and accounting (e.g., Michael Bartel, Martina Cotton, Csilla Csori and Veronica Moran), personnel (Katheryn Ritten), student affairs (e.g., Irene Jacobo, Jorrie

Miller), and seminar programs (Lore Meanley), which are closely related to our research activities.

At the present time, the research activities in the lab are focused on the following projects. Jason Haga studies the interplay between mechanical stretch and growth factor stimulation in regulating Rho family small GTPases signaling in smooth muscle cell. Sepideh Hagvall uses a co-culture system to investigate the interaction between vascular endothelial cells and smooth muscle cells in the regulation of angiogenic factors. Yingli Hu (p. 213) investigates the roles of microtubules and small GTPases in endothelial cell migration. Sung Hur is developing an experimental technique and theoretical modeling for the assessment of cellular viscoelasticity by the manipulation of magnetic particles introduced inside the cell. Roland Kaunas (p. 221) is investigating the various modes of mechanical stimulation in modulating the signaling responses of endothelial cells and smooth muscle cells, especially in relation to focal adhesions and cytoskeleton organization. Susumo Kudo is developing a nanotechnolgical approach to assess membrane dynamics in endothelial cells in response to mechanical shearing. Jane Li is studying the gene expression pattern in endothelial cells in response to different modes of mechanical shearing. Julie Li is investigating the roles of reactive oxygen species in the shear-induced regulation of gene expression via NFκB. Ian Lian, under the joint advisorship of Amy Sung, is studying the gene expression pattern in mice with heterologous deletion of the tropomoduling gene. Hui Miao is using in vivo animal models to investigate the signal transduction mechanisms in arteries. Alex Schuster uses an *ex vivo* system to dissect the effects of perfusion pressure and flow on gene expression in perfused arteries. Shunichi Usami is studying the dynamics of molecular interactions on cell surface by using the laser trap technology. Yingxiao Wang (p. 218) works on the cross-talk among different membrane sensors in initiating the signaling process. Yihua Zhao is using the micropatterning technique to study the effects of cell geometry and boundaries in modulating cell migration.

We are pursuing actively the use of DNA microarray technology for the study of gene expression in response to various types of stimuli, especially mechanical perturbations. The system we have set up will serve as a core facility for WIBE and also for the UCSD Superfund Grant from the National Institute of Environmental Health Sciences (P.I.: Bob Tukey of the Department of Pharmacology). This involves extensive teamwork. Jerry Norwich (p. 205) is responsible for robotic spotting and scanning. Jian-min Lao is responsible for the cDNA libraries and clones. Suli Yuan (p. 236) and Phu Nguyen work on PCR amplification. Jane Li is doing hybridization. Ian Lian and Yihua Zhao are working on bioinformatics for the analysis, archiving and data mining of the microarray results. We are pleased that our first paper on the application of DNA microarray technology to study gene expression profile in endothelial cells subjected to mechanical shearing was published in 2001 in *Physiological Genomics* (Publication A390).

In addition to DNA microarray, another new area of technological development for answering some of the questions in molecular biomechanics is nanotechnology with the use of the laser trapping system. A laser trap — laser tweezers microscope system has been set up with Michael Berns (p. 174), who is Professor at UC Irvine and also an Adjunct Professor in our department. Shunichi Usami, in collaboration with Daniel Ou-Yang of Lehigh University (p. 54), is setting up another laser trap system that will allow the determination of viscoelasticity at subcellular level. We also collaborate with Jeff Price (p. 170) in applying his scanning cytometry and confocal technologies for studies on cellular and subcellular dynamics.

Since coming to San Diego, we have collaborated with many scientists. Within the department, besides the colleagues in my laboratory, I have co-authored publications with Sangeeta Bhatia, Bert Fung, Ghassan Kassab, Andrew McCulloch (and his student Bill Karlon), Jeff Omens, Geert Schmid-Schönbein, Bob Sah (and his students Melissa Kurtis and Bob Schinagl), Dick Skalak, Amy Sung and Paul Sung. Collaborators in other departments in UCSD include Ken Chien, Michael Karin and Anne Hoger. At other institutions in La Jolla, we collaborated with Miguel del Pozo, Velia Fowler, Mark Ginsberg, Jiahuai Han, Eugene Levin, Martin Lotz, Nigel Mackman, Graham C.N. Parry, R.M. Pietrowicz and Martin Schwartz of the Scripps Research Institute; Tony Hunter and David Schlaepfer of the Salk Institute; and Tony

Ratcliff and Mohammad Sotoudeh of Advanced Tissue Sciences, Inc. Outside San Diego, we have collaborated with Junlin Guan of Cornell University; Mony Frojmovic of McGill University, Herb Lipowsky of Pennsylvania State University; Shu Qian Liu and his students at Northwestern University; John Shyy, Michael Stemerman, Lynn Verna and their colleagues at University of California Riverside; Bauer Sumpio of Yale University; and Shelly Weinbaum, Dan Lemons, David Rumschitzky and their students at the City College of New York. Outside the U.S., we collaborated with Gerhard Artmann, Christine Keleman and their colleagues at the Aachen University of Applied Sciences; P.-I. Brånemark (p. 382), Ulf Bagge and MagnusBraide of the University of Gothenburg; and Hans Gregersen of Aalborg University; as well as the scientists in Taiwan and Mainland China who are mentioned in Section II.C.6. of this article (p. 469). All the collaborations with these outstanding colleagues have been very helpful in our research advancement.

We have recently renewed our NIH R01 grant on Studies on Endothelium, the NIH Training Grant and the PPG, with Geert now taking over the responsibility of being the P.I. of the PPG. In 1999, together with Martin Schwartz and Junlin Guan, we received one of the first NIH Bioengineering Research Partnership (BRP) grants. In addition to the funding of the research and training programs in our laboratory by government agencies, I would also like to express my appreciation for the generous support by the Cho Chang Tsung Foundation of Education and Dr. Shi H. Huang of the Chinfon Group

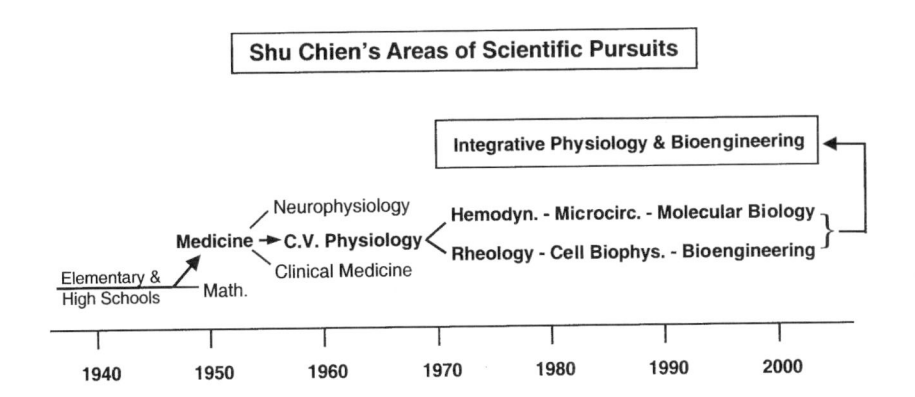

Shu Chien's Areas of Scientific Pursuits

Integrative Physiology & Bioengineering

Neurophysiology
Hemodyn. - Microcirc. - Molecular Biology
Medicine → C.V. Physiology
Rheology - Cell Biophys. - Bioengineering
Clinical Medicine
Elementary & High Schools
Math.

1940 1950 1960 1970 1980 1990 2000

in Taiwan. These valuable supports, together with the Whitaker Foundation Awards to the Department and WIBE for infrastructures, have provided the resources needed to carry out our research and training programs.

I started my training in medicine, and received my Ph.D. training in physiology, working on cardiovascular dynamics at the organ-system level. During the ensuing years in the 70s and 80s at Columbia, I have focused on microcirculation, blood rheology and cell biophysics, as mentioned above (Section II.B.1). Since coming to UCSD in 1988, I have become a faculty in Bioengineering and enhance our research emphasis on molecular biology and genomics. The ultimate goal of my research work is to integrate our knowledge across the biological hierarchy from genes to systems, to combine experiments and theory, to coordinate research and education, and to apply basic research to clinical and industrial application. These integrative approaches to physiology and bioengineering have always been what I advocate not only in my lab, but also in varioius scientific organizations.

Technology Transfer

After my move to San Diego, I have had a much greater degree of interactions with industry because of the rich biotechnology environment and the Industrial Advisory Board of the Whitaker Institute of Biomedical Engineering (Section II.C.1.g of this article). I have served as a consultant or on the Industrial Advisory Board for a number of biotechnology firms, such as Advanced Tissue Sciences (ATS), Alliance Pharmaceuticals Corporation and AVIVA Biosciences. Bert Fung and I have worked with Tony Ratcliffe and his colleagues at ATS, including UCSD graduates Ann Lee-Karlon and Mohammad Soutodeh, on a joint Advanced Technology Program Grant on the development of vascular graft that was supported by the National Institute of Standards and Technology. In the Whitaker Foundation Leadership Award application (Section II.C.1.f of this article), we specifically proposed the establishment of a Technology Transfer and Clinical Development Center then, now Powell-Focht Bioengineering Hall. The von Liebig Center for Entrepreneurism and Technology Development established by the gift from the William J. von Liebig Foundation will make this a reality.

John Shyy and I worked on a research project that involved the use of a negative mutant of the signaling molecule Ras, i.e. RasN17, to block the excessive smooth muscle cell proliferation following endothelial injury due to balloon angioplasty or coronary bypass. Our results on animal experiments have shown that RasN17 can prevent the stenosis induced by balloon injury. Our patent application on "Gene Therapy in Coronary Angioplasty and Bypass" was approved by the U.S. Patent Office in January 2002 (U.S. Patent No. 6,335,010).

Ken Chien, Professor of Medicine with an affiliated appointment in Bioengineering, has developed a new gene therapy technique for treating heart failure by maintaining the calcium concentration in a subcellular compartment (sarcoplasmic reticulum) of the heart muscle cells. Based on this and other technologies, Ken asked me to be a co-founder of a company which is named Celladon. The company is in its formative stage, and this will be a new experience.

2. A Multi-campus Research Unit (MRU) on Bioengineering in the UC System

Following the establishment of the Department of Bioengineering at UCSD and with the rapid advances in the field nationwide, other University of California campuses have also been enhancing their bioengineering programs, including the establishment of departments. Thus, UC Berkeley and UCSF established a joint Department of Biomedical Engineering in 1999, and UC Irvine, UCLA, and UC Davis are also establishing their departments. UC Santa Cruz is forming a Department of Biomolecular Bioengineering. UC Santa Barbara and UC Riverside have formed bioengineering programs, which may evolve into departments in the future. The campuses at Berkeley, Davis, Irvine, San Diego and San Francisco have received the Development Award and/or Leadership Award or their equivalence from the Whitaker Foundation. I have served on the advisory committee of most of these programs and am very impressed by the remarkable progress in bioengineering throughout the UC system and by the complementary strengths among the campuses. Therefore, the time is ripe for forming an intercampus consortium. The opportunity for initiating a multi-campus members. Previously, I had served as President of two professional societies for Chinese Americans, viz., the Chinese American Medical Society (1978–1979) and the Chinese Academic and Professional Association (1979–1980), both being based in Metropolitan New York area. The presidency of the Microcirculatory Society was the first one I ever served in a national scientific society, and I learned a great deal from my colleagues. I learned more than I contributed to the Microcirculatory Society, and I am very grateful for having the opportunity to start at the micro level.

The American Physiological Society

Physiology is my primary area of training and is still an area of my major research and education interests (Johnson, p. 162; Frank, p. 189). I have been a member of the American Physiological Society (APS) for over 30 years. As mentioned above (p. 410), I served on its Council starting in 1985. At the end of my four-year term as council member, I was elected as President-elect to serve as President for the year 1990–1991. Counting the year as President-elect and Past President, I had the pleasure and privilege of serving on APS Council for seven years. APS was founded in 1887 and it has always been an outstanding scientific society. After having been in existence for 100 years, however, there was a sense of malaise among physiologists and by scientists in other biological disciplines ("Long-Range Planning Committee Report," *Physiologist* **33**: 161–180, 1990). There was a need to recruit new members to APS, to enhance the interest and dedication of existing members, and to take leadership in fostering physiology at the frontier of the new biology. Aided by a list of benefits of belonging to the Society prepared by APS Council, I wrote to the chairs of every physiology department, asking them to encourage their faculty to join APS as members. In some cases I had to ask the Chairs themselves to join, since they were not APS members. We also implemented a special recruitment effort for student members, including free membership for one year. These approaches led to a significant increase in membership.

A number of measures were adopted to increase the sense of allegiance of the existing members. APS had over 7000 members, and contacts could be more effectively made through their special areas of scientific interests (the APS Sections) and their regional affiliations. We strengthened the roles of APS Sections in the programming of the annual meetings, formed a Section Advisory Committee (SAC), and instituted frequent meetings between APS Council and SAC to discuss issues of governance. Such direct communications helped to change the relationship from being somewhat adversary to a collaborative and synergistic one. To enhance communications with regional chapters, I attended the regional meetings in New England and Ohio to report APS activities to members and to receive inputs and discuss possible ways of improving the operation of APS.

As mentioned above (p. 409), I promulgated the application of molecular biology to research and education in physiology as part of the integrative approach. This led to a continued evolution of the scientific programs of our annual meetings and the papers published in the journals of the APS such that physiology integrates the whole spectrum of biological hierarchy from genes and molecules to organs and systems. These approaches paved the way for the recent emphasis on physiological genomics by APS under the leadership of Allen Cowley, Jr.

APS also reached out beyond the boundaries of North America. We increased the number of corresponding members from other countries. During my year of Presidency, I visited several countries in Europe (England, Germany and Sweden) and Asia (Japan, Korea and Taiwan) to meet with the leaders in physiology in these countries. Because of my close relationship with the Chinese Physiological Society (CPS) in Taiwan, we were able to hold a Joint APS-CPS meeting in Taipei on 2–5 November 1990 (Frank, p. 189; picture on p. 190). The meeting was attended by nearly 50 American physiologists, including more than one-half of APS Council members, many physiologists from other Asian countries, and nearly all physiologists in Taiwan, most of whom had received training in the U.S. It was a great success both scientifically and socially, leading to fruitful exchanges of scientific progresses and establishment of new collaborations.

It is gratifying that following the APS Long-Range Planning Committee reports in 1990 (see above) and 1996 (Ref. D4 on p. 546), the mood in the physiology community has turned from that of a sense of malaise to the outlook of a bright future.

It was a wonderful and rewarding experience working with the APS. I am most grateful that I was selected as the recipient of the 1999 Ray Daggs Award, which is given annually to a physiologist for distinguished service to the Society and to the science of physiology. Whatever that has been accomplished as described above was the result of cooperative efforts of the members of Council, Sections, various Committees, and the Society. I especially wish to thank the Executive Office of APS, with the wonderful leadership of Martin Frank (p. 189). Marty started his position as Executive Director the year I became a council member and has made wonderful contributions to APS and physiology as a discipline.

One of the most important events during my tenure as President of APS is related to the evolution of the Federation of American Societies of Experimental Biology, which is described in the next section.

The Federation of American Societies for Experimental Biology

The Federation of American Societies of Experimental Biology (FASEB) is an umbrella organization for societies in the fields of biological and medical sciences. In 1989, it was comprised of the APS, American Society for Biochemistry and Molecular Biology (ASBMB), American Society for Pharmacology and Experimental Therapeutics (ASPET), American Society for Investigative Pathology (ASIP), American Institute of Nutrition (AIN; now American Society for Nutritional Sciences, ASNS), and American Association of Immunologists (AAI). These were the same six societies as when AIN and AAI joined in 1943. There was a reluctance on the part of FASEB to add new member societies, and there is also an impediment for other societies to join because of the high assessment of about $60 per society member. The way the financing worked was that the annual FASEB meetings generated a significant amount of revenue from the industrial exhibit program, and the share for each member society was approximately equal to the FASEB assessment. Thus, a society such as the American Society for Cell Biology (ASCB), which has its own annual meeting, would only want to be an associate member society, which paid a small amount of dues rather than the large sum of assessment. The associate member societies, however, did not have the privilege of a full member society such as voting on the FASEB Board.

In the 80s, ASBMB felt that the FASEB meetings were too large and did not address sufficiently its specific needs, and decided to meet separately. Hence, it had to bear the large assessment without a concomitant revenue income from the FASEB meeting. As a result of this and other reasons, ASBMB wanted to withdraw from FASEB, and this caused repercussions through the other member societies, including APS, which also considered the question of withdrawal. Faced with this crisis situation, the FASEB Board held a retreat in Williamsburg in 1989 to discuss the future of the organization. As President-elect of APS, I was one of the participants of that retreat.

The retreat had an in-depth discussion of the past, present and future of FASEB. I, among others, felt strongly that FASEB had an important mission as a unifying organization for experimental biology and that we should do everything to correct what was wrong in the organization and start moving in the right direction. After extensive discussions, the principle evolved that FASEB should serve the needs of its member societies primarily in the area of public affairs. Other FASEB activities, such as placement, publications, scientific meetings and conferences, should be self-sufficient in budget. It was decided that the annual multi-society meetings would no longer be an official FASEB function, and instead would be managed by an Experimental Biology (EB) meeting group formed by the member societies that wish to participate in this joint meeting. It was also recommended that the FASEB Journal be changed from free distribution to paid subscription. As a result of these proposed changes, the large assessment could be replaced by a dues structure in which the societies would pay only $10 per member per year, but this goal would be achieved by a gradual ramping downward over a five-year period. Furthermore, in order to attract new member societies, they would be asked to pay only $10 from the beginning. There was some skepticism as to whether this could work out financially. I made the point that if we did not go this route FASEB would be dissolved anyway; therefore we had no choice but to give this reasonable proposal a real trial. This view of taking a calculated risk received the support of Board Presidents: Howard Schachman (1988–1989), Bill Dewey (1989–1990) and Tom Edgington (1990–1991). The retreat concluded with a general understanding of the principle, but it needed the agreement by each member society, particularly ASBMB.

This issue was discussed extensively in APS Council, and the consensus was to support the recommendations of the Williamsburg retreat. Realizing the importance of getting ASBMB on board, I asked to meet with the ASBMB leadership to discuss this issue. ASBMB Presidents Bob Schimke (1989–1990), Bill Lennarz (1990–1991) and Dan Lane (1991–1992), and Executive Director Chuck Hancock accepted my invitation to meet with APS Council during the APS fall meeting in Orlando in October 1990. I first had a thorough discussion with APS Council in the morning and then invited these ASBMB colleagues to have lunch with me prior to meeting the APS Council as a whole. It was a successful meeting with fruitful, collegial exchange of views. I then invited myself to meet with ASBMB Council in their next meeting in Bethesda. Members of ASBMB Council asked many critical questions regarding the merit of FASEB as an umbrella organization. I stressed the importance of having such an organization to work for the common good for the biomedical research community, so that we would have a voice to represent more than 30,000 scientists. This is extremely important when dealing with the legislative and executive branches of the government, especially in view of the stringent funding situation these days. My meeting with the ASBMB Council lasted over an hour. Later that day, the ASBMB Council passed the resolution of supporting the new FASEB by a single vote! That deciding vote had a very important bearing on the future of biomedical research. During the past decade, as a result of the outstanding leadership by the Presidents and the Board and excellent work by the administrative staff, FASEB has gained increasing recognition on the Hill and has exerted significant influences on many decisions related to biomedical sciences. The coalition has grown from six member societies to 21 and the unified voice now speaks for more than 60,000 scientists (Garrison, p. 194). FASEB now commands the respect and attention of the government and the scientific community and is widely recognized as the leading organization in biomedical research.

Prior to the Williamsburg retreat, the FASEB Presidency was rotated among the six member societies. At the Williamsburg retreat, it was decided that the President would be elected by the FASEB board among the board members. Furthermore, every member society would have two board members serving for four years each, instead of having two or three members (depending on the size of the society) serving for three years each. APS had three positions on the FASEB Board, and they were filled by the Past President, President, and President-elect of APS. The new rule would require filling the two positions with APS presidents elected in alternating years. Seeing what was coming, I proposed that we start with the APS Presidents in 1989–1990 and 1991–1992, thus I would not be in the turn. My colleagues, however, voted instead to have me (1990–1991) and Stan Schultz (1992–1993) as the first two FASEB board members from APS under this new rule. The FASEB Board later voted to have me as its first elected President.

When I took office as President in 1992, ASCB and the Biophysical Society had joined FASEB, and they were followed by the American Association of Anatomists (AAA) to make it a nine-society Federation, but FASEB still was in a relatively precarious situation. I visited nearly every member society during their council meetings (Garrison, p. 194), and had frank and open exchanges on various issues that concerned the societies, especially in relation to FASEB. I found these meetings to be extremely valuable in enhancing communication and avoiding misunderstanding.

Most of the activities of the new FASEB were concerned with public affairs. As FASEB President, I flew frequently from San Diego to Washington area to testify before congress, to give press conferences, to meet with FASEB staff members at the Bethesda headquarters, and to hold consensus conferences regarding biomedical research funding. I am grateful to the FASEB staff for their wonderful cooperation. In particular, I wish to thank the then Executive Director Michael Jackson (p. 192), who did a marvelous job in fostering the remarkable progress of FASEB through a difficult period. I had the pleasure of working with Gar Kaganovich, who led me through the corridors of the senate and house buildings to meet with congressional members and their staff, and with Howard Garrison (p. 194), who has done a terrific job in public policy and policy affairs. Howard has played a major role in the success of the consensus conferences, which are attended by FASEB member societies, other scientific organizations, and government funding agencies. The recommendations made by the consensus conferences have become important references for biomedical research funding. Michael, Gar and Howard helped me to establish the Public Affairs Executive Committee (PAEC), with participation by member society representatives on frequent teleconferences and face-to-face meetings to deal effectively and expeditiously with public affairs matters that require a group decision, including research funding, bioethics, research integrity, etc. When working with FASEB, I learned a great deal about science policies, and also many other aspects of science. For example, Sam Silverstein (p. 121, picture in p. 425) and I wrote an article on the economic impact of application of monoclonal antibodies to medicine and biology, with the help of able FASEB staff members (Publication D1 on p. 546).

Although I made frequent trips to the Washington, D.C. area, my primary job was at UCSD, and most of the communications were made by fax since we did not use e-mail yet. Because of the time difference, there was usually a big pile of fax messages on the machine by the time I got up in the morning (Hsieh, p. 345). Answering these messages was my first order of business every day, sometimes even before brushing my teeth. Although it was a very busy time, it was also a very enjoyable time, because it was such a pleasure to work with all my wonderful colleagues: board members, PAEC representatives, member society leadership and executive officers and FASEB staff. There were many other pleasant duties as well. It was a great pleasure for me to present the 1993 Annual FASEB Public Service Award to Ruth Kirschstein (p. 245), then Director of the National Institute of General Medical Sciences (NIGMS) and later the Acting Director of NIH, in recognition of her tremendous public service through her outstanding contributions at the NIH.

The International Union of Physiological Sciences

The International Union of Physiological Sciences (IUPS) is an international organization that belongs to the International Council of Scientific Unions (ICSU), and is comprised of adhering bodies from more than 60 nations. IUPS has an executive committee composed of the President, two Vice-Presidents, a Secretary General and a Treasurer. Harvey Sparks, who had been APS President in 1987–1988, was IUPS Treasurer from 1990 through 1997. In early 1997, Harvey called to ask me to be a candidate for the Treasurer position succeeding him in January 1998. I told him that I would be quite busy with the Whitaker Awards and a new department, as well as my active research program. He assured me that there would not be that much work except during the International Congress of Physiological Sciences, which is held every four years, with the next one (XXXIIIrd Congress) being in St. Petersburg, Russia in July 1997 and the following one (XXXIVth) in Christchurch, New Zealand in August 2001. As it turned out, the St. Petersburg Congress had a major financial problem. Although the contract stated that the organizer of the Congress was responsible for all financial obligations, many of the obligations were not fulfilled, especially the payment of the travel and local expenses promised to the invited speakers and chairpersons. IUPS President Ewald Weibel and Secretary General Denis Noble made numerous attempts to resolve this issue with the Russian organizers to no avail. Although IUPS was not legally responsible for making these payments, it was morally bound to resolve the matter. There was a very small reserve fund in the IUPS treasury, and we had to settle this financial deficit by asking the adhering bodies to help out and the chairpersons and speakers to take a reduction in reimbursement. Fortunately, several adhering bodies provided generous support and the scientists agreed to a reduced reimbursement. With these valuable help and by using a good portion of the funds available in IUPS, we were able to close this issue after two years of negotiations, communications, processing and accounting. I had not expected to get involved with these complications, and nor did Harvey. I had two excellent administrative staff members helping me with IUPS matters, Martina Cotton (1998–1999) and Elizabeth

Hickman (1999–2001). Elizabeth, who was also my administrative assistant in UCSD Bioengineering, did a great job in assisting me on this difficult financial matter. The IUPS executive committee, composed of Ewald Weibel, Denis Noble, the First Vice-President Ernie Knobil (who unfortunately passed away in 1999) and the Second Vice-President Ramon Latorre, as well as the Executive Secretary Sue Orsoni, was also most helpful. It was indeed a pleasure to work with them on all IUPS matters, including resolving this debacle.

As we were busy closing out the St. Petersburg account, we got started with the New Zealand Congress in 2001. Our colleagues in Christchurch are excellent meeting organizers. It has been a pleasure to work with Tony Macknight, Chair of the Congress, Paul Hill, Chair of the Local Organizing Committee, Bruce Smaill, Congress Treasurer, and other wonderful colleagues. It was an extremely successful IUPS Congress.

As we approached the time of the 2001 Congress, planning was under way for the 2005 Congress in the U.S. The decision to hold this XXXVth Congress in U.S. was made by the IUPS General Assembly in 1997 in response to an invitation from the U.S. National Academy of Sciences President Bruce Alberts. Under the auspices of the U.S. National Committee (USNC) on IUPS of the National Academy of Sciences, a U.S. National Organizing Committee (NOC) for the 2005 Congress was formed. This Committee is composed of representatives from six societies related to physiology, viz. APS, Society for Neurosciences, Society of Comparative Physiology, Society of General Physiology, Biomedical Engineering Society and Microcirculatory Society, with APS being the leading society. Allen Cowley, who had been the Chairman of the NOC, was nominated in May 2001 to be the next IUPS President to succeed Ewald Weibel. On 8 July 2001, I was notified that I had been elected to be the Chair of the NOC to replace Allen. On 10 July, I received from Marty Frank a nicely prepared prospectus for the 2005 Congress in Washington, D.C. in August 2005. I replied saying that "The 2005 Congress will be a very exciting one. We must take advantage of the New Zealand Congress and transmit fully that excitement, both scientifically and socially, including all the attractions in the DC area and vicinity, as you so ably portrayed."

On 15 July, I received an e-mail from Marty to Allen Cowley, Virginia (Ginger) Huxley (Chair of USNC on IUPS), Walter Boron (US Scientific Program Committee Chair for the 2005 Congress), and myself that the APS Council had just decided to move the 2005 IUPS Congress to San Diego in April to coincide with the Experimental Biology meeting (EB 2005), in order to avoid splitting the attendees from the physiology community for the two meetings in the same year less than four months apart. Marty also informed us that APS President John Hall and President-elect Barbara Horwitz wanted to set up a conference call on 18 July to discuss the 2005 IUPS Congress. I realized right away that, while the proposed change is a rational one, its implementation would encounter many difficulties in receiving approval from our international colleagues because of the potential perception that the U.S. would overwhelm the IUPS Congress with the larger EB meeting.

Although this proposal could be presented to the meetings of the IUPS Executive Committee (ExCo), Council and General Assembly in Christchurch, New Zealand in late August 2001, it would not have given sufficient lead time for adequate deliberation and positive reaction. For controversial matters that require discussions and deliberations, I prefer to speak directly with the people involved, rather than using the e-mail or other written modes of communication. Since I was serving on the ExCo and working closely with its members, I phoned them immediately. Fortunately, I was able to reach both President Ewald Weibel and Secretary General Denis Noble by phone on 16 July.

Ewald was surprised by this sudden change. After considerable discussions, he began to feel that there are positive aspects in this proposed change. He sent me an e-mail on 19 July saying that he believed that it was a good strategy to combine the American Physiology meeting with the IUPS Congress as it would ensure the participation of the local physiologists. So in that sense he would approve this in principle. While he had some concerns, Ewald said that if a good scheme could be worked out, it could be successful and that there might be the fringe benefit of having some non-physiologists exposed to physiology. But he wanted it to be clear that IUPS 2005 would not simply be a small corner of the big EB 2005. He said that the change from August to Spring should be no problem, but San Diego versus Washington would have to be justified, because the U.S. invitation, which was accepted by the General Assembly in 1997, was built strongly on the new Washington Convention Center. He did note, however, that UK and NZ also changed the cities. He suggested that we present this as a new option for approval, but not as a fait accompli.

My phone call to Denis on 16 July caught him in the nick of time before his departure for a trip. I presented the proposal of changing the venue and time to him. After some thoughtful discussions, Denis was in favor of holding the meeting in San Diego in April. His main concern was also the identity of IUPS. He would not like to have the international community to have the conception (or misconception) that IUPS would become a subset of American Physiology.

I was not able to reach Ramon until 24 July. Ramon and I also had a fruitful discussion. His reaction was similar to that of Denis. He felt that on the whole, the pluses override the minuses, but he would like to give this more thought and then send me an e-mail about it The major issue he brought up was again the identity of IUPS, and he wanted to be sure that the role of IUPS would not be lost in the process. Thus, the three IUPS ExCo members (besides Ernie Knobil, who had passed away, and myself) all felt that it was feasible to make the change, but, as I expected, there was a great deal of concern about the identity of IUPS in a larger crowd at the EB meeting.

While I was discussing this matter with the IUPS ExCo members, Marty contacted the EB office and the executives of some FASEB Societies who will participate in EB 2005 and sent e-mails to NOC members. Geri Swindle of the EB Meeting Management Office said that it would be possible to hold the Congress along with EB and have a separate headquarter hotel assigned to IUPS. Two of the society executives contacted did not have objections but pointed out some of the potential difficulties. Most of the NOC members were in favor of the shift of IUOS 2005 to San Diego in April, but two members did not agree with these proposed changes. Stan Schultz, Chair Emeritus of NOC, expressed the view that, regardless of whether we make the change, we should not have an APS meeting as such in April 2005, but rather roll everything in Physiology into the IUPS Congress.

During the teleconference call on 18 July, I reported my conversations with Ewald and Denis, and Marty reported his communications with the EB office, FASEB Societies and NOC societies. Following extensive discussions, the consensus was to do everything possible to assure the identity of the IUPS Congress by starting it ahead of and then in conjunction with EB in San Diego in April 2005. It was concluded that, while there would be difficulties, they were not insurmountable. The pluses of the proposed change include a better attendance, the planning of a single meeting in 2005, the attraction of participants from the sister EB societies, a pleasant venue with an amenable climate and easy accessibility to participants from the Pacific Rim, and the availability of a large exhibit. The minuses include the potential impressions by the international physiology community that the EB meeting might overwhelm the IUPS Congress and by the EB sister societies that this would be only a Physiology meeting. Other factors are that schools would be in session and that the West Coast is less accessible from Europe. The results of the teleconference were summarized by an e-mail sent by Ginger to the leadership of the six member societies of the USNC/IUPS for comments and suggestions.

Based on the concerns about the IUPS identity, as expressed by the ExCo and some NOC members, I made the following statement in my 24 July e-mail to NOC. "I think it is time for us to be magnanimous and be willing to say that we will NOT have an APS meeting as such and that we throw all our support to the IUPS Congress, which deals with the same science anyway. We still should start the Congress 1 1/2 to 2 days ahead of the EB to hold the opening ceremony, some special lectures, workshops, etc. and start the regular symposia and poster sessions together with EB. As far as the relationship with EB is concerned, we can proceed in the regular way, but the banners and printed materials must say only IUPS. APS can still have its various functions, including the Section Distinguished Lectures, etc. but APS should be big-hearted enough to label the regular sessions IUPS. APS will not lose from this, but rather it will gain a tremendous amount of goodwill and appreciation from the international community, which is more precious than APS gaining any direct billing." On 25 July, I sent an e-mail to Marty saying that "I understand some of the difficulties involved in emphasizing IUPS and putting APS more in the background, but we must do as much as we can toward that goal. The principle is to give the international physiology community a clear sense that IUPS Congress is not a subset of the EB meeting. I am sure that the IUPS Congress will bring better science and more scientists to the EB meeting in 2005. It is a win-win situation, and EB and IUPS should be partners for the betterment of science."

Following the 18 July teleconference and many e-mail exchanges among the participants, Marty drafted a proposal to the IUPS for the U.S. to host the XXXVth Congress in San Diego in April instead of Washington, D.C. in August 2005. I made a number of revisions to further emphasize the identity of IUPS (while holding the Congress with an overlap with EB) and the benefits of the interactions with sister disciplines. I also added a subheading to indicate that this is an Alternative Plan and stated in the text that this is a possible alternative to the original proposal for the IUPS ExCo, Council and General Assembly to consider for approval. We did not want to have this perceived as a unilateral decision, but rather as a proposal to the IUPS leadership, who clearly has the ultimate authority to make the decision. This proposal received the enthusiastic support of USNC Chair Ginger Huxley and most of its members, but several members expressed their concerns, particularly with respect to the question of the identity of IUPS.

This proposal was submitted to the ExCo, Council and General Assembly of IUPS on 31 July 2001. It met with considerable opposition, with most of the e-mail responses being negative. In response to these messages, Ewald sent an e-mail to the ExCo members on 9 August, saying that "My concerns about the US proposal are rising." Then, with the consent of the ExCo, he sent a message to IUPS Council on 10 August, making the following points regarding the proposal: "(1) It appears as an interesting proposal to expand the spectrum of the physiological sciences… in EB. (2) The chief concern is that the proposal… may not safeguard the identity of the IUPS Congress, for two reasons: (a) The actual IUPS events are limited to two days…., after which the IUPS Congress becomes part of EB; and (b) There will be no specific registration for the IUPS Congress whereas all participants will have to register for the EB meeting." He asked the Council members "to look at these questions carefully and to also consider the consequence this new proposal will have on IUPS as an institution in the long run."

Ewald's message arrived about a week before Allen Cowley and I were to depart for the New Zealand Congress. In the ensuing days, we had several telephone conversations with APS President John Hall, discussing with him the importance of assuring the identity of IUPS if the 2005 Congress were held in conjunction with EB. John made the important decisions of making ALL 2005 San Diego physiology sessions as IUPS instead of APS events and requiring all APS and its guest society members to register through IUPS. Given these new directions, I further revised the proposal to move the 2005 Congress to San Diego in April during my flight to New Zealand. This new revision includes the following key points:

(1) Registration for IUPS Congress will be separate and clearly identified for all participants, with a separate IUPS registration kit and website. APS will require all of its members to register through the IUPS.
(2) The scientific program in physiology during IUPS 2005 (and EB 2005) will be that set up by the IUPS prepared by the International Scientific Program Committee. A separate IUPS 2005 program book/website will be produced.
(3) Meeting
   (a) IUPS will have its headquarters at a top-level hotel (e.g. Marriott, Hyatt, or a new hotel under construction) adjacent to the San Diego Convention Center. On 31 March and 1 April, IUPS will be the only meeting. On 2–5 April 2005, the IUPS Congress will have its own separate site and clear identity, and yet is close enough to others for interdisciplinary exchange.
   (b) The IUPS poster sessions, i.e. all physiology posters, will be positioned in a clearly identified area with clear indication that they are the IUPS posters. They will be set up in such a

way that the international physiological community can meet in that area for exchange of information.

I arrived in Christchurch in the afternoon of 21 August. That evening Ewald invited Denis, Ramon, Allen, Sue Orsoni, K.C. and me to have a drink in his suite in the Parkroyal Hotel, and we had dinner together. Allen, who was sitting next to Ewald and Sue, told me after the dinner that it did not seem likely that we can get the proposed change approved. I arranged to have breakfast with Denis the next morning. I explained to him the advantages of the proposed change for the 2005 Congress. Denis agreed with the rationale, but felt that it was a difficult proposition to get approval and likened it to walking on a tight rope. We needed to do everything right to succeed. I then spoke to Ewald and to Ramon about this. During the ExCo meeting on 22 August, we first discussed many of the other IUPS agenda items, including budget, etc. The 2005 Congress was on the agenda for the next day, but Ewald decided to bring it up that afternoon. I presented the revised proposal that would assure the identity of IUPS, and it was generally well received. The discussion continued on the next day, and by the end of the ExCo meeting there was a consensus that the new proposal represents a workable solution. I was extremely pleased that Ewald asked Denis to draft a memorandum of agreement between IUPS and USNC to be signed by Ewald and Ginger, covering these points. Denis prepared a draft memo based on the document that I had prepared for the ExCo. Ginger arrived in Christchurch on 23 August and I told her about the situation and the memo to be signed. I wanted to have this memorandum signed to indicate Ewald's approval of our suggestions prior to the delibration of this matter by Council at its meeting on 24–25 August.

Originally, the question of the 2005 Congress was going to be discussed at the Council meeting on 25 August in the morning, followed by the 2009 Congress site selection. Ewald felt rightly that, in order to allow sufficient time for Council to deliberate on the proposed changes in venue and time of the 2005 Congress, this issue should be brought up at the Council meeting in the afternoon of 24 August. Ewald invited Ginger, Marty and Wendy White of the National Academy of Sciences to also attend the Council meeting when this 2005 Congress was to be discussed. Marty arrived in Christchurch in the morning of 24 August, and I was able to arrange for Ginger, Allen, Wendy, Marty and myself to meet immediately after his arrival. I went over the recent developments. With Marty's input, we were able to incorporate in the memorandum the financial arrangements, i.e. USNC/APS will allocate 5% of the registration fee to IUPS, as well as $100,000 or 25% of net profit, whichever is greater. I did everything I could, including running back and forth between the meeting room and the business office in the hotel for revisions and printing, to have the memorandum ready for signing prior to the IUPS Council meeting. I was pleased that Ewald approved this memorandum, which was signed by him and Ginger in the morning of 24 August. At lunch time on 24 August, I went over with Ginger, Marty and others a Powerpoint presentation on the alternative proposal, which incorporated all the key points of discussion and pictures on the Convention Center, hotels and other attractive features in San Diego. This presentation was initially prepared by Marty, and I continuously worked on it, prior to leaving San Diego, during the flight, and in Christchurch.

After I made the presentation on the alternative proposal to Council, the members had several questions, mainly focused on the identity of IUPS and the mechanics of having the Congress overlapping partially with EB. They were satisfied with the answers. In the morning of 25 August, Council voted unanimously to approve the alternative plan. This was a critical step that positioned us for the General Assembly meeting to be held the next day.

During the General Assembly meeting, Ewald gave an opening statement that summarized the events leading to the alternative proposal and stated that this proposal had received careful considerations by the ExCo and Council. He distributed the memorandum he signed with Ginger to the members of the General Assembly so that the members could understand the conditions involved. Ginger made an excellent presentation on the alternative proposal, using a further updated version of the Power Point document. There were a few questions from the floor, which Ginger answered very well. Denis made a superb summary of the actions taken by the ExCo and Council on this proposal. A secret ballot was taken, and our alternative proposal received an overwhelming endorsement by the General Assembly with a vote of 71 for and 3 against. This greatly exceeded our expectation!

Having received the IUPS approval, we next needed the approval by the EB Board to implement this joint effort. The Chair of the EB Board is Paul Insel, Professor of Pharmacology at UCSD, and I was able to discuss directly with him this issue. He was supportive of the idea of holding the IUPS Congress in conjunction with the EB meeting in 2005 and invited me to make a presentation to the EB Board on 24 January 2002 at the FASEB headquarters in Bethesda, MD. I contacted the other five members of the Board by phone either directly or through other NOC members prior to the meeting and sent them an e-mail on 2 January outlining the proposal. At the meeting on 24 January, the Board enthusiastically endorsed the holding of the IUPS Congress in conjunction with the EB 2005 meeting in San Diego in 2005. Paul transmitted this endorsement to me in writing, including the Board's view that this is a great opportunity and an excellent precedent that should be a win-win situation. Paul also transmitted the Board's comments and suggestions on the mechanics of the meeting, which was satisfactorily addressed by the NOC at it meeting on 24 April.

Now, we have everything set for this exciting meeting in 2005, when the IUPS Congress will be held in conjunction with the EB meeting. This will provide a wonderful forum for physiological sciences by bringing together physiologists from all parts of the world and an excellent opportunity for the rich interaction of international physiologists with scientists in their sister disciplines. I look forward to the pleasure of working closely with colleagues in IUPS, NOC and everywhere to put together a high-quality congress with the theme of "From Genomics to Function."

4. Professional Organizations in Biomedical Engineering

First World Congress of Biomechanics

Just before we went to San Diego, Dick Skalak and I were asked to co-chair the First World Congress of Biomechanics (WCBM) in San Diego in 1990. It was most appropriate to hold this meeting at UCSD, since Bert Fung is widely regarded as the father of Biomechanics. We started planning for the meeting shortly after our arrival. It was wonderful to have such a great group of colleagues to work with. Bert and Ben Zweifach, both were Honorary Chairmen of this WCBM, had organized the World Congress of Microcirculation and the International Congress of Biorheology a decade earlier. Many of the faculty and staff had a great deal of experience and expertise running international meetings. Geert was the Secretary General, Andrew McCulloch (p. 160) was Treasurer, and Savio Woo (p. 284) was Chair of the Program Committee. Gene Mead was responsible for industrial exhibits and social functions. Frank Delano designed a beautiful logo for the Congress, which is still used today. All faculty, students and staff pitched in and were immensely helpful. The Congress attracted over 1000 participants, which were more than expected. The quality of the meeting was very high, and the facilities on campus were excellent. This Congress set the stage for many more to come.

Biomedical Engineering Society

The Biomedical Engineering Society (BMES) holds its annual spring meetings in conjunction with the FASEB (and later EB) meetings and its fall meetings rotating among different campuses. UCSD was chosen as the site for the 1997 fall meeting to be held on 2–5 October, and Dick and I were appointed as the co-chairs. We could not hold the meeting on campus, because at that time classes were in session. Dick and I, together with other faculty members, looked at several hotels and decided to hold it at Hyatt Islandia and Bahia Hotels in La Jolla. Unfortunately, Dick passed away two months prior to the meeting, and his wisdom, advice and friendship have been sorely missed. Nineteen ninety-seven was a bad year in that BMES lost its key person, the Executive Secretary Rita Schaffer. In the same year UCSD lost Ben Zweifach, Gene Mead, who worked very hard in getting together an excellent exhibits program at this BMES fall meeting, and Pierre Galletti of Brown University, who was the Chair of the Dean's Advisory Committee for UCSD Bioengineering.

The 1997 BMES fall meeting was attended by nearly 800 registrants and the presentations were uniformly high in quality. The eve before the meeting, Bert was invited by the graduate student group to give a talk on his perspectives of biomedical engineering. The session was attended by 200 people, some of them were faculty members. Shelly Weinbaum delivered an outstanding Whitaker Lecture, and Cathy Galbraith, who did her Ph.D. dissertation research with Dick and me at UCSD, was given the Young Investigator Award based on her research at Duke University with Michael Sheetz, who is now at Columbia University.

I served on the BMES Publications Board from 1997 and as the chair from 1999 to 2001. We were fortunate to have Jim Bassingthwaighte as Editor-in-Chief of the Annals of *Biomedical Engineering (ABME)*, which is the official journal of BMES. Under his outstanding leadership, *ABME* has made great strides to become a leading journal in the field. When Jim's term ended in 2001, we needed to search for a successor, and I am extremely pleased that Larry McIntire (p. 181) has agreed to take on this important position. I am grateful to the BMES for selecting me as the recipient of its Distinguished Service Award in 2001.

American Institute for Medical and Biological Engineering

The American Institute for Medical and Biological Engineering (AIMBE) is an umbrella organization in medical and biological engineering, with a combined membership of over 30,000 scientists. The Institute, which was formed ten years ago with Bob Nerem (p. 278) as the Founding President, has the goals of promoting medical and biological engineering by enhancing public awareness, increasing liaison with government and industry, improving intersociety relations and co-operation, and serving the national interest in science, engineering and education. AIMBE is composed of the Council of Societies (15 Member Societies), the College of Fellows (outstanding medical and biological engineers are elected as Fellows, now about 700 in number), an Academic Council (69 educational programs in the U.S.), and an Industrial Council (industrial members related to medical and biological engineering). It has a similar mission for medical and biological engineering as FASEB does for experimental biology, but AIMBE does not have the same financial resources and infrastructure.

I was elected as Chair of the College of Fellows in 1998–1999. One of the responsibilities of the chair is to organize the Annual Meeting. The theme of the symposium during the AIMBE meeting in March 1999 was "Therapeutic Delivery: Opportunities and Challenges for Bioengineering." Thanks to the tremendous help and valuable advice of Rakesh Jain (p. 259), we had a very successful program. The Meeting also had outstanding plenary addresses by Rita Colwell, Director of the National Science Foundation, and Ellie Ehrenfeld, Director of the Center for Scientific Reviews at the National Institutes of Health.

During my service as AIMBE President in 2000–2001, I had the pleasure of working with Past President Jack Linehan, President-elect Peer Portner, Secretary-Treasurer Dan Miller, the Chairs of College of Fellows Don Giddens (p. 183), Academic Council (Ken Lutchen), Council of Societies (Susan Blanchard and Mike Ackerman), Industry Council (Paul Citron), other members of the Board, and Executive Director Kevin O'Connor (p. 197) on matters of importance to medical and biological engineering. The major tasks facing AIMBE were to amplify its public affairs activities, strengthen its administrative infrastructure, build a strong financial base, and enhance its relations with the member societies. In cooperation with the Academy of Radiology Research and other organizations, AIMBE worked with the U.S. Congress to foster the establishment of the new National Institute for Biomedical Imaging and Bioengineering at the NIH. This goal was accomplished at the end of year 2000, as detailed in the next section (Section II.C.5 of this article, p. 298). AIMBE collaborated with the NIH Bioengineering Consortium (BECON) to co-sponsor the 2000 symposium on "Nanoscience and Nanotechnology," the 2001 symposium on "Reparative Medicine: Growing Tissues and Organs," and the 2002 symposium on "Sensors in Biological Research and Medicine." AIMBE co-sponsors the Medical Technology Leadership Forum, which addresses important issues in medical technology development.

AIMBE has benefited from a grant from the Whitaker Foundation to improve its administrative structure, and this allowed the addition of Karen Anderson to its office. Both Kevin and Karen have been doing an excellent job in managing the many administrative and executive functions of AIMBE.

In order to generate additional revenues for AIMBE, I have been working on the possibility of changing the dues structure for the member societies from a fixed rate (which was quite low) to a per-member rate similar to what is done in FASEB. The eventual goal is to have the resources needed to provide the appropriate service to AIMBE Member Societies. Following several valuable discussion sessions, the Council of Societies has agreed to first try this on a voluntary basis for society members to make the contributions. The hope is to eventually go to a more definitive arrangement. We are extremely grateful to AIMBE Founding Fellow Jen-shih Lee, his wife Lian-pin, and the Lee family for their most generous gift of $250,000 given to AIMBE in 2000 as a challenge grant for equal or larger matching to foster public policy. Jen-shih is a Past Chair of AIMBE Council of Societies (1995–1997), and from this experience he feels strongly that AIMBE needs the resources to provide the service to its member societies.

The AIMBE Board has decided to go forward with a development plan to generate the financial resources required to effectively carry out its functions, especially in public affairs. The consulting firm Moore Associates has been engaged to evaluate and establish the mechanisms by which the development efforts should be implemented. Bob Nerem and I have agreed to co-chair the fund-raising committee.

Medical and biological engineering is at a most exciting time. We need to seize the opportunity and meet the challenge to bring these fields to a new level as we enter the new century. The matching of the Lee gift will give AIMBE an excellent start toward funding its public policy efforts. With appropriate planning and collaboration between the NIH and the scientific community, the newly formed National Institute of Biomedical Imaging and Bioengineering will undoubtedly play a leading role in fostering biomedical imaging and bioengineering, with the ultimate goal of enhancing the health and wellbeing of our people. The development of this new institute, from my perspective, is summarized in the next section.

5. The National Institute for Biomedical Imaging and Bioengineering

Under the initiatives of former Director Harold Varmus and Acting Director Ruth Kirschstein (p. 245), the NIH established in 1997 the Bioengineering Consortium (BECON) chaired by Deputy Director Wendy Baldwin (p. 199). Under Wendy's outstanding leadership, BECON has played a major role in fostering inter-institutional activities related to biomedical engineering and has organized annual symposia with the cooperation of intramural bioengineers, e.g. John Watson (p. 177), and the extramural community (see II.C.4c. on AIMBE above). In response to a Congressional directive in the fiscal year 2000 NIH Appropriations Act, the NIH established an Office of Bioengineering, Bioimaging and Bioinformatics. Along with these significant developments in bioengineering at the NIH, there were also efforts to create an NIH Institute that is devoted to bioengineering and bioimaging, but such a goal was not uniformly supported by the existing NIH Institutes.

The bill for establishing the National Institute of Biomedical Imaging and Bioengineering (NIBIB) was passed by the U.S. House of Representatives in September 2000 and the Senate on 15 December 2000. It was signed into law by President Clinton on 29 December 2000. The dedication of leaders in the bioengineering and bioimaging communities to achieve this goal extends back more than 25 years (Publication D8), but the final passage of this law came rather suddenly. It caught the research community and the NIH by surprise.

Realizing the importance of creating a consensus for working on this new institute, I drafted a letter during the Christmas/New Year Holiday for submission to the *Science* journal. This letter stressed the interdisciplinary nature of NIBIB and proposed that "Certain imaging and bioengineering research efforts must be closely integrated with approaches pursued in other NIH institutes and centers. These institutes/centers, therefore, should retain sufficient and appropriate resources to continue efforts in imaging and bioengineering that are integral to their missions. The NIBIB, on the other hand, should be devoted primarily to basic and applied bioimaging and bioengineering research and training that are likely to cut across different organ-systems, diseases and conditions. The new Institute should strengthen and complement (not subtract from or substitute for) research programs in other NIH institutes and centers."

On 2 January 2001, I was called to NIH to meet with Drs. Ruth Kirschstein (p. 245), Wendy Baldwin (p. 199), Donna Dean and other high level NIH officials. The NIH leadership reacted very positively to the principles outlined in the letter I drafted. Later I was told by Ruth and Donna that it served as a base for the Mission Statement developed for NIBIB by the NIH Internal Working Group composed of Steven Hyman (chair), Stephen Katz, Richard Klausner, Claude Lenfant and Lawrence Tabak. I appreciate very much that C. Douglas Maynard, President of Academy of Radiology Research (ARR) co-signed this letter, which was published in the 2 March 2001 issue of Science (Publication D7).

On 4 January 2001, a joint committee was formed between AIMBE (Bill Hendee, Jack Linehan, Peer Portner, Buddy Ratner and myself) and ARR (Stanley Baum, Reed Dunnick, Bruce Hillman, C. Douglas Maynard and Elias Zerhouni) to provide input from the research communities to NIH regarding the new institute. Stan Baum and I served as co-chairs. The joint committee held several meetings, and its representatives met with Ruth Kirschstein and Donna Dean, as well as the NIH Internal Working Group, to exchange views between the extramural communities and the NIH leadership.

As AIMBE President, I had invited Ruth Kirschstein as the plenary speaker at the AIMBE Annual Meeting on 1 March 2001. This turned out to be an excellent choice in view of the establishment of the new institute. Ruth gave a superb speech. In January 2001, I had also invited Donna Dean to give a presentation on NIBIB at the AIMBE Forum on 2 March. The presentation was followed by a discussion panel, with Donna, Stan Baum and Jack Linehan serving as members and myself as a moderator. The plenary lecture, the forum presentation and the panel discussions were very well received. They provided the attendees with excellent, timely information regarding the status and future directions of the NIBIB.

NIBIB made more interactions with the scientific community than any other new institutes at the NIH. It is hoped that these interactions will be continued and enhanced to ascertain that this important new institute can fully achieve its purpose of developing bioengineering and bioimaging to improve the health of people.

6. Organizations related to the Chinese People

Institute of Biomedical Sciences

After Cheng-wen took over as the Director of IBMS, I continued to help him in the development of the institute. In my capacity as Chairman of the Advisory Committee, I was in frequent communication with him and his staff by phone, fax and e-mail. The advisory committee holds its meetings in Taiwan one to two times per year. From 1988 to 1994, prior to the recruitment of a sufficient number of senior members, the advisory committee was responsible for the review of appointments and promotions of the institute. The advisory committee has been responsible for the organization of periodic reviews of the institute by inviting outstanding biomedical scientists from the U.S. to go to IBMS for on-site review. This review system was later adopted by other institutes in Academia Sinica and also by other academic institutions in Taiwan.

The recruitment of scientists from abroad had several difficulties, as already mentioned above (Section II.B.2 of this article, p. 413), and a major problem is salary. As part of Academia Sinica, which is a government institution, the salary scale in IBMS is the same as other government institutions and not competitive with that in the U.S. The National Science Council (NSC) in Taiwan had special fellowships for short-term support (maximum three years) at a rate closer to that in the U.S., but there was no mechanism available for long-term appointees. In the late 1980s, President Ta-You Wu, Paul Yu and I recognized the importance of this issue in recruiting outstanding scientists and in the search for a long-term director for IBMS. We worked on solving this problem by proposing an arrangement in which a limited number of truly outstanding scientists can be appointed with salaries comparable to their current levels of compensation in the U.S. We presented this to government leaders, including President Chiang Ching-Ko, Premier Kuo-Hwa Yu of the Executive Yuan and Minister Kwoh Ting Li, stressing the importance of biomedical research in the health and wellbeing of the people, as well as the economy of the nation. We received positive responses and valuable advice from these top government officials.

These ideas were started when I was the IBMS director, but they were not yet materialized when I left IBMS to return to the U.S. Since Cheng-wen, as the new director, was a potential beneficiary, he had to recuse himself, and I took the initiative of pursuing this matter by making several trips back to Taiwan. During one of the meetings with Premier Yu in January 1989, I finally received the permission to go ahead. Accordingly, a "Plan for the Recruitment of Outstanding Scientists in Basic and Clinical Medical Research by the National Science Council of the Executive Yuan" was drafted, and it was formally submitted by President Wu to the Executive Yuan on 25 February 1989.

The plan proposed three types of recruitment from abroad, with different salary ranges: (a) Special Medical Research Chairs for outstanding senior scientists (e.g. Academy members), (b) Special Medical Researchers for internationally recognized scientists (already at full professor level), and (c) Medical Researchers and Associate Medical Researchers for scientists who have completed their three-year term under the NSC Special Fellowships with outstanding accomplishments and demonstrated potentials.

On 17 April 1989, Premier Yu officially notified President Wu that the proposed plan was approved with ten Special Medical Research Chairs, 20 Special Medical Researchers, and an unspecified number of Medical Researchers and Associate Medical Researchers. The document stated that this was a Special Action to be limited to the recruitment of basic and clinical medical scientists in IBMS and that it was not to be used as a precedent to extend to other cases. The government had the concern that if such a plan were to be applied to fields other than biomedical sciences and to other institutions, it would be very difficult to handle all of them at that time. I fully understood that it would have to start this way at the beginning. In fact I was amazed that this plan would be approved for implementation — it was a miracle! I still feel it is miraculous that we could succeed in this breakthrough. I salute these government leaders for their vision and courage.

I felt all along that it was most important to get this new concept implemented as a breakthrough and that it would be possible to extend this initiative gradually so that research in Taiwan can benefit in a much broader way, despite the limitations placed in the initial approval. Indeed, later Academia Sinica was able to use this plan as a blueprint to establish a system of special chairs across the entire Academy, including all institutes in the natural sciences, humanities, and life sciences sections. Such arrangements have also been implemented in universities by the Ministry of Education. Furthermore, President Yuan T. Lee (p. 129) of Academia Sinica has set up a private Foundation for the Advancement of Outstanding Scholarship to provide salary support for outstanding scientists in Taiwan. All of these arrangements are not limited to biomedical researchers or scientists going to Taiwan from abroad. These developments reinforced my belief that temporary inequity can bring about a broader benefit and that we need to take certain actions that may not be popular to a large sector initially, as long as we can follow through to eventually achieve the greater goal.

With the implementation of the Special Medical Research Chair system in 1989, Academia Sinica was able to offer such a chair to Cheng-wen following a thorough review. It was fortunate for IBMS and for health research in Taiwan that Cheng-wen decided to accept the offer and stay as its full-time, long-term director. His outstanding, sustained leadership from 1988 to 1996 was very important for the development of the institute. As stated in the next section on the National Health Research Institute, Cheng-wen became president of the institute in 1996. From 1996 to 1999, Kenneth K. Wu, Chair Professor in the Division of Hematology and Oncology at the University of Texas, Houston, served as the IBMS director with distinction. Between

1999 and 2001, Te-chang Lee and Sho-tong Lee provided excellent service as acting directors prior to the recruitment of Yuan-Tsong Chen from Duke University as a long-term director in July 2001. Yuan-Tsong received his Ph.D. degree from the Department of Human Genetics and Development at Columbia and went to Duke University to become a Professor of Pediatrics and Human Genetics. His appointment is wonderful not only for the continued development of IBMS, but also for the implementation of the new initiative on genomics program promulgated by President Lee and Vice-President Sonny Chan in Academia Sinica.

National Health Research Institutes

In the early 1990s, a group of Academia Sinica members proposed to establish the National Health Research Institutes (NHRI), which is patterned after the NIH of the U.S. Its missions are to enhance and coordinate health research, investigate important diseases, develop novel medical technologies, improve health policy and preventive medicine, and train health research personnel, with the ultimate goal of improving the health and wellbeing of people. This proposal received the support of President Ta-You Wu, who helped to move it forward.

In order to attain a greater degree of flexibility, it was proposed that NHRI be established as a Foundation, although initially the bulk of its funding would have to come from the national government. The sole source of biomedical research funding in Taiwan had been the National Research Council (NSC) and represented less than 20% of the total research funding, in contrast to the much higher percentages at countries such as the U.S. The proposal of the formation of NHRI as an additional funding agency, as well as an organization to coordinate health science research in Taiwan, received broad support in the biomedical community, but it ran into difficulties during government review. There were concerns of duplication of NSC functions, and by early 1993 it seemed almost certain that the plan would be dropped. At that time I was going to Taipei to attend the NHRI advisory committee meeting. On the trans-Pacific flight, I realized that we needed to go to the top government leadership to rescue this important organization. We were very fortunate to get the appointments for the advisory committee to see President Teng-hui Lee and Premier Chan Lien on very short notices. The Premier had taken office only a few days before our visit; Cheng-wen's secretary Annie Lin worked very hard to obtain this difficult appointment. After hearing our explanations of the importance of NHRI, the President and the Premier agreed to move ahead with the establishment of the institute. The newspapers reported the next day that "NHRI has been resurrected from its death," or "NHRI has come back to life from the Losing Bracket" (i.e. using an analogy to a double-elimination sports tournament).

This turning point was followed by continued efforts by Cheng-wen, who had been appointed as Director for the NHRI Preparatory Office, Chung-Fu Lan, Ming Yang Yeh, Winston Yu, and others in the office, members of the advisory committee, the Department of Health Secretary General Po-Ya Chang and Councilor Hsueh Yun Chi, and the biomedical research community in Taiwan. After the passage of the law by the Legislative Yuan for the formation of NHRI, the institute was formally established in January 1996, with Cheng-wen as its founding president; I have been serving as chairman of the advisory committee. Cheng-wen has done an excellent job. The institute has ten divisions, viz. Environmental Health and Occupational Medicine, Health Policy Research, Medical Engineering, Mental Health and Drug Abuse, Gerontology Research, Molecular and Genomic Medicine, Biotechnology and Pharmaceutical Research, Biostatistics and Bioinformatics, Cancer Research and Clinical Research. Eight excellent division heads have already been recruited. I had the pleasure of participating in a review of NHRI in August 2000; the review committee was very impressed by the progress made by the institute during the first four years of its existence.

NHRI, in addition to having its intramural program, is a granting agency that supports biomedical research in public and private institutions in Taiwan. Prior to the formal establishment of the institute, I set up in 1992 a review system for NHRI, which was modeled after that of the NIH. We started with 15 outstanding Chinese-American scientists serving as reviewers. Most of them had their undergraduate education in Taiwan and received their graduate training in the U.S. The first NHRI study section meeting was held in a hotel in Newark Airport. Cheng-wen and the extremely capable and dedicated staff of Shiu-Ling Chao, Daisy S.F. Tsai and Susan S.C. Tsai flew in from Taiwan. Kung-ming Jan drove down from his Tenafly, NJ home with his computers and printers. Their tremendous teamwork made possible the compilation and tabulation of the reviewers' scores almost as soon as they became available. During the three days of review, everyone worked late into the night and even to the wee hours in the morning. The devotion is truly admirable and the effectiveness is astounding!

The number of reviewers has increased over the years to over 50. The reviews were conducted in the U.S. in the first few years and then moved to Taiwan in recent years. The review process is very rigorous and efficient. We took the strengths of the NIH review system and added a few new approaches so that we can do all the reviews within one week — both the initial review by the study section and the final review by the council. This is faster than what is done elsewhere, including the NIH. The NHRI review system is now recognized as the best in Taiwan. One of the factors contributing to the success of this system is that it has the support of a large number of scientists who received their early education in Taiwan and are now in the States. These people have a special understanding of health sciences in Taiwan and a strong desire to serve, and none of the reviewers is an applicant. It is very rewarding that virtually every scientist asked has agreed to serve in this review system.

Military Training for College Graduates in Taiwan

When I graduated from National Taiwan University College of Medicine, it was the second year that the ROC Government required all male college and technical graduates to have a one-year Reserve Officer Training (ROT). The year of training had its positive aspects in that it strengthened the body and mind of the youth, but it took the graduates away from their intellectual and professional pursuits. Since it is important to have the reserve officers for national defense, on balance it was a worthy experience. A few years later, the rule was changed to lengthen the training to two years. That was too long a period at a critical time of the development of a young college graduate. In 1998, I really felt the impact of the loss of the continuity of professional development of a tremendously promising young scientist. Kuang-Den (Dennis) Chen, a graduate from National Tsinghua University in Hsinchu, Taiwan, told me that he had to serve the two-year ROT as soon as he returned to Taiwan after his completion of his research training in my laboratory. Kuang-Den had done superb research work in my laboratory and published a very original article (paper A#370, in my publication list in this book) identifying some of the molecules the endothelial cells lining the blood vessels may use to sense mechanical forces in modulating its intracellular signaling. I had high hopes that he would be able to return to Tsinghua and make important contributions to scientific development in Taiwan. The two-year ROT would take him away from his laboratory research so long that he would not be able to keep up with the rapid progress in the field and might lose the momentum he had built while working in San Diego.

In July 1998, I attended the biennial membership meeting of Academia Sinica in Taiwan. In the evening on 11 July the ROC Vice-Premier Chao-Shiuan Liu, who had been President of National Tsinghua University, met with the academy members and presented the government's plan for the sustained development of science and technology in the new century. At the end of his excellent talk, I stood up and said the following: "There is no doubt that it is the duty of all citizens to perform military service, which is essential for the strengthening of national defense and assurance of national security. National defense today, however, is not limited to the military, and economics, science and technology are unseparable from the military as the foundation of national strength. Therefore, national defense training should include many types of personnel that can contribute to the total strength of the nation. Such an approach is essential for the build-up of the best, state-of-the-art team of national defense. By putting every college graduate through the same type of military training, the country does not optimize the usage of its talents and will suffer seriously from the loss of valuable young human resources that could otherwise contribute importantly to national defense through their services in science and technology. Therefore, I sincerely wish the government would re-evaluate its ROT policy and make the best use of the nation's human resources." Several other academy members also spoke and agreed with me. Dr. Liu, whom I admire very much as a scientist, an educator and a statesman, said that he understood very well my points and promised to look into this matter. I seized the opportunity and, together with Dr. P.C. Huang of Johns Hopkins University, drafted a letter to Dr. Liu, summarizing these points and asked him to form a cross-ministry task force (including Academia Sinica) to discuss thoroughly the scope and duration of current national defense service needs, with the goal of building up the strongest possible national strength and the best possible national defense. We were able to get 60 academy members to co-sign this letter, including academy president Yuan-Tseh Lee, within an hour. There were press reporters at that meeting and the newspapers had a great deal of coverage of this issue.

Shortly after my return to San Diego, I received a letter from Dr. Chao-Shiuan Liu that he had already formed the Task Force and thanked me for bringing up this important issue. Several months later, I was very pleased to learn that a new regulation was implemented in which college graduates can work on scientific and/or technological research for four years in place of the two-year ROT in the military. I was happy not because I was able to achieve what I proposed, but rather for the country and for the young graduates from college.

Universities and Hospitals in China (Mainland)

I received my education from kindergarten to premedical college in Shanghai and Beijing, where I was born. I owed a great deal of what I know from the teachers and the schools there. After leaving Mainland China in late 1948, I went back to China for the first time in August 1995, after more than 46 years. I accepted the invitation by Dr. Yong-de Shi of the First Shanghai Medical University to give a plenary speech at the International Conference on Biophysics in Shanghai.

Xiang-Chi Jin of the Beijing Overseas Association kindly organized the trip for K.C. and me.

We first went to Beijing to visit my Alma Mater Peking University (Beida) and Peking Medical University, as well as other medical institutions and hospitals. I gave lectures and participated in seminars organized for my visit. President De-Bing Wang kindly appointed me as an Honorary Professor of the Peking Medical University and. Feng-Yuan Zhuang kindly arranged to have me appointed as an Honorary Professor at the China-Japan Friendship Hospital. K.C. and I had a wonderful reunion with my Beida classmates at the famous Fang Shan Restaurant in Bei Hai (North Sea), which was arranged by Wen-Yen Cai and other classmates. Most of them were medical leaders in Beida and elsewhere, and we talked happily about the days when we were premedical students nearly half a century ago. K.C. and I also enjoyed very much our tours to the Great Wall (accompanied by Xiao-Nian Chang) and the Royal Palace (accompanied by Dr. and Mrs. Zhuang). It was a pleasure to visit the house my family lived in before leaving Beijing in 1948; it has become an office for the local government.

At the International Conference in Shanghai, the First Chien-Fung Young Investigator Award (named in honor of Bert Fung and me) was selected by an international committee from among the papers submitted by many outstanding young investigators. The winner was Dr. Hui Miao, who was a scientist in Dr. Zhuang's lab at the China-Japan Friendship Hospital and is now working in my lab in UCSD. I was made an Honorary Member of the Shanghai Biophysical Society. I was very happy to be able to visit the townhouse off Avenue Foch in Shanghai where my family lived. It was a pleasant surprise that the townhouses in that area still remained the same despite a lot of constructions in Shanghai.

We went to Chungqing at the invitation of President and Mrs. Yun Peng Wu of Chungqing University, where I gave lectures and was appointed an honorary professor. K.C. went to Nankai Middle School to visit her Alma Mater and showed me the classrooms and the fields on which she used to play and run. The Wus arranged for us to take a boat trip down the Yangtze River when the famous Three Gorges were not yet altered. It was a marvelous experience.

I was impressed by the scientific accomplishments by the Chinese scientists under difficult circumstances. This trip not only renewed our acquaintance with China, but also established several collaborative efforts in science.

Through Dr. Wen's initiative, a collaborative arrangement was made between his Laboratory of Hemorheology in Peking Medical University and our lab at UCSD. A formal ceremony was held in Beijing in 1997. We have since published several papers together and have ongoing research projects. During the 1997 trip, we also visited the Microcirculatory Laboratory of the Chinese Academy of Medical Sciences (Director: Rui Juin Xiu), and I was appointed as an Honorary Professor of the Academy.

Other Organizations Related to the Chinese People

While I appreciate the opportunity of having my higher education and postdoctoral learning in Columbia (p. 401), I am also indebted for my early education in Mainland China and Taiwan and I would like to be able to help to further research and education in my homeland. In addition to my participation in the activities described above (Sections II.C.6.a-c of this article), I am on the advisory committees of the Institute of Molecular Biology (IMB) at Academia Sinica and the Industrial Technological Research Institute (ITRI), and a member of the Board of the Foundation for the Advancement of Outstanding Scholarship in Taiwan. During the past decade, I was appointed by President Yuan T. Lee (p. 129) to chair the search committees for the Directors of IBMS and IMB in Academia Sinica and by Director General Po-Ya Chang to chair the search committee for the Director of NHRI. I frequently review grant applications for the National Science Council of R.O.C. in Taiwan. I continued to collaborate with scientists in Taiwan, including JengJiann Chiu at the National Health Research Institutes, Konan Peck and Danny Wang at IBMS, Jui-Sung Hung and Jackson Wu at China Medical College, and Simon Mao at the Pig Research Institute. I am a board member of the American Bureau of Medical Advancement in China (Pierson, p. 62), which is an organization with offices in New York and Taipei aiming at advancing medical research and education for the Chinese people.

In Mainland China, I have collaborative projects with Drs. Zongyao Wen and Zongyi Yan of Peking University and I am on the advisory board of the Biochip Laboratories headed by Dr. Jing Cheng of Tsinghua University. In San Diego, I served from 1996 to 1997 as Chairman of the Board of the Chinese Historical Museum, which later had had the excellent leaderships of Shao-Chi Lin and Lilly Cheng (p. 226). I am a member of the San Diego Chinese American Science and Engineering Association and the Phi Lambda Fraternity, which I joined some 30 years ago in New Jersey through the introduction of David Cheng (p. 116).

7. Family

During our years in San Diego, our two daughters May and Ann each gave birth to three lovely daughters. They are Kristen (1990), Renee (1992) and Natalie (1996) Busch (Picture on p. 24); and Katie (1992), Jenny (1993) and Laura (1996) Guidera (Picture on p. 28). It is always a great joy to be with them, either visiting May, Len and their girls in London; or Ann, Steve and their girls in New York (they have moved to the Philadelphia area), or having them in La Jolla. Every girl is lovely and they are all different. They often surprise us by how much they know and what they say. For example, once when I was working on a large pile of papers, I was told by my five-year-old granddaughter, "Grandpa, you have to get organized. Do a little bit every day!"

We had several family reunions in the last few years. In July 1995, my brothers Robert and Fred and their families all came to San Diego from Taiwan, Hong Kong and New York (Picture on p. 12). Robert and Ruth brought their daughters Debra and her husband Frank, and Kathy and her husband David Chou and son Alexander. Fred and Julie brought their son Carl and his wife Jenny, and their daughter Carol and her husband David Sun and daughter Alexandra. We had a marvelous time together.

Since then, May and Ann each has another daughter, Carl and Jenny have three sons (Justin, Darren and Fred, III), and Carol and David have one more daughter (Katherine) and one son (William). So the next reunion will have seven more children. Most of K.C.'s siblings are in California, so it is a little easier to have reunions, e.g., the picture on p. 31 shows her siblings Katherine, Charles, Robert, George and David (p. 30) and their spouses.

In August 2000 K.C. and I, together with our daughters' families, had a marvelous trip to Taiwan and Mainland China. In Taiwan, we were able to get together with Fred and Julie and their children and grandchildren, Robert's son-in-law Frank, and K.C.'s brother Charles and his family. We had a wonderful time touring Taiwan, Beijing, the Great Wall, and Xi'an. This journey meant a lot to us beyond the beautiful artwork in the Palace Museum in Taipei, the historic Great Wall outside Beijing, and the elegant terra cotta warriors in Xi'an. This trip gave the next two generations of our family a chance to see their roots where K.C. and I came from and for us to renew our memories of the past.

In April of this year, May and Len and Ann and Steve brought their two families to San Diego to celebrate K.C.'s 70th Birthday and our 44th wedding anniversary. The pictures on pages 24, 28 and ix were taken at that time.

The following diagram summarizes my families from my parents to my grandchildren, spanning four generations over almost one century. At this time of Thanksgiving in 2002, when I did the final updating of

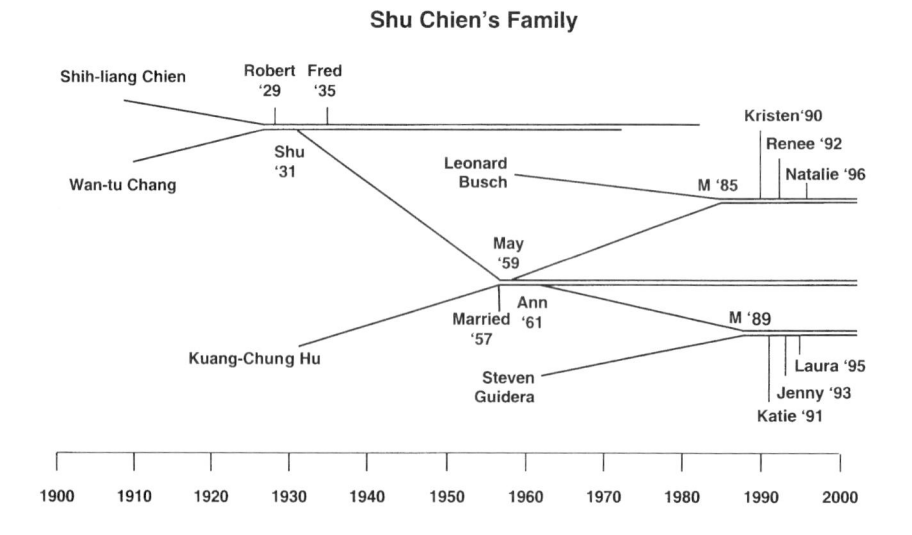

Shu Chien's Family

this article, warmest thanks and gratitude swelled in my heart for being blessed by having such wonderful families.

III. RETROSPECTS AND PROSPECTIVES

It is difficult to believe that 70 years have passed since I was born. Time goes by so quickly when you are having fun. I have been very fortunate in finding life to be so enjoyable. I am very grateful for what I have, especially my lovely family and wonderful friends.

The Chinese saying is that a person reaching 70 can do whatever one wants, because by then he/she knows the limits. I believe by that definition I was 70 quite a while back because I have known my limits for some time. I always try to do the best I can, but only within my limits, though I do try to reach my upper limit. Within these boundaries, I am very happy to have been able to do whatever I want. Life has been really good to me. I have excellent health, a marvelous family and wonderful friends. I have been able to achieve what I would like to do. When I do not succeed the first time, I try harder to succeed on the next, and that makes it even sweeter. A close friend recently told K.C. and me that there is nothing more we can ask from God, and I

In baby carriage with brother Robert, 1931.

A picture taken at 70 with K.C., 2001.

fully agree. I do not want to ask for anything. I want to express my gratitude for what I have, and that is plenty. My last name Chien means money in Chinese. We do not have that much money, but we are extremely rich in our heart, just as K.C. said in her article (p. 16).

Looking back at my life, I could have chosen many different paths at several bifurcation points (see diagram in Section II.C.1.g of this article). If I had gone into mathematics, I probably would not have met K.C. and I would have missed out on our happy lives together. If my parents did not decide to take that plane to fly out of Beijing in 1948, my life would have been totally different. There are many such "if's." What if I choose clinical medicine instead of physiology and bioengineering? What if I took some of the job offers and left Columbia earlier? What if I chose to stay at Columbia instead of coming to La Jolla? There is no need to answer these questions, and there is no need to ask them. What is important is what we have and who we are. We look back only to see where our path has been, and not to wonder what would have happened if we had taken a different path. I am extremely happy with the way things have been. I am totally at peace with myself and with everyone around me.

The path we take and where we end up sometimes seem to be pre-destined. Throughout my career, there were many happenings that seemed to put me in a critical position at a critical time. For example,

I got involved in the development of biomedical sciences in Taiwan after my election to Academia Sinica, and that allowed me to play a role in the establishment of IBMS and NHRI. I was elected as Presidents of APS and FASEB when FASEB was on the verge of collapse, giving me the challenge and opportunity to work with others to bring FASEB to a new height. I moved to UCSD in time for the establishment of the infrastructure of bioengineering in the university and for positioning ourselves for the Whitaker Foundation Awards. I was elected to Presidency of AIMBE in time to work with the NIH and the bioimaging and bioengineering communities on the establishment of the new NIBIB. As I stepped down as the Chair of Bioengineering at UCSD, the opportunity arose for the establishment of the MRU in the UC System. After I decided not to continue as IUPS Treasurer for a second term, I have been drafted to chair the next IUPS Congress in 2005.

Even the writing of this article was the result of events that I had not expected. First, I was so moved by the wonderful tributes written by my friends and relatives that I would like to respond to and resonant with. I was going to write a short note of appreciation and gratitude. It so happened that Frederik Nebeker of the IEEE History Center interviewed me earlier this year for the purpose of writing the oral histories of a number of bioengineers. Subsequently, his colleague David Morton sent me a transcript of that interview, and I was impressed by the amount of information Fred took down during the interview that lasted only a little over an hour. This stimulated me to think about my life and made me write this article. With each iteration of the Tribute Book, I have added more materials, and it is now at least ten times longer than the few pages I had told Amy at the onset.

The various events in life seem to be connected to each other. The training in society leadership at the Chinese American Medical Society, the Chinese Academic and Professional Association, and the Microcirculatory Society prepared me for the jobs at APS and FASEB. The experience with PPG application at Columbia allowed me to do the same at IBMS and UCSD, and eventually the preparation of the applications for the Whitaker Foundation Development and Leadership Awards. The preparation of brochures at IBMS in Taiwan provided me with the experience to produce the brochures for IBME at San Diego.

The experience with grant review at the NIH gave me the know-how to establish a new review system at NHRI in Taiwan. It is important that we can apply what we learn to different situations, transcending time and space.

Because of the continued growth and development of my experience and expertise, everything worked out when the situation arose. I have to attribute my ability to do the various tasks to what was given to me by my parents both in my genetic inheritance and my upbringing, as I mentioned earlier, as well as the education and help by my teachers, colleagues and students. I am grateful that I have been given the courage and ability to deal with these challenges. I appreciate immensely the warm, kind and generous words of my family, friends, colleagues and students in these birthday celebrations. I sincerely feel that there is no reason for me to be proud of what I have accomplished, but that I have every reason to be grateful for the abilities installed in me and the support given to me to handle difficult situations with positive outcomes. As I have this deep gratitude to my parents and my grandparents, I am very gratified to know that they were pleased with what I did, as indicated by the calligraphies written by my maternal grandfather in 1957 (p. 490) and by my father in 1981 (p. 492). The support by my family, especially the loving and unwavering encouragement, support and help by K.C., is a major factor for my having the courage, determination and concentration to pursue the various tasks wholeheartedly without reservation.

The marvelous tributes in this book are generous in that they point to my strengths rather than weaknesses, and I appreciate very much these kind words. I thank everyone for telling me that they were pleased with what I have done, even though I know the words are more than what I deserve.

In recent years, I have been asked by several organizations to speak on the important elements in working as a scientist and for scientific organizations. I have used the seven C's outlined below as my motto, and I would like to take this opportunity to share these with my younger friends.

1. *Compassion*: In order to succeed in anything, we must have compassion. If we have the compassion for what we plan to do, we

will be able to do virtually anything. Work and service then become a joy rather than a chore.

Everything we do starts from our heart and our mind. Confucius taught us that in life we should start with sincerity in our heart, set our mind straight, improve ourselves and harmonize the family; then we can govern the country and make peace in the world. Compassion allows us to start from our heart and expand our love to work, people and society.

2. *Commitment*: In order to accomplish our goals and to excel in any endeavor, we must have serious commitment. As scientists, we are committed to do our best for the pursuit of new knowledge and for the education and training of the next generation of scientists. As a member of the profession, we need to be committed to work with our professional societies to advance our fields. At the same time, we must be committed to our family and friends, and also to ourselves, including the health of our body and mind. We need to balance all these commitments.

The effective management of our time is the key to fulfill all of our commitments. Time is the fairest thing in the world; nobody gets one second more or less than the others. The effectiveness with which we utilize time, however, varies widely. We need to know ourselves and be realistic, so that we will commit to the things we can do, not only in terms of our ability, but also in terms of the time availability. Once the commitments are made, we need to balance our priorities, and give priorities to different responsibilities at different times. In the end, we need to accomplish all of these, and do the best we can, i.e. with a commitment to excellence. Although we can do only one thing at a given time, we must also be able to manage several things at the same time. These seem to be contradictory, but really not. In the overall planning of our activities, we must take all of our commitments into account. While implementing a particular task, we should focus on the work at hand, and we need to bring all of our knowledge and experience to bear on that task.

We should welcome challenges, because that is where opportunity lies. We should never be afraid of making commitment because of fear of failure. We need to prepare ourselves fully for success, but we must be able to accept failure and make improvements accordingly. Failure is the mother of success.

3. *Comprehension* (Learning): In order to enhance our contributions to the progress of science, we must learn and comprehend what is known and has been done. We need to organize what we have learned into a systematic body of knowledge that is our own, i.e. to build a framework to put new knowledge together with the old, so that our knowledge base grows continuously in a systematic manner.

Learning is a life-long process. We can and must learn at all times and all occasions. We learn from good examples, i.e. to follow them, and we learn from bad examples, i.e. how not to do things. We can learn at every moment and from every person.

We need to learn broadly, i.e. beyond our own specific discipline. We need to learn and comprehend beyond the sciences. We need to learn about our society, humanities, arts, politics, economics, etc. especially in matters related to science, e.g. ethics, scientific integrity, animal research, human investigation and funding. The breadth of knowledge is important for us to maximize our contributions to professional organizations and the society.

4. *Creation*: Creativity is the key element in research and also in working with a professional organization. In order to be creative, we must know what has been done before and have a clear understanding of what is available (as discussed under "Comprehension.") We should determine what are the important problems that need be solved and focus on them. We need to create new ideas to solve these critical problems, and the solution should direct at the root of the problem and be as definitive as possible.

We need to know new developments not only in our specific field, but also in all related fields. In this way, we can leverage on new approaches that have been made available in other disciplines. We need to be resourceful in applying these to test our innovative ideas for the creation of new knowledge.

In a professional organization, we also need to provide innovative ideas and be creative, for the advancement of the profession and for the good of the organization as a whole. The approaches are actually the same as in doing research, e.g. to comprehend what is going on in the organization, what are the missions of the organization, and what needs to be done in order to make further advancements.

5. *Communication*: The skill to communicate with other people is an extremely important part of human behavior. This includes both written and oral communications. As scientists, we need to write research manuscripts and grant proposals. In industry, it is also necessary to write various types of reports. Everyday we write notes, letters, e-mails, and other types of documents. We need to orally present our findings and thoughts to other individuals and groups.

In order to improve our presentation, we must keep on practicing. One way to improve our oral presentation is to tape it and listen to our own talk (Section II.B.1 of this article). Then we can understand how the other people perceive what we say. This kind of feedback is the key to improving ourselves. We can learn from others when they make their presentations. From good speakers, we learn the right way to do it, whereas from poor speakers, we learn what to avoid (as mentioned under (Comprehension.")

6. *Cooperation*: No matter what we do, we must cooperate with others. Modern science is becoming increasingly more interdisciplinary, and scientific advancements depend on teamwork among people with common interest and different expertise. Effective and successful collaboration involves give and take. Everyone must be able to balance one's small self and the large self of the team. As a leader, one must provide effective leadership by formulating visions and focussing on matters for the group and never on personal glory. As members of the team, we must work together with the others for the good of the group. In all cases, we must be considerate, have mutual respect and foster a team spirit.

Before serving as a leader in any organization, we must first participate as a member of the team and be truly devoted to the success of the group. We must be willing and able to serve. To paraphrase John F. Kennedy: "Ask not what your organization can do for you. Ask what you can do for your organization." By becoming knowledgeable about how to work for the team and by working sincerely for the team, one gets gradually recognized as someone who can serve the organization as a leader. It is valuable to work in a relatively small organization first, which may be rather limited in either the geographical location or the nature of the group. Through our experience in working with a small organization, we can be prepared to serve the larger ones.

No person is an island. Friendship and collegiality are not only the keys to success, they are the sources of joy. Bert's excellent Foreword in this tribute book quotes the "Book of Poetry" about gift-giving and working together with mutual respect. It is important to give more and take less (Shyy, p. 290). If everyone can do that, there will be world peace.

7. *Consummation*: In every endeavor, the most important part is the completion of the task. In sports, it does not matter how beautifully one dribbles the basketball or how intricate the footwork is in soccer; all the efforts would be in vain if one cannot score. There is the saying that "At the 90-mile point of a 100-mile walk, you are only half way there." This illustrates the importance of the completion the task. The other six C's culminate in Consumption.

In carrying out these mottos, the most important element is people. In working with people, the most important element lies in our heart. If we care about people from our heart and treat every one with consideration and dignity, then there will be no problem in communication and cooperation, and there would be no prejudice or discrimination in the world. In this regard, the word "heart" in the English language is used interchangeably with "mind," and it is interesting that these two words are one and the same in the Chinese language. No wonder I had difficulty deciding whether to go into cardiovascular physiology or neurophysiology! Our heart/mind is also the key in performing any task. We need to think through the plan carefully, to pursue it with determination, confidence and devotion, and to enjoy what we do. The enjoyment is the greatest when we can benefit other people, enhancing their health and happiness. I am fortunate to have the opportunity of working on research and education that can use my small self to benefit the large self, i.e. people and society. Looking back at my career, it is obvious that I enjoy working with other people for the good of people, and that is what makes life so enjoyable.

I list below my Chinese translation of these seven C's, with some extensions. It is to be noted that the second character of each line is the word "heart (mind)." In parentheses are the literal translations of these Chinese characters back to English. Everything starts from our "heart/mind." We can succeed in what we do only if we put our "heart/mind" to it. In interacting with people and matters, we got to have "heart."

Compassion	熱心誠意	(warm and sincere in heart and mind)
Commitment	決心承擔	(resolved in one's mind to be responsible)
Comprehension	用心了解	(application of one's mind to understand)
Creation	精心創新	(focusing one's mind for innovation)
Cooperation	同心協力	(sharing of minds to work together)
Communication	推心交流	(expressing one's heart in interaction)
Consummation	盡心完成	(whole-hearted for task completion)

Looking ahead, I will do the same as I have been doing, as long as I can continue to do them well. I will always do my best to reach my capability, which has always been changing and will continue to be. I will heed the advice by my brother Fred (p. 13) and my friend Ernie Huang and his family (p. 228) to play more golf, and the advice by Savio Woo to spend some time on orthopaedic research so that I can improve the biomechanics of golfing (p. 284). However, I do know my limits. For example, I will not be able to score my age (p. 13) or to rule the world of golf (p. 284). Nevertheless, I do intend to fulfill my promise to May and Ann that I will play golf three times a week (but I did not specify how many holes for each round). One thing I can promise without hedging is that together with K.C., my family and my friends, I will continue to enjoy life and contribute to the happiness of others, whatever my age. I sincerely thank everyone for giving me 70 years of an extremely happy life and such a wonderfully happy time when I reached 70. My heart is filled with appreciation and gratitude!

Grandfather's Calligraphy, 1957

Chao-ching Chang[*]

Grandfather Chao-Ching Chang, Taipei 1952.

This calligraphy was written by Shu's maternal grandfather Mr. Chao-ching Chang (1874–1962), who sent it from Taiwan to New York. The body of the calligraphy was taken from a discussion of the Buddhist Bible. The annotation in the last line says:

Grandson Shu left some five years ago. Everyday his virtues grow and his knowledge become richer. Hence, I am sending this as a memento. December 1957. Maternal grandfather Chao-ching Chang at the age of 84 (83 by Western counting).

---

[*]Mr. Chao-ching Chang, who was the Magistrate of DaMing County, Hebei Province, when Shu's paternal grandfather was a Judge in DaMing. The two families arranged the marriage of Shu's parents.

昔者分形分蹟之時言未馳而成化當常現常之世人仰德而知遵及乎晦影歸真遷儀越世金容掩色不鏡三千之光麗象開圖空端四八之相於是微言廣被拯含類於三途遺訓宣導群生於十地然而真教難仰莫能一其指歸曲學易遵邪正於焉紛糺所以空有之論或習俗而是非大小之乘作訟時而隆替有玄奘法師者法門之領袖也幼懷貞敏早悟三空之心長契神情先苞四忍之行松風水月未足比其清華仙露明珠詎能方其朗潤故以智通無累神測未形超六塵而迥出隻千古而無對凝心內境悲正法之陵遲栖慮玄門慨慨深文之訛謬思欲分條析理廣破前聞

熙孫別巳五年德日進而學日豐寄此存念　丁酉臘月朔外祖張昭芹時年八十有四

Father's Calligraphy, 1981

Shih-liang Chien, Ph.D.*

English translation of calligraphy by Shu's father Dr. Shih-liang Chien (1908–1983):

> Thinking about the past events in Beijing,
> I recall the unbound happiness while holding the infant.
>
> Being warm as the spring sun, Shu should be able to heal the world.
> Kuang being supportive and caring, two hearts work as one.
>
> They have both reached 50 and have been wedded for half of that time.
> I cherish the spreading of Shu's reputation across the ocean.*
>
> You should note that my mind at this age still works rather well,
> As I am enjoying the flourishing of orchids and magnolia in my garden.

Written by Father
August 1981, Taipei

Note by Shu: Father rarely wrote poems or calligraphy. When I was in Taipei during one of my visits home, he told me that he had been thinking of writing a poem for K.C. (Kuang-Chung) and me. He put

---

*Referring to Shu's recent reception of the First Fåhraeus Medal (p. 494) at the European Conference of Clinical Hemorheology.

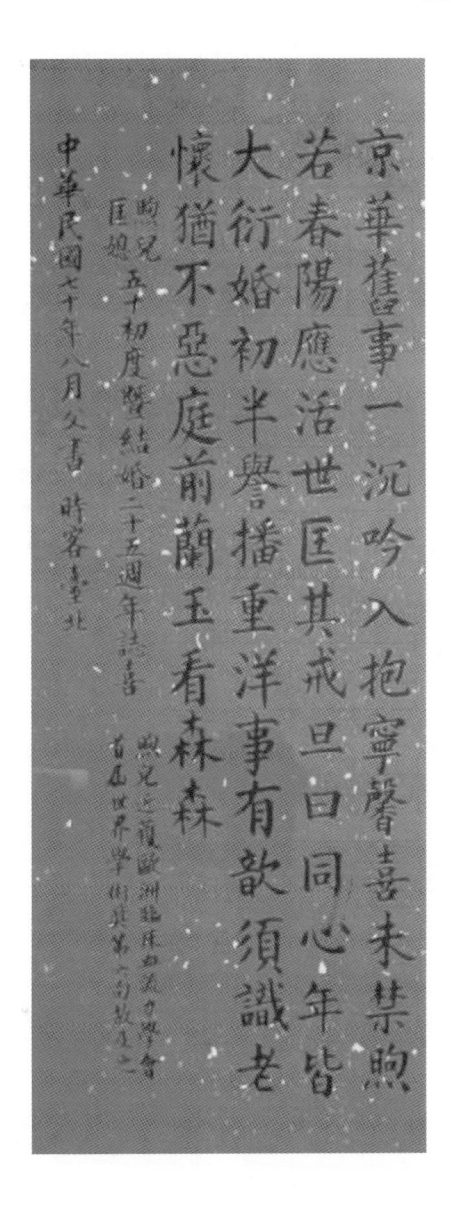

京華舊事一沉吟 入抱寧馨喜未禁 煦
若春陽應活世 匡其戒旦日同心 年皆
大衍婚姻初半譽播重洋事有歆 須識老
懷猶不惡庭前蘭玉看森森

煦兄
匡姐 五十初度暨結婚二十五週年誌喜

煦兄迄荷歐洲瑞德荷瑞四國講學命
匡姐世界學術選第六句教授之一

中華民國七十年八月久書 時客臺北

this piece of red calligraphy paper on his desk. I grounded the calligraphy ink and watched him writing this beautifully composed and elegantly written calligraphy. Twenty years later, I still can vividly recall his sense of joy and this precious moment.

Medals

Medals 495

National Health Medal, Department of Health R.O.C. (1998).

Curriculum Vitae of Shu Chien

Address

Office: Department of Bioengineering, Mail Code 0412, Powell-Focht Bioengineering Hall (PFBH), University of California, San Diego, 9500 Gilman Drive, La Jolla, CA 92093-0412, USA
Tel: 858-534-5195.   Fax: 858-534-5453.
E-mail: shuchien@ucsd.edu

Birth Date: 23 June 1931; Beijing, China. U.S. Citizen.
Status: Married 7 April 1957.   Wife: Kuang-Chung Hu Chien, M.D.
Two daughters: May Chien Busch, M.B.A. and Ann Chien Guidera, M.D.

Education

National Peking University (Premed), 1947–1948.
National Taiwan University (College of Medicine), 1949–1953, M.D.
Columbia University (Physiology), 1954–1957, Ph.D.
Oak Ridge Institute of Nuclear Studies, Radioisotope Training Course, February 1958.
Jackson Laboratory, Bar Harbor, Maine, Genetics Training Course, August 1983.

Appointments

Rotating Intern, National Taiwan University Hospital, 1952–1953.
Assistant in Physiology, Columbia University College of Physicians & Surgeons, New York City, 1954–1956.

Instructor in Physiology, Columbia University College of Physicians & Surgeons, 1956–1958.
Assistant Professor of Physiology, Columbia University College of Physicians & Surgeons, 1958–1964.
Associate Professor of Physiology, Columbia University College of Physicians & Surgeons, 1964–1969.
Professor of Physiology and Director of Laboratory of Hemorheology, Columbia University College of Physicians & Surgeons, 1969–1988.
Director, Division of Circulatory Physiology and Biophysics, Columbia University College of Physicians & Surgeons, 1974–1988.
Director, Institute of Biomedical Sciences, Academia Sinica, Taipei, Taiwan, ROC, 1987–1988.
Professor of Bioengineering & Medicine, and Member of Center for Molecular Genetics, University of California, San Diego, 1988–Present.
Coordinator, Bioengineering Group, University of California, San Diego, 1989–1994.
Chair, Department of Bioengineering, University of California, San Diego, 1994–1999.
Director, Whitaker Institute of Biomedical Engineering, University of California, San Diego, 1991–present.

Areas of Research Interest

Effects of mechanical forces on gene expression and signal transduction (current major interest).
Biomechanical properties and molecular organization of cell membranes.
Energy balance and molecular mechanisms of cell-cell interactions.
Transport of macromolecules across vascular endothelium.
Blood rheology and microcirculatory dynamics in health and disease.
Molecular, cellular and tissue bioengineering.

Honors and Awards

Li Foundation Fellow, 1954–1956.
Golden Tooth Award, School of Oral and Dental Surgery, Columbia University, 1971.

Teacher of the Year, Columbia University College of Physicians & Surgeons, 1972, 1974, 1981, 1983.
Member, Academia Sinica, ROC, 1976.
Research Achievement Award, American Chinese Medical Society, 1979.
Distinguished Visiting Scientist, Albany Medical College, 1980.
Nanci Medal for Research in Blood Rheology, 1980.
The First Fåhraeus Medal in Clinical Haemorheology, 1981.
Nathanson Lectureship, University of Southern California, 1983.
Landis Award, Microcirculatory Society, 1983.
Outstanding Research Award, Chinese Academic & Professional Association (Eastern America), 1985.
NSF Special Creativity Grant Award (with Dr. Sheldon Weinbaum et al.), 1985–1988.
The First Honorary Member, Chinese American Medical Society of New Jersey, 1986.
Achievement Award, Chinese-American Engineers & Scientists Association of Southern California, 1988.
NIH MERIT Award (National Institute of Heart, Lung and Blood), 1989–1998.
Best Paper in the *Journal of Biomechanical Engineering*, American Society of Mechanical Engineers, 1989.
Distinguished Visiting Scientist, University of South Alabama, 1990.
Distinguished Achievement Award, The Chinese Hospital, San Francisco, 1990.
Melville Medal (with Dr. Cheng Dong et al.), American Society of Mechanical Engineers, 1990.
Zweifach Award, Fifth World Congress for Microcirculation, 1991.
Founding Fellow, American Institute of Medical and Biological Engineering, 1992.
Vaishnav Lectureship, The Catholic University of America, 1992.
Research Achievement Award, San Diego Chinese American Scientists & Engineers Association, 1992.
ALZA Award, Biomedical Engineering Society, 1993.
Distinguished Accomplishment Award, Mid-America Chinese Science & Technology Association, 1993.
SCBA Lecturer, Fifth International SCBA Symposium and Workshops, 1993.

Distinguished Faculty Lecturer, University of California, San Diego, 1994.
Merck Distinguished Lecturer, Rutgers University, 1994.
Member, Institute of Medicine, National Academy of Sciences, 1994.
Wellcome Professorship, Federation of American Societies for Experimental Biology, 1995.
Honorable Visiting Scientist, National Science Council, ROC, 1995.
Honorary Professor, Peking Medical University, 1995.
Honorary Member, Shanghai Biophysical Society, 1995.
Honorary Professor, Chongqing University, 1995.
Chien-Fung Young Investigator Award in Biorheology, established by Chinese Society of Biorheology and International Society of Biorheology, 1995.
Best Paper in the *Journal of Biomechanical Engineering*, American Society of Mechanical Engineers, 1995.
Donald M. Nelson Lectureship, University of Missouri-Columbia, 1995.
Joseph Mather Smith Prize for Distinguished Alumni Research, Columbia University, 1996.
Borun Lecturer, University of California, Los Angeles, 1996.
Melville Medal (with Dr. Yaqi Huang *et al.*), American Society of Mechanical Engineers, 1996.
Molecular, Cellular and Physiological Biology Distinguished Lecturer, University of Illinois, 1997.
Honorary Professor, Chinese Academy of Medical Sciences, 1997.
Member, National Academy of Engineering, 1997.
Lifetime Achievement Award, Chinese-American Engineers & Scientists Association of Southern California, 1998.
National Health Medal, Department of Health, Taiwan, ROC, 1998.
Ray Daggs Award, American Physiological Society, 1999.
Whitaker Distinguished Lecturer, University of Pennsylvania, 2000.
Yehia Habib Honor Lecture, International Congress of African Association of Physiological Sciences, 2000.
Distinguished Keynote Lecturer, Inauguration of Department of Biomedical Engineering, Columbia University, 2000.
Distinguished Lecturer in Biomedical Engineering, Stanford University, 2001.

Services in Scientific Organizations

European Society for Microcirculation: Executive Committee, 1971–1989.
The Microcirculatory Society: Councillor, 1975–1978; President, 1980–1981.
National Institutes of Health: NHLBI Program Project Grant Review Committee B, 1977–1981; Hypertension Task Force, 1977–1979; Co-chair, Panel on Functional Genomics, BECON, 1998; Member, Center for Scientific Review Advisory Committee, 1998– .
American Chinese Medical Society: President, 1978–1979.
Society for Experimental Biology and Medicine: Membership Committee, 1978–1981.
Chinese Academic and Professional Association: President, 1979–1980.
International Society of Biorheology: Councillor, 1981–1983, Vice President, 1983–1989.
National Committee on Biomechanics: Executive Committee, 1982–1986.
North American Society of Biorheology: Chairman, Steering Committee, 1985–1986.
American Physiological Society: Councillor, 1985–1989; President, 1990–1991; Chairman, Long Range Planning Committee, 1994–1996, Member, International Physiology Committee, 1998–2001.
Federation of American Societies for Experimental Biology: Board Member, 1989–1993, President, 1992–1993, Chair, Public Affairs Executive Committee, 1993–1994.
Institute of Biomedical Sciences, Academia Sinica, Taiwan, ROC: Chair, Advisory Committee, 1991– .
National Health Research Institute (Analogous to NIH), Department of Health, Taiwan, ROC: Chairman, Advisory Committee, 1991– .
Biomedical Engineering Society: Publications Committee, 1994–2001 (Chair 1998–2001).
International Union of Physiological Sciences: Treasurer, 1998–2001.
National Academy of Engineering: Chair, Section 2 Peer Committee, 2001–2002.
American Institute for Medical and Biological Engineering: Chair, College of Fellows, 1998-1999; President, 2000–2001.
Chair, Organizing Committee of 2005 International Congress of Physiological Sciences, Washington, D.C., 2001–2005.

Shu Chien's Publications

A. Full-length Journal Papers

A1. Chien, S., Lukin, L., Holt, A.P., Cherry, S.H., Root, W.S. and Gregersen, M.I. The effect of total-body x-irradiation on the circulation of splenectomized dogs. *Radiat. Res.* **7**: 277–287, 1957.

A2. Chien, S. Quantitative evaluation of circulatory adjustment of splenectomized dogs to hemorrhage. *Am. J. Physiol.* **193**: 605–614, 1958.

A3. Peng, M.T., Chien, S. and Gregersen, M.I. Effect of large doses of head irradiation in dogs. *Am. J. Physiol.* **194**: 344–350, 1958.

A4. Rawson, R.A., Chien, S., Peng, M.T. and Dellenback, R.J. Determination of the residual blood volume required for survival in rapidly hemorrhaged splenectomized dogs. *Am. J. Physiol.* **196**: 179–183, 1959.

A5. Gregersen, M.I., Sear, H., Rawson, R.A., Chien, S. and Saiger, G.L. Cell volume, plasma volume, total blood volume and F cells factor in the rhesus monkey. *Am. J. Physiol.* **196**: 184–187, 1959.

A6. Gregersen, M.I., Chien, S., Rawson, R.A. and Muelheims, G. Effect of histamine on F cells value in splenectomized dogs. *Proc. Soc. Exp. Biol. Med.* **100**: 872–875, 1959.

A7. Chien, S. Supersensitivity of denervated superior cervical ganglia to acetylcholine. *Am. J. Physiol.* **198**: 949–954, 1960.

A8. Chien, S. Blood volume in sympathectomized-splenectomized dogs. *Proc. Soc. Exp. Biol. Med.* **104**: 733–736, 1960.

A9. Chien, S. and Billig, S. F cells value in sympathectomized-splenectomized dogs after hemorrhage. *Proc. Soc. Exp. Biol. Med.* **104**: 737–739, 1960.

A10. Chien, S. and Billig, S. The effect of hemorrhage on the cardiac output of sympathectomized dogs. *Am. J. Physiol.* **201**: 475–479, 1961.

A11. Dische, Z., Pallavicini, C., Kavasaki, H., Smirnow, N., Cizek, L.J. and Chien, S. Influence of the nature of the secretory stimulus on the composition of the carbohydrate moiety of glycoproteins of the submaxillary saliva. *Arch. Biochem. Biophys.* **97**: 459–469, 1962.

A12. Gregersen, M.I., Pallavicini, C. and Chien, S. Studies on the chemical composition of the central nervous system in relation to the effects of x-irradiation and of disturbances in water and salt balance. I. Chemical composition of various specific areas and structures of the brain in the dog and monkey. *Radiat. Res.* **17**: 209–225, 1962.

A13. Gregersen, M.I., Pallavicini, C. and Chien, S. Studies on the chemical composition of the central nervous system in relation to the effects of x-irradiation and of disturbances in water and salt balance. II. Effects of x-irradiation on the chemical composition of brain tissues in dogs. *Radiat. Res.* **17**: 226–233, 1962.

A14. Chien, S., Pallavicini, C., Cizek, L.J. and Gregersen, M.I. Studies on the chemical composition of the central nervous system in relation to the effects of x-irradiation and of disturbances in water and salt balance. III. Effect of disturbances in water and electrolyte balance on the chemical composition of brain tissues in dogs. *Radiat. Res.* **17**: 234–243, 1962.

A15. Dische, Z., Pallavicini, C., Cizek, L.J. and Chien, S. Changes in the control of the secretion of mucus glycoproteins as possible pathogenic factor in cystic fibrosis of the pancreas. *Ann. N.Y. Acad. Sci.* **93**: 526–540, 1962.

A16. Usami, S., Peric, B. and Chien, S. Release of antidiuretic hormone due to common carotid occlusion and its relation with vagus nerve. *Proc. Soc. Exp. Biol. Med.* **111**: 189–193, 1962.

A17. Chien, S., Peric, B. and Usami, S. The reflex nature of the release of antidiuretic hormone upon common carotid occlusion in vagotomized dogs. *Proc. Soc. Exp. Biol. Med.* **111**: 193–196, 1962.

A18. Chien, S. Cell volume, plasma volume and cell percentage in splanchnic circulation of splenectomized dogs. *Circ. Res.* **12**: 22–28, 1963.

A19. Chien, S. and Krakoff, L. Hemodynamics of dogs in histamine shock, with special reference to splanchnic blood volume and flow. *Circ. Res.* **12**: 29–30, 1963.

A20. Gregersen, M.I., Peric, B., Usami, S. and Chien, S. Relation of molecular size of dextran to its effects on the rheological properties of blood. *Proc. Soc. Exp. Biol. Med.* **112**: 883–887, 1963.

A21. Usami, S. and Chien, S. Role of hepatic blood flow in regulation plasma concentration of antidiuretic hormone after hemorrhage. *Proc. Soc. Exp. Biol. Med.* **113**: 606–609, 1963.

A22. Gregersen, M.I., Usami, S., Peric, B., Chang, C., Sinclair, D. and Chien, S. Blood viscosity at low shear rates. Effects of low and high molecular dextrans. *Biorheology* **1**: 247–253, 1963.

A23. Chien, S. Role of sympathetic nervous system in surviving acute hemorrhage. *Am. J. Physiol.* **206**: 21–24, 1964.

A24. Gregersen, M.I., Peric, B., Usami, S. and Chien, S. Relation of molecular weight of dextran to its effects on viscosity and sedimentation rate of blood. *Bibl. Anat.* **4**: 58–61, 1964.

A25. Chien, S., Sinclair, D.G., Chang, C., Peric, B. and Dellenback, R.J. Simultaneous study of capillary permeability to several macromolecules. *Am. J. Physiol.* **207**: 513–517, 1964.

A26. Chien, S., Sinclair, D.G., Dellenback, R.J., Chang, C., Peric, B., Usami, S. and Gregersen, M.I. Effect of endotoxin on capillary permeability to macromolecules. *Am. J. Physiol.* **207**: 518–522, 1964.

A27. Chien, S. and Usami, S. Circulation of sympathectomized dogs under pentobarbital anesthesia. *Am. J. Physiol.* **208**: 790–794, 1965.

A28. Chien, S., Dellenback, R.J., Usami, S. and Gregersen, M.I. Hematocrit changes in endotoxin shock. *Proc. Soc. Exp. Biol. Med.* **18**: 1182–1187, 1965.

A29. Chien, S., Dellenback, R.J. and Usami, S. Effect of endotoxin on the transfer of fibrinogen from plasma to lymph. *Proc. Soc. Exp. Biol. Med.* **118**: 1187–1190, 1965.

A30. Taylor, H.M., Chien, S., Gregersen, M.I. and Lundberg, J.L. Comparison of viscometric behavior of suspensions of polystyrene latex and human blood cells. *Nature* **207**: 77–78, 1965.

A31. Chien, S., Dellenback, R.J., Usami, S. and Gregersen, M.I. Plasma trapping in hematocrit determination: differences among animal species. *Proc. Soc. Exp. Biol. Med.* **119**: 1155–1158, 1965.

A32. Gregersen, M.I., Usami, S., Chien, S. and Swank, R.L. Comparison of screen filtration pressure and low-shear viscosity of blood. *J. Appl. Physiol.* **20**: 1362–1364, 1965.

A33. Chien, S., Usami, S., Taylor, H.M., Lundberg, J.L. and Gregersen, M.I. Effects of hematocrit and plasma proteins on human blood rheology at low shear rates. *J. Appl. Physiol.* **21**: 81–87, 1966.

A34. Chien, S., Usami, S., Simmons, R.L., McAllister, F.F. and Gregersen, M.I. Blood volume and age: repeated measurements on normal men after 17 years. *J. Appl. Physiol.* **21**: 583–588, 1966.

A35. Chien, S., Chang, C., Dellenback, R.J., Usami, S. and Gregersen, M.I. Hemodynamic changes in endotoxin shock. *Am. J. Physiol.* **210**: 1401–1410, 1966.

A36. Chien, S., Dellenback, R.J., Usami, S., Treitel, K., Chang, C. and Gregersen, M.I. Blood volume and its distribution in endotoxin shock. *Am. J. Physiol.* **210**: 1411–1418, 1966.

A37. Dellenback, R.J., Chien, S., Usami, S., Potter, R.T. and Gregersen, M.I. Hemodynamic effects of pericardial tamponade. *Proc. Soc. Exp. Biol. Med.* **123**: 623–627, 1966.

A38. Chien, S. Role of the sympathetic nervous system in hemorrhage. *Physiol. Rev.* **47**: 214–288, 1967.

A39. Gregersen, M.I., Usami, S., Chien, S. and Dellenback, R.J. Characteristics of torque-time records on heparinized and defibrinated elephant, human and goat blood at low shear rates (0.01 sec^{-1}): effects of fibrinogen and dextran (Dx 375). *Bibl. Anat.* **9**: 276–281, 1967.

A40. Gregersen, M.I., Bryant, C.A., Hammerle, W., Usami, S. and Chien, S. Flow characteristics of human erythrocytes through polycarbonate sieves. *Science* **157**: 825–827, 1967.

A41. Chien, S., Usami, S., Dellenback, R.J. and Gregersen, M.I. Blood viscosity: influence of erythrocyte deformation. *Science* **157**: 827–829, 1967.

A42. Chien, S., Usami, S., Dellenback, R.J., Gregersen, M.I., Nanninga, L.B. and Guest, M.M. Blood viscosity: influence of erythrocyte aggregation. *Science* **157**: 829–831, 1967.

A43. Chien, S., Dellenback, R.J., Usami, S., Seaman, G.V.F. and Gregersen, M.I. Centrifugal packing of suspensions of erythrocytes hardened with acetaldehyde. *Proc. Soc. Exp. Biol. Med.* **127**: 982–985, 1968.

A44. Chien, S. and Usami, S. Effects of hemorrhage on anesthetized dogs: role of the sympathetic system. *Am. J. Physiol.* **216**: 1322–1329, 1969.

A45. Dellenback, R.J. and Chien, S. The extinction coefficient of canine fibrin in alkaline urea at 282 mm. *Proc. Soc. Exp. Biol. Med.* **131**: 599–600, 1969.

A46. Chien, S. Blood rheology and its relation to flow resistance and transcapillary exchange, with special reference to shock. *Adv. Microcirc.* **2**: 89–103, 1969.

A47. Usami, S., Chien, S. and Gregersen, M.I. Viscometric characteristics of blood of the elephant, man, sheep and goat. *Am. J. Physiol.* **217**: 884–890, 1969.

A48. Dellenback, R.J., Usami, S., Chien, S. and Gregersen, M.I. Effect of splenectomy on blood picture, blood volume, and plasma proteins in beagles. *Am. J. Physiol.* **217**: 891–897, 1969.

A49. Gregersen, M.I., Bryant, C.A., Chien, S., Dellenback, R.J., Magazinovic, V. and Usami, S. Species differences in the flexibility and deformation of erythrocytes (RBC). *Bibl. Anat.* **10**: 104–108, 1969.

A50. Chien, S., Usami, W. and Bertles, J.F. Abnormal rheology of oxygenated blood in sickle cell anemia. *J. Clin. Invest.* **49**: 623–634, 1970.

A51. Chien, S. Shear dependence of effective cell volume as a determinant of blood viscosity. *Science* **68**: 977–978, 1970.

A52. Dellenback, R.J. and Chien, S. The extinction coefficient of fibrinogen from man, dog, elephant, sheep, and goat at 280 nm. *Proc. Soc. Exp. Biol. Med.* **134**: 353–355, 1970.

A53. Chien, S., Usami, S., Dellenback, R.J. and Gregersen, M.I. Shear-dependent deformation of erythrocytes in rheology of human blood. *Am. J. Physiol.* **219**: 136–142, 1970.

A54. Chien, S., Usami, S., Dellenback, R.J. and Gregersen, M.I. Shear-dependent interaction of plasma proteins with erythrocytes in blood rheology. *Am. J. Physiol.* **219**: 143–153, 1970.

A55. Benis, A.M., Usami, S. and Chien, S. Effect of hematocrit and inertial losses on pressure-flow relations in the isolated hindpaw of the dog. *Circ. Res.* **27**: 1047–1068, 1970.

A56. Usami, S., Magazinovic, V., Chien, S. and Gregersen, M.I. Viscosity of turkey blood: rheology of nucleated erythrocytes. *Microvasc. Res.* **2**: 489–499, 1970.

A57. Chien, S. Hemodynamics in hemorrhage: influences of sympathetic nerves and pentobarbital anesthesia. *Proc. Soc. Exp. Biol. Med.* **136**: 271–275, 1971.

A58. Dellenback, R.J., Usami, S. and Chien, S. Metabolism and distribution of homologous and autologous fibrinogen in beagle dogs. *Am. J. Physiol.* **220**: 828–836, 1971.

A59. Usami, S., Chien, S. and Gregersen, M.I. Hemogolobin solution as a plasma expander: effects on blood viscosity. *Proc. Soc. Exp. Biol. Med.* **136**: 1232–1235, 1971.

A60. Chien, S., Luse, S.A. and Bryant, C.A. Hemolysis during filtration through micropores: a scanning electron microscopic and hemorheologic correlation. *Microvasc. Res.* **3**: 183–203, 1971.

A61. Miller, L.H. and Chien, S. Density distribution of red cells infected by *Plasmodium knowlesi* and *Plasmodium coatneyi*. *Exp. Parasitol.* **29**: 451–456, 1971.

A62. Chien, S., Usami, S., Dellenback, R.J. and Bryant, C.A. Comparative hemorheology — hematological implications of species differences in blood viscosity. *Biorheology* **8**: 35–57, 1971.

A63. Chien, S., Usami, S. and Rowe, A.W. Rheologic properties of erythrocytes preserved in liquid nitrogen. *J. Lab. Clin. Med.* **78**: 175–180, 1971.

A64. Miller, L.H., Usami, S. and Chien, S. Alteration in the rheologic properties of *Plasmodium knowlesi*-infected red cells. A possible mechanism for capillary obstruction. *J. Clin. Invest.* **50**: 1451–1455, 1971.

A65. Dellenback, R.J. and Chien, S. The quantitation of isotope loss with tyrosine loss in the canine fibrinogen-fibrin transformation. *Proc. Soc. Exp. Biol. Med.* **137**: 751–758, 1971.

A66. Dellenback, R.J. and Chien, S. The coagulability of ammonium sulfate precipitated canine fibrinogen. *Proc. Soc. Exp. Biol. Med.* **137**: 759–762, 1971.

A67. Chien, S. A theory for the quantification of transcapillary exchange in the presence of shunt flow. *Circ. Res.* **29**: 173–180, 1971.

A68. Benis, A.M., Usami, S. and Chien, S. Determination of the shear stress-shear rate relation for blood for Couette viscometry. *Biorheology* **8**: 65–69, 1971.

A69. Benis, A.M., Usami, S. and Chien, S. Evaluation of viscous and inertial pressure losses in isolated tissue with a simple mathematical model. *Microvasc. Res.* **4**: 81–93, 1972.

A70. Miller, L., Chien, S. and Usami, S. Decreased deformability of *Plasmodium coatneyi*-infected red cells and its possible relation to cerebral malaria. *Am. J. Trop. Med. Hyg.* **21**: 133–137, 1972.

A71. Skalak, R., Chen, P.H. and Chien, S. Effect of hematocrit and rouleaux on apparent viscosity in capillaries. *Biorheology* **9**: 67–82, 1972.

A72. Miller, L.H., Cooper, G.W., Chien, S. and Fremount, H.N. Surface charge on *Plasmodium knowlesi* and *P. coatneyi*-infected red cells of macaca mullata. *Exptl. Parasitol.* **32**: 86–95, 1972.

A73. Benis, A.M., Chien, S., Usami, S. and Jan, K.-M. Inertial pressure losses in perfused hindlimb: a reinterpretation of the results of Whittaker and Winton. *J. Appl. Physiol.* **34**: 383–389, 1973.

A74. Chien, S. and Jan, K.-M. Ultrastructural basis of the mechanism of rouleaux formation. *Microvasc. Res.* **5**: 155–166, 1973.

A75. Skalak, R., Tözeren, A., Zarda, R.P. and Chien, S. Strain energy function of red blood cell membranes. *Biophys. J.* **13**: 245–264, 1973.

A76. Jan, K.-M. and Chien, S. Role of surface electric charge in red blood cell interactions. *J. Gen. Physiol.* **61**: 638–654, 1973.

A77. Jan, K.-M. and Chien, S. Influence of ionic composition of fluid medium on red cell aggregation. *J. Gen Physiol.* **61**: 655–668, 1973.

A78. Usami, S. and Chien, S. Optical reflectometry of red cell aggregation under shear flow. *Bibl. Anat.* **11**: 91–97, 1973.

A79. Chien, S. Electrochemical and ultrastructural aspects of red cell aggregation. *Bibl. Anat.* **11**: 244–250, 1973.

A80. Jan, K.-M. and Chien, S. Role of the electrostatic repulsive force in red cell interactions. *Bibl. Anat.* **11**: 281–288, 1973.

A81. Chien, S. Force balance at the surfaces of aggregating cells. *Bibl. Anat.* **11**: 303–309, 1973.

A82. Chien, S., Dellenback, R.J., Usami, S., Burton, D., Gustavson, P. and Magazinovic, V. Blood volume, hemodynamic and metabolic changes in hemorrhagic shock in normal and splenectomized dogs. *Am. J. Physiol.* **225**: 866–879, 1973.

A83. Usami, S. and Chien, S. Shear deformation of red cell ghosts. *Biorheology* **10**: 425–430, 1973.

A84. Chien, S. and Jan, K.-M. Red cell aggregation by macromolecules: roles of surface adsorption and electrostatic repulsion. *J. Supramol. Struct.* **1**: 385–409, 1973.

A85. Chien, S., Cooper, G.W., Jr., Jan, K.-M., Miller, L.H., Howe, C., Usami, S. and Lalezari, P. N-acetylneuraminic acid deficiency in erythrocyte membranes: biophysical and biochemical correlates. *Blood* **43**: 445–460, 1974.

A86. Chien, S. and Usami, S. Rate and mechanism of release of antidiuretic hormone after hemorrhage. *Circ. Shock* **1**: 71–80, 1974.

A87. Benis, A.M., Chien, S., Usami, S. and Jan, K.-M. A reappraisal of Whittaker and Winton's results on the basis of inertial losses. *Biorheology* **11**: 153–161, 1974.

A88. Scholz, P.M., Chien, S., Gump, F.E., Karis, J.H. and Kinney, J.M. Use of blood viscosity measurements to assess vascular tone in patients with altered peripheral resistance. *Surg. Forum* **25**: 239–241, 1974.

A89. Usami, S., Chien, S., Scholz, P.M. and Bertles, J.F. Effect of deoxygenation on blood rheology in sickle cell disease. *Microvasc. Res.* **9**: 324–334, 1975.

A90. Scholz, P.M., Kinney, J.M. and Chien, S. Effects of major abdominal operations on human blood rheology. *Surgery* **77**: 351–359, 1975.

A91. Jan, K.M., Chien, S. and Bigger, J.T., Jr. Observations on blood viscosity changes after acute myocardial infarction. *Circulation* **51**: 1079–1084, 1975.

A92. Tietjen, G.W., Chien, S., Scholz, P.M., Gump, F.E. and Kinnery, J.M. Changes in blood viscosity and plasma proteins in carcinoma. *Surg. Forum* **26**: 166–168, 1975.

A93. Usami, S., Chien, S. and Bertles, J.F. Deformability of sickle cells as studied by microsieving. *J. Lab. Clin. Med.* **86**: 274–282, 1975.

A94. Copley, A.L., King, R.G., Chien, S., Usami, S., Skalak, R. and Huang, C.R. Microscopic observations of viscoelasticity of human blood in steady and oscillatory shear. *Biorheology* **12**: 257–263, 1975.

A95. Usami, S., King, R.G., Chien, S., Skalak, R., Huang, C.R. and Copley, A.L. Microcinephotographic studies on red cell aggregation in steady and oscillatory shear — a note. *Biorheology* **12**: 323–325, 1975.

A96. Chien, S., Peng, M.T., Chen, K.P., Huang, T.F., Chang, C. and Fang, H.S. Longitudinal measurements of blood volume and essential body mass in human subjects. *J. Appl. Physiol.* **39**: 818–824, 1975.

A97. Chien, S., Peng, M.T., Chen, K.P., Huang, T.F., Chang, C. and Fang, H.S. Longitudinal studies on adipose tissue and its distribution in human subjects. *J. Appl. Physiol.* **39**: 825–830, 1975.

A98. Scholz, P.M., Karis, J.H., Gump, F.E., Kinney, J.M. and Chien, S. Correlation of blood rheology with vascular resistance in critically ill patients. *J. Appl. Physiol.* **39**: 1008–1011, 1975.

A99. Chien, S., King, R.G., Skalak, R., Usami, S. and Copley, A.L. Viscoelastic properties of human blood and red cell suspensions. *Biorheology* **12**: 341–346, 1975.

A100. Tietjen, G.W., Chien, S., Leroy, E.C., Gavras, I., Gavras, H. and Gump, F.E. Blood viscosity, plasma proteins, and Raynaud syndrome. *Arch. Surg.* **110**: 1343–1346, 1975.

A101. Chien, S. Electrochemical interactions between erythrocyte surfaces. *Thrombosis Res.* **8**: 189–202, 1976.

A102. Chien, S. Significance of macrorheology and microrheology in atherogenesis. *Ann. N.Y. Acad. Sci.* **275**: 10–27, 1976.

A103. Tietjen, G.W., Chien, S., Scholz, P., Gump, F. and Kinnery, J.M. Changes in blood viscosity and plasma proteins in carcinoma. *J. Surg. Oncol.* **9**: 53–59, 1977.

A104. Carlin, R. and Chien, S. Partition of xenon and iodo-antipyrine among erythrocytes, plasma and myocardium. *Circ. Res.* **40**: 497–504, 1977.

A105. Carlin, R. and Chien, S. Effect of hematocrit on the washout of xenon and iodo-antipyrine from dog myocardium. *Circ. Res.* **40**: 505–509, 1977.

A106. Zarda, R.P., Skalak, R. and Chien, S. Elastic deformation of red blood cells. *J. Biomech.* **10**: 211–221, 1977.

A107. Chien, S. Blood rheology in hypertension and cardiovascular diseases. *Cardiovasc. Med.* **2**: 356–360, 1977.

A108. Chien, S., Sung, L.A., Kim, S., Burke, A.M. and Usami, S. Determination of aggregation force in rouleaux by fluid mechanical technique. *Microvasc. Res.* **13**: 327–333, 1977.

A109. Chien, S. Principles and techniques for assessing erythrocyte deformability. *Blood Cells* **3**: 71–99, 1977.

A110. Chien, S. Rheology of sickle cells and erythrocyte content. *Blood Cells* **3**: 283–296, 1977.

A111. Jan, K.M. and Chien, S. Effect of hematocrit variations on coronary hemodynamics and oxygen utilization. *Am. J. Physiol.* **233**: H106–H133, 1977.

A112. Chien, S., Simchon, S., Abbott R.E. and Jan, K.M. Surface adsorption of dextrans on human red cell membrane. *J. Colloid Interface Sci.* **62**: 461–470, 1977.

A113. Chen, R.Y.Z. and Chien, S. Plasma volume, red cell volume and thoracic duct lymph flow in hypothermia. *Am. J. Physiol.* **2**: H605–H612, 1977.

A114. Chien, S. Clinical rheology in cardiovascular disease. *Bibl. Anat.* **16**: 472–474, 1977.

A115. Lee, W.M., Lebwohl, O. and Chien, S. Hyperviscosity syndrome attributable to hyperglobulinemia in chronic active hepatitis. *Gastroenterology* **74**: 918–922, 1978.

A116. Chen, R.Y.Z. and Chien, S. Hemodynamic functions and blood viscosity in surface hypothermia. *Am. J. Physiol.* **235**: H136–H143, 1978.

A117. Chien, S., Sung, K.-L.P., Skalak, R., Usami, S. and Tözeren, A. Theoretical and experimental studies on viscoelastic properties of red cell membrane. *Biophys. J.* **24**: 463–487, 1978.

A118. Somer, H., Sung, L.A., Thurn, A. and Chien, S. Erythrocytes in Duchenne dystrophy: osmotic fragility and membrane deformability. *Neurology* **29**: 519–522, 1979.

A119. Lee, M.M.L. and Chien, S. Morphologic effects of pressure changes on canine carotid artery endothelium as observed by scanning electron microscopy. *Anat. Rec.* **194**: 1–14, 1979.

A120. Fan, F.-C., Chen, R.Y.Z., Schuessler, G.B. and Chien, S. Comparison between the ^{133}Xe clearance method and the microsphere technique in cerebral blood flow determinations. *Circ. Res.* **44**: 653–659, 1979.

A121. Fan, F.-C., Schuessler, G.B., Chen, R.Y.Z. and Chien, S. Determinations of blood flow and shunting of 9- and 15 μm microspheres in regional vascular beds. *Am. J. Physiol.* **234**: H25–H33, 1979.

A122. Burke, A.M., Chien, S., McMurtry, J.G. III and Quest, D.O. Effects of low molecular weight dextran on blood viscosity after craniotomy for intracranial aneurysms. *Surg. Gynecol. Obstet.* **148**: 9–15, 1979.

A123. Blank, M., King, R.G., Soo, L., Abbott, R.E. and Chien, S. The viscoelastic properties of monolayers of red cell membrane proteins. *J. Colloid Interface Sci.* **69**: 67–73, 1979.

A124. Chen, R.Y.Z., Lee, M.M.L. and Chien, S. Local anesthetics and the rheologic behavior of red cell suspensions. *Anesthesiology* **51**: 245–250, 1979.

A125. Kaperonis, A.A., Bertles, J.F. and Chien, S. Variability of intracellular pH within individual populations of SS and AA erythrocytes. *Br. J. Haematol.* **43**: 391–400, 1979.

A126. Meltzer, J.I., Keim, H.J., Laragh, J.H., Sealey, J.E., Jan, K.-M. and Chien, S. Nephrotic syndrome: vasoconstriction and hypervolemic types indicated by renin-sodium profiling. *Ann. Int. Med.* **91**: 688–696, 1979.

A127. Letcher, R.L., Chien, S. and Laragh, J.H. Changes in blood viscosity accompanying the response to Prazosin in patients with essential hypertension. *J. Cardiovasc. Pharmacol.* **1** (Suppl.): S8–S20, 1979.

A128. Kim, S., Fan, F.-C., Chen, R.Y.Z., Simchon, S., Schuessler, G.B. and Chien, S. Effects of changes in systemic hemodynamic parameters on pulpal hemodynamics. *J. Endod.* **6**: 394–399, 1980.

A129. Schmid-Schöenbein, G.W., Usami, S., Skalak, R. and Chien, S. Cell distribution in capillary networks. *Microvasc. Res.* **19**: 18–44, 1980.

A130. Schmid-Schöenbein, G.W., Usami, S., Skalak, R. and Chien, S. The interaction of leukocytes and erythrocytes in capillary and postcapillary vessels. *Microvasc. Res.* **19**: 45–70, 1980.

A131. Fan, F.-C., Schuessler, G.B., Chen, R.Y.Z. and Chien, S. Effect of hematocrit alteration on the regional hemodynamics and oxygen transport. *Am. J. Physiol.* **238**: H545–H552, 1980.

A132. Lipowsky, H.H., Usami, S. and Chien, S. *In vivo* measurements of "apparent viscosity" and microvessel hematocrit in the mesentery of the cat. *Microvasc. Res.* **19**: 297–319, 1980.

A133. Lipowsky, H.H., Usami, S. and Chien, S. Hematocrit determination in small bore tubes from optical density measurements under white light illumination. *Microvasc. Res.* **20**: 51–70, 1980.

A134. Jan, K.-M., Heldman, J. and Chien, S. Coronary hemodynamics and oxygen utilization after hematocrit variations in hemorrhage. *Am. J. Physiol.* **239**: H326–H332, 1980.

A135. Chen, R.Y.Z., Fan, F.-C., Kim, S., Jan, K.-M., Usami, S. and Chien, S. Tissue-blood partition coefficient for xenon: temperature and hematocrit dependence. *J. Appl. Physiol.* **49**: 178–183, 1980.

A136. Chen, R.Y.Z., Wicks, A.E. and Chien, S. Hemoconcentration induced by surface hypothermia in infants. *J. Thorac. Cardiovasc. Surg.* **80**: 236–241, 1980.

A137. Fan, F.-C., Kim, S., Simchon, S., Chen, R.Y.Z., Schuessler, G.B. and Chien, S. Effects of sodium nitroprusside on systemic and regional hemodynamics and oxygen utilization in the dog. *Anesthesiology* **53**: 113–120, 1980.

A138. Schmid-Schöenbein, G.W., Shih, Y. and Chien, S. Morphometry of human leukocytes. *Blood* **56**: 866–875, 1980.

A139. Handley, D.A., Arbeeny, C.M., Witte, L.D. and Chien, S. Colloidal gold-low density lipoprotein conjugates as membrane receptor probes. *Proc. Nat. Acad. Sci. U.S.A.* **78**: 92–95, 1980.

A140. Zawicki, D.F., Jain, R.K., Schmid-Schöenbein, G.W. and Chien, S. Dynamics of neovascularization in normal tissue. *Microvasc. Res.* **21**: 27–47, 1981.

A141. Seplowitz, A.H., Chien, S. and Smith, F.R. Effects of lipoproteins on plasma viscosity. *Atherosclerosis* **38**: 89–95, 1981.

A142. Chien, S., Sung, L.A. and Skalak, R. Energy balance in red cell aggregation. *Bibl. Anat.* **20**: 207–210, 1981.

A143. Handley, D.A. and Chien, S. Oxidation of ruthenium red for use as an intercellular tracer. *Histochemistry* **71**: 249–258, 1981.

A144. Rakow, A., Simchon, S., Sung, L.A. and Chien, S. Aggregation of red cells with membrane altered by heat treatment. *Biorheology* **18**: 3–8, 1981.

A145. Skalak, R. and Chien, S. Capillary flow: history, experiments and theory. *Biorheology* **18**: 307–330, 1981.

A146. Chien, S. and Weinbaum, S. Vesicle transport in arterial endothelium and the influence of mechanical factors on macromolecular permeability. *J. Biomech. Eng.* **103**: 186–196, 1981.

A147. Chien, S. Determinants of blood viscosity and red cell deformability. *Scand. J. Lab. Clin. Med.* **41** (Suppl. 156): 7–12, 1981.

A148. Skalak, R., Zarda, P.R., Jan, K.M. and Chien, S. Mechanics of rouleau formation. *Biophys. J.* **35**: 771–781, 1981.

A149. Schmid-Schöenbein, G.W., Sung, K.-L.P., Tözeren, H., Skalak, R. and Chien, S. Passive mechanical properties of human leukocytes. *Biophys. J.* **36**: 243–256, 1981.

A150. Fan, F.C., Chen, R.Y.Z., Chien, S. and Correll, J.W. Bypass blood flow during carotid endarterectomy. *Anesthesiology* **55**: 305–310, 1981.

A151. Chien, S., Lee, M.M.L., Laufer, L.S., Handley, D.A., Weinbaum, S., Caro, C.G. and Usami, S. Effects of oscillatory mechanical disturbance on macromolecular uptake by arterial wall. *Arteriosclerosis* **1**: 326–336, 1981.

A152. Letcher, R.L., Chien, S., Pickering, T.G., Sealey, J.E. and Laragh, J.H. Direct relationship between blood pressure and blood viscosity in normal and hypertensive subjects: role of fibrinogen and concentration. *Am. J. Med.* **70**: 1195–1202, 1981.

A153. Handley, D.A., Arbeeny, C.M., Eder, H.A. and Chien, S. Hepatic binding and internalization of low density lipoprotein-gold conjugates in rats treated with 17a-ethinylestradiol. *J. Cell Biol.* **90**: 778–787, 1981.

A154. Letcher, R.L., Pickering, T.G., Chien, S. and Laragh, J.H. Effects of exercise on plasma viscosity in athletes and sedentary normal subjects. *Clin. Cardiol.* **4**: 172–179, 1981.

A155. Handley, D.A., Alexander, J.T. and Chien, S. The design and use of a simple device for rapid quench-freezing of biological samples. *J. Microsc.* **121**: 273–282, 1981.

A156. Chien, S. The Fahraeus Lecture: hemorheology in disease, pathophysiological significance and therapeutic implications. *Clin. Hemorheol.* **1**: 419–442, 1981.

A157. Chien, S. Hemorheology in clinical medicine. *Clin. Hemorheol.* **2**: 137–142, 1982.

A158. Simchon, S. and Chien, S. A new method for kidney perfusion *in situ*: application to dynamics of autoregulation. *Am. J. Physiol.* **242**: F86–F94, 1982.

A159. Chen, R.Y.Z., Matteo, R.S., Fan, F.C., Schuessler, G.B. and Chien, S. Resetting of baroreflex sensitivity after induced hypotension. *Anesthesiology* **56**: 29–35, 1982.

A160. Chien, S. Rheology in the microcirculation in normal and low flow states. *Adv. Shock Res.* **8**: 71–90, 1982.

A161. Skalak, R. and Chien, S. Rheology of blood cells as soft tissues. *Biorheology* **19**: 453–461, 1982.

A162. Jan, K.M., Usami, S. and Chien, S. The disaggregation effect of dextran 40 on red cell aggregation in macromolecular suspensions. *Biorheology* **19**: 543–554, 1982.

A163. Chien, S., King, R.G., Kaperonis, A.A. and Usami, S. Viscoelastic properties of sickle cells and hemoglobin. *Blood Cells* **8**: 53–64, 1982.

A164. Lipowsky, H.H., Usami, S. and Chien, S. Human SS red cell rheological behavior in the microcirculation of cremaster muscle. *Blood Cells* **8**: 113–126, 1982.

A165. Chien, S., Laufer, L. and Handley, D.A. Vesicle distribution in the arterial endothelium determined with ruthenium as an extracellular marker. *J. Ultrastruct. Res.* **79**: 198–206, 1982.

A166. Chen, R.Y.Z., Fan, F.-C., Schuessler, G.B. and Chien, S. Baroreflex control of heart rate in humans during nitroprusside-induced hypotension. *Am. J. Physiol.* **243**: R18–R24, 1982.

A167. Lipowsky, H.H., Usami, S., Chien, S. and Pittman, R.N. Hematocrit determination in small bore tubes by differential spectrophotometry. *Microvasc. Res.* **24**: 42–55, 1982.

A168. Sung, K.-L.P., Schmid-Schöenbein, G.W., Skalak, R. Schuessler, G.B., Usami, S. and Chien, S. Influence of physicochemical factors on rheology of human neutrophils. *Biophys. J.* **39**: 101–106, 1982.

A169. Tözeren, A., Skalak, R., Sung, K.-L.P. and Chien, S. Viscoelastic behavior of erythrocyte membrane. *Biophys. J.* **39**: 23–32, 1982.

A170. Chien, S., Schmalzer, E.A., Lee, M.M.L., Impelluso, T. and Skalak, R. Role of white blood cells in filtration of blood cell suspensions. *Biorheology* **20**: 11–27, 1982.

A171. Schmalzer, E.A., Skalak, R., Usami, S., Vayo, M. and Chien, S. Influence of red cell concentration on filtration of blood cell suspensions. *Biorheology* **20**: 29–40, 1982.

A172. Skalak, R., Impelluso, T., Schmalzer, E.A. and Chien, S. Theoretical modeling of filtration of blood cell suspensions. *Biorheology* **20**: 41–56, 1982.

A173. Arminski, L., Weinbaum, S. and Chien, S. Theory for vesicle-vesicle and vesicle-plasmalemma interactions. *J. Colloid Interface Sci.* **90**: 390–408, 1982.

A174. Chien, S. and Lipowsky, H.H. Correlation of hemodynamics in macrocirculation and microcirculation. *Int. J. Microcirc. Clin. Exp.* **1**: 351–365, 1983.

A175. Kim, S., Schuessler, G. and Chien, S. Measurement of blood flow in the dental pulp of dogs with the ^{133}Xenon washout method. *Arch. Oral Biol.* **28**: 501–505, 1983.

A176. Handley, D.A., Arbeeny, C.M., Witte, L.D., Goodman, D.S. and Chien, S. Ultrastructural visualization of low-density lipoproteins during receptor binding and cellular endocytosis. *J. Ultrastruct. Res.* **83**: 43–47, 1983.

A177. Schmalzer, E., Chien, S. and Brown, A.K. Transfusion therapy in sickle cell disease. *Am. J. Ped. Hematol./Oncol.* **4**: 395–406, 1983.

A178. Secomb, T.W., Chien, S., Jan, K.M. and Skalak, S. The bulk rheology of close-packed red blood cells in shear flow. *Biorheology* **20**: 295–309, 1983.

A179. Skalak, R., Hanss, M. and Chien, S. Indices of filterability of red blood cell suspensions. *Biorheology* **20**: 311–316, 1983.

A180. Copley, A.L., King, R.G. and Chien, S. On the antithrombogenic action of low molecular weight heparins and of chondroitins A, B and C. *Biorheology* **20**: 697–704, 1983.

A181. Chen, R.Y.Z., Fan, F.-C., Schuessler, G.B., Usami, S. and Chien, S. Effects of sphere size and injection site on regional cerebral blood flow measurements. *Stroke* **14**: 769–776, 1983.

A182. Simchon, S. and Chien, S. Effect of variations in renal hemodynamics on the time course of renin secretion rate. *Am. J. Physiol.* **245**: F784–F791, 1983.

A183. Letcher, R.L., Chien, S., Pickering, T.G. and Laragh, J.H. Elevated blood viscosity in patients with borderline essential hypertension. *Hypertension* **5**: 757–762, 1983.

A184. Handley, D.A., Arbeeny, C.M. and Chien, S. Sinusoidal endothelial endocytosis of low density lipoprotein-gold conjugates in perfused livers of ethinyl-estradiol treated rats. *Eur. J. Cell Biol.* **30**: 266–271, 1983.

A185. Chien, S. Aggregation of red blood cells: energy balance at the interface. *Ann. N.Y. Acad. Sci.* **404**: 103–104, 1983.

A186. Handley, D.A., Chien, S. and Arbeeny, C.M. Ultrastructure of hepatic cholesterol crystals in the hypercholesterolemic-diabetic rat. *Path. Res. Pract.* **177**: 13–21, 1983.

A187. Handley, D.A. and Chien, S. Colloidal gold: a pluripotent receptor probe. *Proc. Soc. Exp. Biol. Med.* **174**: 1–11, 1983.

A188. Chabanel, A., Flamm, M., Sung, K.-L.P., Lee, M.M.L., Schachter, D. and Chien, S. Influence of cholesterol content on red cell membrane viscoelasticity and fluidity. *Biophys. J.* **44**: 171–176, 1983.

A189. Chien, S. Hemorheology: its clinical implications. *Vasc. Med.* **1**: 123–143, 1983.

A190. Chien, S. Hemorheology: normal physiology and pathophysiological changes in blood viscosity and blood flow. *Symp. Front. Pharmacol.* **3**: 15–33, 1983.

A191. Kim, S., Lipowsky, H.H., Usami, S. and Chien, S. Arteriovenous distribution of hemodynamic parameters in the rat dental pulp. *Microvasc. Res.* **27**: 28–38, 1984.

A192. Chien, S., Sung, L.A., Simchon, S., Lee, M.M.L. and Skalak, R. Energy balance in red cell interactions. *Ann. N.Y. Acad. Sci.* **416**: 138–148, 1984.

A193. Skalak, R. and Chien, S. Theoretical models of rouleau formation and disaggregation. *Ann. N.Y. Acad. Sci.* **416**: 190–206, 1984.

A194. Tözeren, A., Skalak, R., Fedorciw, B., Sung, K.-L.P. and Chien, S. Constitutive equations of erythrocyte membrane incorporating evolving preferred configuration. *Biophys. J.* **45**: 541–549, 1984.

A195. Chen, R.Y.Z., Fan, F.-C., Schuessler, G.B., Simchon, S., Kim, S. and Chien, S. Regional blood flow and oxygen consumption of the canine brain during hemorrhagic hypotension. *Stroke* **15**: 343–350, 1984.

A196. Tözeren, H., Chien, S. and Tözeren, A. Estimation of viscous dissipation inside an erythrocyte during aspirational entry into a micropipette. *Biophys. J.* **45**: 1179–1184, 1984.

A197. Baldwin, A.L. and Chien, S. Endothelial transport of anionized and cationized ferritin in the rabbit thoracic aorta and vasa vasorum. *Arterioclerosis* **4**: 372–382, 1984.

A198. Chien, S., Fan, F.C., Lee, M.M.L. and Handley, D.A. Effects of arterial pressure on endothelial transport of macromolecules. *Biorheology* **21**: 631–641, 1984.

A199. Schmalzer, E.A. and Chien, S. Filterability of subpopulations of leukocytes: effects of pentoxifylline. *Blood* **2**: 542–546, 1984.

A200. Kim, S., Chen, R.Y.Z., Wasserman, H., Usami, S. and Chien, S. Determination of partition coefficient of ^{133}Xenon between oral tissues and blood. *Arch. Oral Biol.* **29**: 721–727, 1984.

A201. Braide, M., Amundson, B., Chien, S. and Bagge, U. Quantitative studies on the influence of leukocytes on the vascular resistance in a skeletal muscle preparation. *Microvasc. Res.* **27**: 331–352, 1984.

A202. Firrell, J.C., Lipowsky, H.H., Usami, S. and Chien, S. Segmental and modular resistances of individual microvascular networks during hemorrhagic hypotension. *Am. J. Physiol.* **247**: H361–H370, 1984.

A203. Chien, S. and Sung, K.-L.P. Effect of colchicine on viscoelastic properties of neutrophils. *Biophys. J.* **46**: 383–386, 1984.

A204. Reinhart, W.H., Usami, S., Schmalzer, E.A., Lee, M.M.L. and Chien, S. Evaluation of red blood cell filterability test: influences of pore size, hematocrit and flow rate. *J. Lab. Clin. Med.* **104**: 501–516, 1984.

A205. King, R.G., Chien, S., Usami, S. and Copley, A.L. Biorheological methods employing the Weissenberg Rheogoniometer. *Biorheology* **I** (Suppl.): 23–24, 1984.

A206. Usami, S., Reinhart, W., Danoff, S. and Chien, S. A new capillary viscometer for measurements on small blood samples over a wide shear rate range. *Clin. Hemorheol.* **4**: 295–297, 1984.

A207. Chien, S. Rheology of sickle cells and the microcirculation. *New Engl. J. Med.* **311**: 1567–1569, 1984.

A208. Kim, S., Edwall, L., Trowbridge, H. and Chien, S. Effects of local anesthetics on pulpal blood flow in dogs. *J. Dent. Res.* **63**: 650–652, 1984.

A209. Devereaux, R., Drayer, J., Chien, S., Pickering, T., Letcher, R., DeYoung, J., Sealey, J. and Laragh, J. Whole blood viscosity as a determinant of cardiac hypertrophy in systemic hypertension. *Am. J. Cardiol.* **54**: 592–595, 1984.

A210. Reinhart, W.H., Danoff, S.J., King, R.G. and Chien, S. Rheology of fetal and maternal blood. *Pediat. Res.* **19**: 147–153, 1985.

A211. Schmid-Schöenbein, G.W., Jan, K.-M., Skalak, R. and Chien, S. Deformation of leukocytes on a hematological blood film. *Biorheology* **21**: 767–781, 1984.

A212. Solomon, R.A., Antunes, J.L., Chen, R.Y.Z., Bland, L. and Chien, S. Decrease in cerebral blood flow in rats after experimental subarachnoid hemorrhage: a new animal model. *Stroke* **16**: 58–64, 1985.

A213. Chien, S. Role of blood cells in microcirculatory regulation. *Microvasc. Res.* **29**: 129–151, 1985.

A214. Baldwin, A.L. and Chien, S. Regulation of aortic endothelial vesicular uptake of cationized ferritin by plasmalemma binding. *Atherosclerosis* **55**: 233–245, 1985.

A215. Simchon, S., Fan, F.C., Chen, R.Y.Z., Kim, S. and Chien, S. Effects of experimental hypotension on hemodynamics and renin secretion rate. *Circ. Shock* **15**: 27–36, 1985.

A216. Chien, S., Tvetenstrand, C.D., Farrell-Epstein, M.A. and Schmid-Schöenbein. Model studies on distributions of white and red blood cells at microvascular bifurcations. *Am. J. Physiol.* **248**: H568–H576, 1985.

A217. Schmukler, R. and Chien, S. Rapid deoxygenation of red cells and hemoglobin solution using hollow capillary fibers. *Biorheology* **22**: 21–29, 1985.

A218. Reinhart, W.H. and Chien, S. Roles of cell geometry and cytoplasmic viscosity in red cell passage through narrow pores. *Am. J. Physiol.* **248**: C473–C479, 1985.

A219. Reinhart, W.H. and Chien, S. Stomatocytic transformation of red blood cells after marathon running. *Am. J. Hematol.* **19**: 201–204, 1985.

A220. Chien, S. Hemodynamics of the Dental Pulp. *J. Dent. Res.* **64**: 602–606, 1985.

A221. Appel, F.B., Blum, C.B., Chien, S., Kunis, C.L. and Appel, A.S. The hyperlipidemia of the nephrotic syndrome: relation to plasma albumin concentration, oncotic pressure and viscosity. *New Engl. J. Med.* **312**: 1544–1548, 1985.

A222. Liu, C.Y., Handley, D.A. and Chien, S. Gold-labeling of thrombin and ultrastructural studies of thrombin-gold conjugate binding by fibrin. *Anal. Chem.* **147**: 49–56, 1985.

A223. Chabanel, A., Abbott, R.E., Chien, S. and Schachter, D. Effects of benzyl alcohol on erythrocyte shape, membrane hemileaflet fluidity and membrane viscoelasticity. *Biochim. Biophys. Acta* **816**: 142–152, 1985.

A224. Sung, L.A., Kabat, E.A. and Chien, S. Interaction of lectins with membrane receptors on red cell surfaces. *J. Cell Biol.* **101**: 646–651, 1985.

A225. Sung, L.A., Kabat, E.A. and Chien, S. Interaction energies in lectin-induced red cell aggregation. *J. Cell Biol.* **101**: 652–659, 1985.

A226. Vayo, M.M., Lipowsky, H.H., Karp, N., Schmalzer, E.A. and Chien, S. A model of oxygen transport in sickle cell disease. *Microvasc. Res.* **30**: 195–206, 1985.

A227. Weinbaum, S., Tzeghai, G., Ganatos, P., Pfeffer, R. and Chien, S. Effect of cell turnover and leaky junctions on arterial macromolecular transport. *Am. J. Physiol.* **248**: H945–H960, 1985.

A228. Tözeren, A. and Chien, S. Modeling of time-variant coupling between left ventricle and aorta in cardiac cycle. *Am. J. Physiol.* **249**: H560–H569, 1985.

A229. Reinhart, W.H., Danoff, S.J., Usami, S. and Chien, S. Rheological measurements on small samples with a new capillary viscometer. *J. Lab. Clin. Med.* **104**: 921–931, 1985.

A230. Sung, K.-L.P., Chabanel, A., Freedman, J. and Chien, S. Effect of complement on the viscoelastic properties of human erythrocyte membrane. *Br. J. Haematol.* **61**: 455–466, 1985.

A231. Baldwin, A.L. and Chien, S. Effects of plasma proteins on endothelial binding and vesicle loading of anionized ferritin in the rabbit aorta. *Arteriosclerosis* **5**: 451–458, 1985.

A232. Kaperonis, A.A., Handley, D.A. and Chien, S. Fibers, crystals and other forms of HbS polymers in deoxygenated sickle erythrocytes. *Am. J. Hematol.* **21**: 269–275, 1986.

A233. Simchon, S., Chen, R.Y.Z., Carlin, R.D., Fan, F.-C., Jan, K.-M., and Chien, S. Effects of blood viscosity on plasma renin activity and renal hemodynamics. *Am. J. Physiol.* **250**: F40–F46, 1986.

A234. Handley, D.A. and Chien, S. Ultrastructural studies of endothelial and platelet receptor binding of thrombin-colloidal gold probes. *Eur. J. Cell Biol.* **39**: 291–398, 1985.

A235. Reinhart, W.H. and Chien, S. Red cell rheology in stomatocyte-echinocyte transformation: roles of cell geometry and cell shape. *Blood* **67**: 1110–1118, 1986.

A236. Chabanel, A., Spiro, A., Schachter, D. and Chien, S. Some biophysical properties of the erythrocyte membrane in Duchenne Muscular Dystrophy. *J. Neurol. Sci.* **76**: 131–142, 1986.

A237. Reinhart, W., Sung, L.A. and Chien, S. Quantitative relationship between Heinz body formation and red blood cell deformability. *Blood* **68**: 1376–1383, 1986.

A238. Sung, K.-L.P., Sung, L.A., Crimmins, M., Burakoff, S.J. and Chien, S. Determination of junction avidity of cytotoxic T cell and target cell. *Science* **234**: 1405–1408, 1986.

A239. Reinhart, W., Sung, L.A., Schuessler, G. and Chien, S. Membrane protein phosphorylation during stomatocyte-echinocyte transformation of human erythrocytes. *Biochim. Biophys. Acta* **862**: 1–7, 1986.

A240. Chien, S. Blood rheology in myocardial infarction and hypertension. *Biorheology* **23**: 633–653, 1986.

A241. Chabanel, A., Reinhart, W. and Chien, S. Increased resistance to membrane deformation of shape-transformed human red blood cells. *Blood* **69**: 739–743, 1987.

A242. Chien, S. Red cell deformability and its relevance to blood flow. *Ann. Rev. Physiol.* **49**: 177–192, 1987.

A243. Skalak, R., Soslowsky, L., Schmalzer, E., Impelluso, T. and Chien, S. Theory of filtration of mixed blood suspensions. *Biorheology* **24**: 35–52, 1987.

A244. Schmalzer, E.A., Lee, J.O., Brown, A.K., Usami, S. and Chien, S. Viscosity of mixtures of sickle and normal red cells at varying hematocrit levels. *Transfusion* **27**: 228–233, 1987.

A245. Simchon, S., Jan, K.M. and Chien, S. Influence of reduced red cell deformability on regional blood flow. *Am. J. Physiol.* **253**: H898–H903, 1987.

A246. Chien, S. and Gargus, J.J. Molecular biology in physiology. *FASEB J.* **1**: 97–102, 1987.

A247. Chien, S. and Sung, L.A. Physicochemical basis and clinical implications of red cell aggregation. *Clin. Hemorheol.* **7**: 71–92, 1987.

A248. Chien, S. Effects of inflammatory agents on leukocyte rheology and microcirculation. *Prog. Appl. Microcirc.* **12**: 67–78, 1987.

A249. Reinhart, W. and Chien, S. The time course of filtration test as a model for microvascular plugging by white cells and hardened red cells. *Microvasc. Res.* **34**: 1–2, 1987.

A250. Chien, S., Kaperonis, A., King, R.G., Lipowsky, H.H., Schmalzer, E.A., Sung, L.A., Sung, K.-L.P. and Usami, S. Rheology of sickle cells and its role in microcirculatory dynamics. *Prog. Clin. Biol. Res.* **240**: 151–165, 1987.

A251. Chabanel, A., Schachter, D. and Chien, S. Increased rigidity of RBC membrane in young spontaneously hypertensive rats. *Hypertension* **10**: 603–607, 1987.

A252. Reinhart, W.H., Chabanel, A., Vayo, M. and Chien, S. Evaluation of a filter aspiration technique to determine membrane deformability. *J. Lab. Clin. Med.* **110**: 483–494, 1987.

A253. Chien, S., Sung, K.-L.P., Schmid-Schöenbein, G.W., Skalak, R., Schmalzer, E.A. and Usami, S. Rheology of leukocytes. *Ann. N.Y. Acad. Sci.* **516**: 333–361, 1987.

A254. Reinhart, W.H. and Chien, S. Echinocyte-stomatocyte transformation and shape control of human red blood cells: morphological aspects. *Am. J. Hematol.* **24**: 1–14, 1987.

A255. Devereux, R.B., Pickering, T.G., Alderman, M.G., Chien, S., Borer, J.S. and Laragh, J.H. Left ventricular hypertrophy in hypertension. Prevalence and relationship to pathophysiologic variable. *Hypertension* **9**: 53–60, 1987.

A256. Handley, D.H. and Chien, S. Colloidal gold labeling studies related to vascular and endothelial function, hemostasis and receptor-mediated processing of plasma macromolecules. *Eur. J. Cell Biol.* **43**: 163–174, 1987.

A257. Solomon, R.A., Lovitz, R.L., Hegemann, M.T., Schuessler, G.B., Young, W.L. and Chien, S. Regional cerebral metabolic activity in the rat following experimental subarachnoid hemorrhage. *J. Cereb. Blood Flow Metab.* **7**: 193–198, 1987.

A258. Young, W.L., Josovitz, K., Morales, O. and Chien, S. The effect of nimodipine on post-ischemic cerebral glucose utilization and blood flow in the rat. *Anesthesiology* **67**: 54–59, 1987.

A259. Lemons, D.E., Chien, S., Crawshaw, L.I., Weinbaum, S. and Jiji, L.M. Significance of vessel size and type in vascular heat transfer. *Am. J. Physiol.* **253**: R128–R135, 1987.

A260. Reinhart, W.H. and Chien, S. Red cell vacuoles: their size and distribution under normal conditions and after splenectomy. *Am. J. Hematol.* **27**: 265–271, 1987.

A261. Starc, T.J., Schmalzer, E.A. and Chien, S. Effect of beta adrenergic agents on granulocyte deformability. *Clin. Hemorheol.* **7**: 699–706, 1987.

A262. Kaperonis, A.A., Michelson, C.B., Askanazi, J., Kinney, J.M. and Chien, S. Effects of total hip replacement and bed rest on blood rheology and red cell metabolism. *J. Trauma* **28**: 453–457, 1988.

A263. Seldin, D.W., Simchon, S., Jan, K.M., Chien, S. and Alderson, P.O. Dependence of technetium-99m red blood cell labeleing efficiency on red cell surface charge. *J. Nucl. Med.* **29**: 1710–1713, 1988.

A264. Lin, S.J., Jan, K.M., Schuessler, G.B., Weinbaum, S. and Chien, S. Enhanced macromolecular permeability of aortic endothelial cells in association with mitosis. *Atherosclerosis* **73**: 223–232, 1988.

A265. Reinhart, W.H., Sung, L.A., Sung, K.-L.P., Bernstein, S.E. and Chien, S. Impaired echinocytic transformation of ankyrin- and spectrin-deficient erythrocytes in mice. *Am. J. Hematol.* **29**: 195–200, 1988.

A266. Kuo, C.D., Bai, J.J., Chang, I.T., Wang, J.H. and Chien, S. Continuous monitoring of erythrocyte sedimentation process: a new possible mechanism of erythrocyte sedimentation. *J. Biomech. Eng.* **110**: 392–395, 1988.

A267. King, R.G., Procyk, R., Chien, S., Blomback, B. and Copley, A.L. The effect of factor XIII and fibronectin on the viscoelasticity of fibrinogen surface layers. *Thromb. Res.* **49**: 139–144, 1988.

A268. Lee, M.M.L., Schuessler, G. and Chien S. Time-dependent effects of endotoxin on the ultrastructure of aortic endothelium. *Artery* **15**: 71–89, 1988.

A269. Chien, S. Molecular biology in biorheology. *Biorheology* **24**: 659–573, 1988.

A270. Dong, C., Skalak, R., Sung, K.-L.P., Schmid-Schönbein, G.W. and Chien, S. Passive deformation analysis of human leukocytes. *J. Biomech. Eng.* **110**: 27–36, 1988.

A271. Baldwin, A.L. and Chien, S. Effects of dextran 40 on endothelial binding and vesicle loading of ferritin in rabbit aorta. *Arteriosclerosis* **8**: 140–146, 1988.

A272. Simchon, S., Jan, K.M. and Chien, S. Studies of sequestration of neuraminidase-treated red blood cells. *Am. J. Physiol.* **254**: H1167–H1171, 1988.

A273. Sung, K.-L.P., Dong, C., Schmid-Schönbein, G.W., Chien, S. and Skalak, R. Leukocyte relaxation properties. *Biophys. J.* **54**: 331–336, 1988.

A274. Chien, S., Lin, S.J., Weinbaum, S., Lee, M.M. and Jan, K.M. The role of arterial endothelial cell mitosis in macromolecular permeability. *Adv. Exp. Med. Biol.* **242**: 99–109, 1988.

A275. Sung, K.-L.P., Sung, L.A., Crimmins, M., Burakoff, S.J. and Chien, S. Dynamic changes in viscoelastic properties in cytotoxic T-lymphocyte-mediated killing. *J. Cell Sci.* **91**: 179–189, 1988.

A276. Weinbaum, S., Ganatos, P., Pfeffer, R., Wen, G.B., Lee, M. and Chien, S. On the time-dependent diffusion of macromolecules through transient open junctions and their subendothelial spread. I. Short-time model for cleft exit region. *J. Theor. Biol.* **135**: 1–30, 1988.

A277. Wen, G.B., Weinbaum, S., Ganatos, P., Pfeffer, R. and Chien, S. On the time-dependent diffusion of macromolecules through transient open junctions and their subendothelial spread. 2 Long-time model for interaction between leaky sites. *J. Theor. Biol.* **135**: 219–253, 1988.

A278. Chen, R.Y.Z., Carlin, R.D., Simchon, S., Jan, K.M. and Chien, S. Effects of dextran-induced hyperviscosity on regional blood flow and hemodynamics in dogs. *Am. J. Physiol.* **256**: H898–H905, 1989.

A279. Tözeren, A., Sung, K.-L.P. and Chien, S. Analysis of the strength of adhesion between a pair of cytotoxic T-cell and target cell. *Biophys. J.* **55**: 479–493, 1989.

A280. Young, W.L. and Chien, S. Effect of nimodipine on cerebral blood flow and metabolism in rats during hyperventilation. *Stroke* **20**: 275–280, 1989.

A281. Chabanel, A., Sung, K.-L.P., Rapiejko, J., Prchal, J.T., Palek, J., Liu, S.C. and Chien, S. Viscoelastic properties of red cell membrane in hereditary elliptocytosis. *Blood* **73**: 592–595, 1989.

A282. Lin, S.J., Jan, K.M., Weinbaum, S. and Chien, S. Transendothelial transport of low density lipoprotein in association with cell mitosis in rat aorta. *Arteriosclerosis* **9**: 230–236, 1989.

A283. Kaperonis, A.A. and Chien, S. Effects of centrifugation on transmembrane water loss from normal and pathologic erythrocytes. *Proc. Soc. Exp. Biol. Med.* **190**: 174–178, 1989.

A284. Schmid-Schönbein, G.W. and Chien, S. Morphometry of human leukocyte granules. *Biorheology* **26**: 331–343, 1989.

A285. Gallik, S., Usami, S., Jan, K.M. and Chien, S. Shear stress-induced detachment of human polymorphonuclear leukocyte from endothelial cell monolayers. *Biorheology* **26**: 823–834, 1989.

A286. Schmalzer, E.A. and Chien, S. Effects of methylxanthines, cytochalasin B and FMLP on neutrophil filtrability in a constant flow system. *Clin. Hemorheol.* **9**: 69–80, 1989.

A287. Lipowsky, H.H. and Chien, S. Role of leukocyte-endothelium adhesion in affecting recovery from ischemic episodes. *Ann. N.Y. Acad. Sci.* **565**: 308–315, 1989.

A288. Kaperonis, A.A., King, R.G., Smith, J.A. and Chien, S. Electron microscopic quantitation of hemoglobin S polymer in SS red blood cells and rheological correlation. *Ann. N.Y. Acad. Sci.* **565**: 463–465, 1989.

A289. Simchon, S., Manger, W.M., Carlin, R.D., Peeters, L.L., Rodriguiz, J., Batista, D., Brown, T., Merchant, N.B., Jan, K.M. and Chien, S. Salt-induced hypertension in Dahl salt-sensitive rats. Hemodynamis and renal responses. *Hypertension* **13**: 612–621, 1989.

A290. Schmalzer, E.A., Manning, R.S. and Chien, S. Filtration of sickle cells: recruitment into a rigid fraction as a function of density and oxygen tension. *J. Lab. Clin. Med.* **113**: 727–734, 1989.

A291. Lai, C.S., Yang, C.M., Tsai, Y.T., Chen, H.I., Simchon, S. and Chien, S. An animal study of portal hypertension. *Chin. Med. J.* **44**: 19–24, 1989.

A292. Kuo, C.D., Bai, J.J. and Chien, S. Erythrocyte sedimentation realizable in terms of population dynamics. *Biorheology* **26**: 1003–1010, 1989.

A293. Feng, S.S., Skalak, R. and Chien, S. Velocity distribution on the membrane of a tank-treading red blood cell. *Bull. Math. Biol.* **51**: 449–465, 1989.

A294. Lipowsky, H.H. and Chien, S. Role of leukocyte-endothelium adhesion in affecting recovery from ischemic episodes. *Ann. N.Y. Acad. Sci.* **565**: 308–315, 1989.

A295. Sung, L.A., Chien, S., Chang, L.-S., Lambert, K., Bliss, S.A., Bouhassira, E., Nagel, R.L., Schwartz, R.S. and Rybicki, A.C. Molecular cloning of human protein 4.2: A major component of the red cell membrane. *Proc. Nat. Acad. Sci. U.S.A.* **87**: 955–959, 1990.

A296. DeSimone, G., Devereux, R.B., Roman, M.J., Ganau, A., Chien, S., Alderman, M.H., Atlas, S.A. and Laragh, J.H. Increasing age accentuates gender differences in volume regulatory hormonal profile, blood viscosity and left ventricular anatomy in normal adults. *Circulation* **81**: 107–117, 1990.

A297. Chuang, P.T., Cheng, H.-J., Lin, S.-J., Jan, K.-M., Lee, M.M.L. and Chien, S. Macromolecular transport across arterial and venous endothelium in rats: studies with Evans blue-albumin and horseradish peroxidase. *Arteriosclerosis* **10**: 188–197, 1990.

A298. Lin, S.J., Jan, K.M. and Chien, S. The role of dying endothelial cells in transendothelial macromolecular transport. *Arterioclerosis* **10**: 703–709, 1990.

A299. Chien, S., Feng, S.S., Vayo, M., Sung, L.A., Usami, S. and Skalak, R. The dynamics of shear disaggregation of red blood cells in a flow channel. *Biorheology* **27**: 135–147, 1990.

A300. Simchon, S., Carlin, R.D., Jan, K.M. and Chien, S. A double isotope technique to determine regional albumin permeability: effects of anesthesia. *Proc. Soc. Exp. Biol. Med.* **195**: 114–118, 1990.

A301. Wu, C.H., Chi, J.C., Jerng, J.S., Lin, S.J., Jan, K.M., Wang, D.L. and Chien, S. Transendothelial macromolecular transport in the aorta of spontaneously hypertensive rats. *Hypertension* **16**: 154–161, 1990.

A302. Chien, S. and Sung, L.A. Molecular basis of red cell membrane rheology. *Biorheology* **27**: 589–597, 1990.

A303. Chi, L.M., Wu, W.G., Sung, K.-L.P. and Chien, S. Biophysical correlates of lysophosphatidylcholine- and ethanol-mediated shape transformation and hemolysis of human erythrocytes. Membrane viscoelasticity and NMR measurement. *Biochim. Biophys. Acta* **1027**: 163–171, 1990.

A304. Lin, S.J., Jan, K.M. and Chien, S. Temporal and spatial changes in macromolecular uptake in rat thoracic aorta and relation to thymidine uptake. *Atherosclerosis* **85**: 229–238, 1990.

A305. Young, T.H., Cuttner, J., Meyer, R. and Chien, S. Leukapheresis reduced blood viscosity in chronic myelogenous leukemia patients with hyperleukocytosis. *Clin. Hemorheol.* **10**: 191–203, 1990.

A306. Yuan, F., Weinbaum, S., Pfeffer, R. and Chien, S. A mathematical model for the receptor mediated cellular regulation of the low density lipoprotein metabolism. *J. Biomech. Eng.* **113**: 1–10, 1991.

A307. Yuan, F., Chien, S. and Weinbaum, S. A new view of convective-diffusive transport processes in the arterial intima. *J. Biomech. Eng.* **113**: 314–329, 1991.

A308. Tang, T.K., Tam, K.-B. and Chien, S. Two RFLPs in the human protein 4.1 gene (EL1). *Nucleic Acids Res.* **19**: 6057, 1991.

A309. DeSimone, G., Devereux, R.B., Roman, M.J., Ganau, A., Chien, S., Alderman, M.H., Atlas, S.A. and Laragh, J.H. Gender differences in left ventricular anatomy, blood viscosity and volume regulatory hormones in normal adults. *Am. J. Cardiol.* **68**: 1704–1708, 1991.

A310. Chien, K.R., Knowlton, K.U., Zhu, H. and Chien, S. Regulation of cardiac gene expression during myocardial growth and hypertrophy: molecular studies of an adaptive physiologic response. *FASEB J.* **5**: 3037–3046, 1991.

A311. Su, C.Y., Tsai, A.I., Liu, H.P., Chen, L.T. and Chien, S. Ultrastructural studies on splenic microcirculation of the rat. *Chin. J. Physiol.* **34**: 223–234, 1991.

A312. Reinhart, W.H., Huang, C., Vayo, M., Norwich, G., Chien, S. and Skalak, R. Folding of red blood cells in capillaries and narrow pores. *Biorheology* **28**: 537–549, 1991.

A313. Sung, L.A., Fowler, V.M., Lambert, K. and Chien, S. Molecular cloning and characterization of human erythrocyte tropomodulin: A tropomyosin-binding protein. *J. Biol. Chem.* **267**: 2616–2621, 1992.

A314. Tözeren, A., Sung, K.-L.P., Sung, L.A., Dustin, M.L., Chan, P.Y., Springer, T.A. and Chien, S. Micromanipulation of adhesion of a Jurkat cell to a planar membrane containing lymphocyte function-associated antigen 3 molecules. *J. Cell Biol.* **116**: 997–1006, 1992.

A315. Itoh, T., Chien, S. and Usami, S. Deformability measurements on individual sickle cells using a new system with pO2 and temperature control. *Blood* **79**: 2141–2147, 1992.

A316. Sung, L.A., Chien, S., Fan, Y.-S., Lin, C.C., Lambert, K., Zhu, L., Lam, J.S. and Chang, L.-S. Human erythrocyte protein 4.2: Isoform expression, differential splicing and chromosomal assignment. *Blood* **79**: 2763–2770, 1992.

A317. Wang, D.L., Chang, Y.N., Hsu, H.T., Usami, S. and Chien, S. Prostaglandin E1 and dibutyryl cyclic AMP enhance platelet resistance to deformation. *Thromb. Res.* **65**: 757–768, 1992.

A318. Chien, S. Blood cell deformability and interactions: from molecules to micromechanics and microcirculation. *Microvasc. Res.* **44**: 243–254, 1992.

A319. Sung, K.-L.P., Kuhlman, P., Maldonado, F., Lollo, B.A., Chien, S. and Brian, A.A. Force contribution of the LFA-1/ICAM-1 complex to T cell adhesion. *J. Cell Sci.* **103**: 259–266, 1992.

A320. Huang, A.I., Jan, K.M. and Chien, S. Role of intercellular junctions in the passage of horseradish peroxidase across aortic endothelium. *Lab. Invest.* **67**: 201–209, 1992.

A321. Usami, S., Wung, S.L., Skierczynski, B.A., Skalak, R. and Chien, S. Locomotion forces generated by a polymorphonuclear leukocyte. *Biophys. J.* **63**: 1663–1666, 1992.

A322. Chien, S., Sung, L.A., Lee, M.M.L. and Skalak, R. Red cell membrane elasticity as determined by flow channel technique. *Biorheology* **29**: 467–478, 1992.

A323. Chang, R.R.K., Chien, N.T.Y., Chen, C.-H., Jan, K.-M., Schmid-Schönbein, G.W. and Chien, S. Spontaneous activation of circulating granulocytes in patients with acute myocardial and cerebral diseases. *Biorheology* **29**: 549–561, 1992.

A324. Lin, S.J., Hong, C.Y., Chang, M.S., Chiang, B.N. and Chien, S. Long-term nicotine exposure increases aortic endothelial cell death and enhances transendothelial macromolecular transport in rats. *Arterioscler. Thromb.* **12**: 1305–1312, 1992.

A325. Sung, K.-L.P. and Chien, S. Influence of temperature on rheology of human erythrocytes. *Chin. J. Physiol.* **35**: 81–94, 1992.

A326. Sung, K.-L.P., Frojmovic, M.M., O'Toole, T.E., Zhu, C., Ginsberg, M.H. and Chien, S. Determination of adhesion force between single cell pairs generated by activated GpIIb-IIIa receptors. *Blood* **81**: 419–423, 1993.

A327. Usami, S., Chien, H.H., Zhao, Y., Chien, S. and Skalak, R. Design and construction of a linear shear stress flow chamber. *Ann. Biomed. Eng.* **21**: 77–83, 1993.

A328. Shyy, Y.J., Wickham, L.L., Hagan, J.P., Hsieh, H.J., Hu, Y.L., Telian, S.H., Valente, A.J., Sung, K.-L.P. and Chien, S. Human monocyte colony-stimulating factor stimulates the gene expression of monocyte chemotactic protein-1 and increases the adhesion of monocytes to endothelial monolayers. *J. Clin. Invest.* **92**: 1745–1751, 1993.

A329. Lin, S.J., Hong, C.Y., Chang, M.S., Chiang, B.N. and Chien, S. Increased aortic endothelial death and enhanced transendothelial macromolecular transport in streptozotocin-diabetic rats. *Diabetologia* **36**: 926–930, 1993.

A330. Skalak, R., Skierczynski, B.A., Wung, S.L., Chien, S. and Usami, S. Mechanical models of pseudopod formation. *Blood Cells* **19**: 389–397, 1993.

A331. Weinbaum, S. and Chien, S. Lipid transport aspects of atherogenesis. *J. Biomech. Eng.* **115**: 602–610, 1993.

A332. Skierczynski, B.A., Usami, S., Chien, S. and Skalak, R. Active motion of polymorphonuclear leukocytes in response to chemoattractant in a micropipette. *J. Biomech. Eng.* **115**: 503–509, 1993.

A333. Tang, T.K., Hong, T.-M., Lin, C., Lai, M.-L., Liu, C.H.L., Lo, H.-J., Wang, M.E., Chen, L.B., Chen, W.-T., Ip, W., Lin, D.C., Lin, J.J.C., Lin, S., Sun, T.-T., Wang, E., Wang, J.L., Wu, R., Wu, C.-W. and Chien, S. Nuclear proteins of the bovine esophageal epithelium: I. monoclonal antibody *w*2 specifically reacts with condensed nuclei or differentiated superficial cells. *J. Cell Sci.* **104**: 237–247, 1993.

A334. Kuo, C.D., Bai, J.J. and Chien, S. A fractal model for erythrocyte sedimentation. *Biorheology* **31**: 77–89, 1994.

A335. Shyy, Y.J., Hsieh, H.J., Usami, S. and Chien, S. Fluid shear stress induces a biphasic response of human monocyte chemotactic protein 1 gene expression in vascular endothelium. *Proc. Nat. Acad. Sci. U.S.A.* **91**: 4678–4682, 1994.

A336. Itoh, T., Chien, S. and Usami, S. Effects of hemoglobin concentration on deformability of individual sickle cells after deoxygenation. *Blood* **85**: 2245–2253, 1995.

A337. Schmid-Schönbein, G.W., Kosawada, T., Skalak, R. and Chien, S. Membrane model of endothelial cells and leukocytes. A proposal for the origin of a cortical stress. *J. Biomech. Eng.* **117**: 171–178, 1995.

A338. Shyy, Y.J., Lin, M.C., Han, J., Lu, Y., Petrime, M. and Chien, S. The cis-acting phorbol ester "12-O-tetradecanoylphorbol 13-acetate"-responsive element is involved in shear stress-induced monocyte chemotactic protein 1 gene expression. *Proc. Nat. Acad. Sci. U.S.A.* **92**: 8069–8073, 1995.

A339. Huang, Y., Rumschitzki, D., Chien, S. and Weinbaum, S. A fiber matrix model for the growth of macromolecular leakage spots in the arterial intima. *J. Biomech. Eng.* **116**: 430–445, 1995.

A340. Zhao, Y., Chien, S. and Skalak, R. A stochastic model of leukocyte rolling. *Biophys. J.* **69**: 1309–1320, 1995.

A341. Shyy, Y.J., Li, Y.S., Lin, M.C., Chen, W., Yuan, S., Usami, S. and Chien, S. The shear stress-mediated gene expression and the cis-elements involved. *J. Biomech.* **28**: 1451–1457, 1995.

A342. Wang, D.L., Wung, B.S., Shyy, Y.J., Lin, C.F., Chao, Y.J, Usami, S. and Chien, S. Mechanical strain induces monocyte chemotactic protein-1 gene expression in endothelial cells. Effects of mechanical strain on monocyte adhesion to endothelial cells. *Circ. Res.* **77**: 294–302, 1995.

A343. Skierczynski, B.A., Skalak, R. and Chien, S. Modeling of molecular mechanisms of cell adhesion. *Biochem. Cell Biol.* **73**: 399–409, 1995.

A344. Chen, Y.L., Jan, K.M., Lin, H.S. and Chien, S. Ultrastructural studies on macromolecular permebililty in relation to endothelial cell turnover. *Atherosclerosis* **118**: 89–104, 1996.

A345. Hansen, J.C., Skalak, R., Chien, S. and Hoger, A. An elastic network model based on the structure of the red blood cell membrane skeleton. *Biophys. J.* **70**: 146–166, 1996.

A346. Li, Y.S., Shyy, J.Y.J., Li, S., Lee, J.D., Su, B., Karin, M. and Chien, S. The Ras/JNK pathway is involved in shear-induced gene expression. *Molec. Cell. Biol.* **16**: 5947–5954, 1996.

A347. Li, S., Pietrowicz, R.S., Levin, E.G., Shyy, Y.J. and Chien, S. Fluid shear stress induces the phosphorylation of small heat shock proteins in vascular endothelial cells. *Am. J. Physiol., Cell Physiol.* **271**: C994–C1000, 1996.

A348. Hu, Y.L. and Chien, S. Effects of shear stress on protein kinase C distribution in endothelial cells. *J. Histochem. Cytochem.* **45**: 237–249, 1997.

A349. Lin, M.C., Almus-Jacobs, F., Chen, H.-H., Parry, G.C.N., Mackman, N., Shyy, J.Y.-J. and Chien, S. Shear stress induction of tissue factor gene expression and phosphorylation of Sp1. *J. Clin. Invest.* **99**: 737–744, 1997.

A350. Yin, Y., Lim, K.-H., Weinbaum, S., Chien, S. and Rumschitzki, D. A model for the initiation and growth of extracellular lipid liposomes in arterial intima. *Am. J. Physiol., Model. Physiol.* **272**: H1033–H1046, 1997.

A351. Artmann, G.M., Sung, K.-L.P., Horn, T., Whittemore, D., Norwich, G. and Chien, S. Micropipette aspiration of human erythrocytes induces echinocytes via membrane phospholipid translocation. *Biophys. J.* **72**: 1434–1441, 1997.

A352. Huang, Y., Rumschitzki, D.S., Chien, S. and Weinbaum, S. A fiber matrix model for the filtration through fenestral pores in a compressible arterial intima. *Am. J. Physiol., Model. Physiol.* **272**: H2023–H2039, 1997.

A353. Hansen, J.C., Skalak, R., Chien, S. and Hoger, A. Influence of network topology on the elasticity of the red blood cell membrane skeleton. *Biophys. J.* **72**: 2369–2381, 1997.

A354. Chen, Y.L., Jan, K.M., Lin, H.S. and Chien, S. Relationship between endothelial cell turnover and permeability to horseradish peroxidase. *Atherosclerosis* **133**: 7–14, 1997.

A355. Shyy, J.Y.J. and Chien, S. Role of integrins in cellular responses to mechanical stress and adhesion. *Curr. Opin. Cell Biol.* **9**: 707–713, 1997.

A356. Gregersen, H., Kassab, G., Pallencaoe, E., Lee, C., Chien, S., Skalak, R. and Fung, Y.C. Morphometry and strain distribution in guinea pig duodenum with reference to the zero-stress state. *Am. J. Physiol.* **273** (*Gastrointest. Liver Physiol.* **36**): G865–G874, 1997.

A357. Li, S., Kim, M., Schlaepfer, D.D., Hunter, T., Chien, S. and Shyy, J.Y.-J. The fluid shear stress induction of JNK pathway is mediated through FAK-Grb2-sos. *J. Biol. Chem.* **272**: 30455–30622, 1997.

A358. Jalali, S., Sotoudeh, M., Yuan, S., Chien, S. and Shyy, J.Y.-J. Shear stress activates p60src-Ras-MAPK signaling pathways in vascular endothelial cells. *Arterioscler. Thromb. Vasc. Biol.* **18**: 227–234, 1997.

A359. Hansen, J., Skalak, R., Chien, S. and Hoger, A. Spectrin properties and the elasticity of the red blood cell membrane skeleton. *Biorheology* **34**: 327–348, 1997.

A360. Chien, S., Li, S. and Shyy, J.Y.J. Effects of mechanical forces on signal transduction and gene expression in endothelial cells. *Hypertension* **31**(Part 2): 162–169, 1998.

A361. Chiu, J.J., Wang, D.L., Chien, S., Skalak, R. and Usami, S. Effects of disturbed flows on endothelial cells. *J. Biomech. Eng.* **120**: 2–8, 1998.

A362 Sotoudeh, M., Jalali, S., Usami, S., Shyy, J.Y.J. and Chien, S. A strain device imposing dynamic and uniform equi-biaxial strain to cultured cells. *Ann. Biomed. Eng.* **26**: 181–189, 1998.

A363. Chien, S. and Shyy, J.Y.J. Effects of hemodynamic forces on gene expression and signal transduction in endothelial cells. *Biol. Bull.* **194**: 390–393, 1998.

A364. Galbraith, C.G., Skalak, R. and Chien, S. Shear stress produces three-dimensional remodeling in the endothelial cell cytoskeleton. *Cell Motil. Cytoskeleton* **40**: 317–330, 1998.

A365. Wen, Z.Y., Song, L.C., Yan, Z.Y., Lu, Z.H., Sun, D.G. and Chien, S. An animal model to study erythrocyte senescence with a narrow time window of erythrocyte production. *Clin. Hemorheol. Microcirc.* **18**: 75-84, 1998.

A366. Bhullar, I.S., Li, Y.-S., Miao, H., Zandi, E., Kim, M., Shyy, J.Y-J. and Chien, S. Fluid shear stress activation of I?B kinase is integrin-dependent. *J. Biol. Chem.* **273**: 30544-30549, 1998.

A367. Wen, Z.Y., Song, L.C., Yan, Z.Y., Lu, Z.H., Sun, D.G., Shi, Y. and Chien, S. An animal model to study erythrocyte senescence with a narrow time window of erythrocyte production: alterations in osmotic fragility and deformability of erythrocytes during their life span. *Clin. Hemorheol. Microcirc.* **19**: 299-306, 1998.

A368. Artmann, G.M., Kelemen, C., Porst, D., Idt, B. and Chien, S. Temperature transitions of protein properties in human red blood cells. *Biophys. J.* **75**: 3179-3183, 1998.

A369. Li, S., Chen, P.-C.B., Azuma, N., Hu, Y.-L., Wu, Z.S., Sumpio, B.E., Shyy, J.Y.-J. and Chien, S. Distinct roles for the small GTPases Cdc42 and Rho in endothelial responses to shear stress. *J. Clin. Invest.* **103**: 1141-1150, 1999.

A370. Chen, K.-D., Li, Y.-S., Kim, M., Li, S., Chien, S. and Shyy, J.Y.-J. Mechanotransduction in response to shear stress: roles of receptor tyrosine kinases, integrins, and Shc. *J. Biol. Chem.* **274**: 18393-18400, 1999.

A371. Schinagl, R.M., Kurtis, M.S., Ellis, K.D., Chien, S. and Sah, R.L. Effects of seeding duration on the strength of chondrocyte adhesion to articular cartilage. *J. Orthop. Res.* **17**: 121-129, 1999.

A372. Hu, Y.L., Li, S., Shyy, J.Y.J. and Chien, S. Sustained JNK activation induces endothelial apoptosis: studies with colchicine and shear stress. *Am. J. Physiol. (Heart Circ.)* **277**: H1593-H1599, 1999.

A373. Li, S., Chien, S. and Brånemark, P.I. Heat shock-induced necrosis and apoptosis in osteoblasts. *J. Orthop. Res.* **17**: 891-899, 1999.

A374. Karlon, W.J., Hsu, P.P., Li, S., Chien, S., McCulloch, A.D. and Omens, J.H. Measurement of orientation and distribution of

cellular alignment and cytoskeletal organization. *Ann. Biomed. Eng.* **27**: 712–720, 1999.
A375. Gregersen, H., Lee, T.C., Chien, S., Skalak, R. and Fung, Y.C. Strain distribution in the layered wall of the esophagus. *J. Biomech. Eng.* **121**: 442–448, 1999.
A376. Devereaux, R.B., Case, D.B., Alderman, M.H., Pickering, T.G., Chien, S. and Laragh, J.H. Possible role of increased blood viscosity in the hemodynamics of systemic hypertension. *Am. J. Cardiol.* **85**: 1265–1268, 2000.
A377. Lin, K., Hsu, P.P., Chen, B.P., Yuan, S., Usami, S., Shyy, J.Y.J. and Chien, S. Molecular mechanism of endothelial growth arrest by laminar shear stress. *Proc. Nat. Acad. Sci. U.S.A.* **97**: 9385–9389, 2000.
A378. Jin, G., Sah, R.L., Li, Y.S., Lotz, M., Shyy, J.Y.J. and Chien, S. Biomechanical regulation of matrix metalloproteinase-9 in chondrocytes. *J. Orthop. Res.* **18**: 899–908, 2000.
A379. Jin, G., Wu, C.H., Li, Y.H., Hu, Y.L., Shyy, J.Y.J. and Chien, S. Effects of active and negative mutants of Ras on rat arterial neointima formation. *J. Surg. Res.* **94**: 124–132, 2000.
A380. Butler, P.J., Weinbaum, S., Chien, S. and Lemons, D.E. Endothelium-dependent, shear-induced vasodilation is rate-sensitive. *Microcirculation* **7**: 53–65, 2000.
A381. Jalali, S., del Pozo, M.A., Chen, K.D., Miao, H., Li, Y.S., Schwartz, M.A., Shyy, J.Y.J. and Chien, S. Integrin-mediated mechanotransduction requires its dynamic interaction with specific ECM ligands. *Proc. Nat. Acad. Sci. U.S.A.* **98**: 1042–1046, 2001.
A382. Zhao, Y.H., Chien, S. and Weinbaum, S. Dynamic contact forces on leukocyte microvilli and their penetration of the endothelial glycocalyx. *Biophys. J.* **80**: 1124–1140, 2001.
A383. Moore, M.M., Goldman, J., Patel, A.R., Chien, S. and Liu, S.Q. Role of mechanical stretch in the induction of smooth muscle cell death in experimental vein grafts. *J. Biomech.* **34**: 289–297, 2001.
A384. Zhao, Y.H., Lipowsky, H.H., Skalak, R. and Chien, S. Leukocyte rolling in rat mesenteric venules. *Ann. Biomed. Eng.* **29**: 360–372, 2001.

A385. Butler, P.J., Norwich, G., Weinbaum, S. and Chien, S. Shear stress induces a time- and position-dependent increase in endothelial cell membrane fluidity. *Am. J. Physiol. Cell Physiol.* **280**: C962–C969, 2001.

A386. Wu, C.H., Lin, C.S., Hung, J.S., Wu, C.J., Lo, P.H., Jin, G., Shyy, J.Y.J., Mao, S.J.T. and Chien, S. Inhibition of neointimal formation in porcine coronary artery by a Ras mutant. *J. Surg. Res.* **99**: 100–106, 2001.

A387. Zhu, Y., Liao, H.L., Wang, N., Ma, K.-S., Verna, L., Shyy, J.Y., Chien, S. and Stemerman, M.B. LDL-Activated p38 in endothelial cells is mediated by Ras. *Arterioscler. Thromb. Vasc. Biol.* **21**: 1159–1164, 2001.

A388. Kelemen, C., Artmann, G.M. and Chien, S. Temperature transitions of human hemoglobin at body temperature: effects of calcium. *Biophys. J.* **80**: 2622–2630, 2001.

A389. Hsu, P.P., Li, S., Usami, S., Ratcliffe, A., Wang, X. and Chien, S. Effects of flow patterns on endothelial cell migration into a zone of mechanical denudation. *Biochem. Biophys. Res. Comm.* **285**: 751–759, 2001.

A390. Chen, B.P.C., Li, Y.-S., Zhao, Y., Chen, K.-D., Li, S., Lao, J., Yuan, S., Shyy, J.Y.-J. and Chien, S. DNA microarray analysis of gene expression in endothelial cells in response to 24-hour shear stress. *Physiol. Genomics* **7**: 55–63, 2001.

A391. Li, S., Bhatia, S., Hu, Y.L., Shiu, Y.-T., Li, Y.-S., Usami, S. and Chien, S. Effects of morphological patterning on endothelial cell migration. *Biorheology* **38**: 101–108, 2001.

A392. Fisher, A.B., Chien, S., Barakat, A.I. and Nerem, R.M. Endothelial cellular response to altered shear stress. *Am. J. Physiol. Lung* **C281**: L529–L533, 2001.

A393. Wen, Z., Xie, L., Yan, Z., Yao, W., Chen, K., Ke, W. and Chien, S. Effect of ^{60}Co irradiation on characteristics of hemorheology in rabbits. *Clin. Hemorheol.* **25**: 75–81, 2001.

A394. Chen, B.P.C., Li, Y.-S., Zhao, Y., Chen, K.-D., Li, S., Lao, J., Yuan, S., Shyy, J.Y.-J. and Chien, S. DNA microarray analysis of gene expression in endothelial cells in response to 24-hour shear stress. *Physiol. Genomics* **7**: 55–63, 2001.

A395. Butler, P.J., Tsou, T.C., Li, J.Y.S., Usami, S. and Chien, S. Rate sensitivity of shear-induced changes in the lateral diffusion of endothelial cell membrane lipids: a role for membrane perturbation in shear-induced MAPK activation. *FASEB J.* **16**: 216–218, 2002 (http://www.fasebj.org/cgi/doi/10.1096/fj.01-0434fje).

A396. Li, S., Butler, P., Wang, Y.X., Shiu, Y.T., Hu, Y.L., Han, D.C., Usami, S., Guan, J.L. and Chien, S. The role of the dynamics of focal adhesion kinase in the mechanotaxis of endothelial cells. *Proc. Nat. Acad. Sci. U.S.A.* **99**: 3546–3551, 2002.

A397. Liu, Y., Chen, B.P., Lu, M., Zhu, Y., Stemerman, M.B., Chien, S. and Shyy, J.Y. Shear stress activation of SREBP1 in endothelial cells is mediated by integrins. *Arterioscler. Thromb. Vasc. Biol.* **22**: 76–81, 2002.

A398. Sotoudeh, M., Li, Y.-S., Chang, C.C., Tsou, T.C., Wang, Y.B., Usami, S., Lee, A., Ratcliffe, A., Chien, S. and Shyy, J.Y. Induction of apoptosis in vascular smooth muscle cells by mechanical stretch. *Am. J. Physiol. (Cell Physiol.)* **282**: H1709–H1716, 2002.

A399. Miao, H., Li, S., Hu,Y.L., Yuan, S., Chen, B.P.C., Puzon-McLaughlin, W., Tarui, T., Shyy, J.Y.-J., Takada, Y., Usami, S. and Chien, S. Differential regulation of Rho GTPases by α1 and β3 integrins: the role of an extracellular domain of integrins in intracellular signaling. *J. Cell Sci.* **115**: 2199–2206, 2002.

A400. Miao, H., Yuan, S., Wang, Y., Tsygankov, A. and Chien, S. Role of Cbl in shear-activation of PI 3-kinase and JNK in endothelial cells. *Biochem. Biophys. Res. Commun.* **292**: 892–899, 2002.

A401. Tzima, E., del Pozo, M.A., Shattil, S.J., Chien, S. and Schwartz, M.A. Activation of integrins in endothelial cells by fluid shear stress mediates Rho-dependent cytoskeletal alignment. *EMBO J.* **20**: 4639–4647, 2002.

A402. Katsumi, A., Milanini, J., Kiosses, W.B., del Pozo, M.A., Kaunas, R., Chien, S., Hahn, K.M. and Schwartz, M.A. Effects of cell tension on membrane protrusion through regulation of the GTPase Rac. *J. Cell Biol.* **158**: 153–164, 2002.

A403. Yajima, N., Masuda, M., Miyazaki, M., Nakajima, N., Chien, S. and Shyy, J.Y. Oxidative stress is involved in the development of experimental abdominal aortic aneurysm: a study of the transcription profile by cDNA microarray. *J. Vasc. Surg.* **36**: 379–385, 2002.

A404. Wang, Y., Miao, H., Li, S., Chen, K.D., Li, Y.-S., Yuan, S., Shyy, J.Y.-J. and Chien, S. Interplay between integrins and Flk-1 in shear stress-induced signaling. *Am. J. Physiol. (Cell Physiol.)* **283**: C1540–1547, 2002.

A405. Weyts, F.A.A., Li, Y.-S., van Leeuwen, J., Weinans, H. and Chien, S. ERK activation and alphavbeta3 integrin signaling through Shc recruitment in response to mechanical stimulation in human osteoblasts. *J. Cell Biochem.* **87**(1): 85–92, 2002.

A406. Zhao, Y., Chen, B.P.C., Miao, H., Yuan, S., Li, Y.-S., Hu, Y., Rocke, D.M. and Chien, S. Improved significance test for DNA microarray data: temporal modulation of gene expression by shear stress. *Physiol. Genomics* **12**: 1–11, 2002.

A407. Hu, Y.L., Li, S., Miao, H., Tsou, T.-C., del Pozo, M.A. and Chien, S. Roles of microtubule dynamics and small GTPase Rac in endothelial cell migration and lamellipodium formation under flow. *J. Vasc. Sci.* **39**: 465–476, 2002.

A408. Shyy, J.Y.-J. and Chien, S. Role of integrins in endothelial mechanosensing of shear stress. *Circ. Res.* **91**: 769–776, 2002.

B. Books

B1. Chien, S. and Ho, C. (eds.). *Nuclear Magnetic Resonance in Biology and Medicine.* Raven Press, New York, 1986, 248 pp.

B2. Skalak, R. and Chien, S. (eds.). *Handbook of Bioengineering.* McGraw-Hill, New York, 1986, 932 pp.

B3. Chien, S., Dormandy, J.A., Ernest, E. and Matrai, A. (eds.). *Clinical Hemorheology, Applications in Cardiovascular and Hematological Disease, Diabetes, Surgery and Gynecology.* Martinus Nijhoff, Dordrecht, The Netherlands, 1987, 387 pp.

B4. Chien, S. (ed.). *Vascular Endothelium in Health and Disease.* Plenum Press, New York, 1988, 234 pp.

B5. Chien, S. (ed.). *Molecular Biology in Physiology.* Raven Press, New York, 1989, 167 pp.
B6. Chien, S. (ed.). *Molecular Biology of Cardiovascular System.* Lea and Febiger, Philadelphia, 1990, 241 pp.
B7. Liu, C.Y. and Chien, S. (ed.). *Fibrinogen, Thrombosis, Coagulation and Fibrinolysis.* Plenum Press, New York, 1990.
B8. Wang, E., Wang, J., Chien, S., Cheung, W.Y. and Wu, C.W. (eds.). *Biochemical and Structural Dynamics of the Cell Nucleus.* Academic Press, Orlando, Florida, 1990.
B9. Liu, C.Y. and Chien, S. (ed.). *Recent Advances in Haemostasis and Cardiovascular Research.* Institute of Biomedical Sciences, Academia Sinica, Taipei, Taiwan, ROC, 1990, 137 pp.

C. Chapters in Books, Reviews, and Conference Publications

C1. Chien, S. and Gregersen, M.I. Determination of body fluid volumes, in *Physical Techniques in Biological Research, Vol. IV*, ed. Nastuk, W.L. Academic Press, 1962, pp. 1–105.
C2. Chien, S. Review of neuroanatomy and neurophysiology, in *Approaches to the Treatment of Patients with Neuromuscular Dysfunction*, ed. Sattely, C. W.C. Brown, Dubuque, Iowa, 1964, pp. 6–16.
C3. Gregersen, M.I., Peric, B., Chien, S., Sinclair, D., Chang, C. and Taylor, H. Viscosity of blood at low shear rates: observations on its relation to volume concentration and size of the red cells, in *Proceedings of the 4th International Congress on Rheology, Symposium on Biorheology.* John Wiley, New York, 1965, pp. 613–628.
C4. Chien, S., Dellenback, R.J., Usami, S. and Gregersen, M.I. Capillary permeability to macromolecules in endotoxin shock, in *Third Conference on Artificial Colloidal Agents.* National Academy of Sciences-National Research Council, Washington, D.C., 1965, pp. 56–73.
C5. Gregersen, M.I. and Chien, S. Blood volume, in *Medical Physiology, 12th edn.*, ed. Mountcastle, V.B. C.V. Mosby, St. Louis, 1968, pp. 244–261.

C6. Chien, S. and Gregersen, M.I. Hemorrhage and shock, in *Medical Physiology, 12th edn.*, ed. Mountcastle, V.B. C.V. Mosby, St. Louis, 1968, pp. 262–282.

C7. Usami, S., Chien, S. and Gregersen, M.I. Viscometric behavior of hardened erythrocytes in relation to deformability and size, in *Proceedings of the Fifth International Congress on Rheology, Vol. 2*, ed. Onogi, S. University Park Press, Baltimore, 1970, pp. 99–108.

C8. Gregersen, M.I., Usami, S., Bryant, C.A. Magazinovic, V. and Chien, S. Evidence on species differences in erythrocyte deformability by filtration through micropores (polycarbonate sieves), in *Proceedings of the Fifth International Congress on Rheology, Vol. 2*, ed. Onogi, S. University Park Press, Baltimore, 1970, pp. 109–113.

C9. Chien, S., Usami, S., Dellenback, R.J., Bryant, C.A. and Gregersen, M.I. Change of erythrocyte deformability during fixation in acetaldehyde, in *Theoretical and Clinical Hemorheology*, eds. Hartert, H.H. and Copley, A.L. Springer-Verlag, Berlin, 1971, pp. 136–143.

C10. Chien, S., Usami, S., Dellenback, R.J. and Gregersen, M.I. Influence of fibrinogen and globulins on blood rheology at low shear rates: comparison among elephant, dog and man, in *Theoretical and Clinical Hemorheology*, eds. Hartert, H.H. and Copley, A.L. Springer-Verlag, Berlin, 1971, pp. 144–153.

C11. Usami, S., Chien S. and Gregersen, M.I. Viscometric behavior or young and erythrocytes, in *Theoretical and Clinical Hemorheology*, eds. Hartert, H.H. and Copley, A.L. Springer-Verlag, Berlin, 1971, pp. 266–270.

C12. Chien, S., Luse, S.A., Jan, K.-M., Usami, S., Miller, L.H. and Fremount, H. Effects of macromolecules on the rheology and ultrastructure of red cell suspensions, in *Proceedings of the 6th European Conference on Microcirculation*. Karger, Basel, 1971, pp. 29–34.

C13. Chien, S., Usami, S., Dellenback, R.J. and Gregersen, M.I. Hepatic and intestinal lymph in endotoxin shock, in *Proceedings of the 6th European Conference on Microcirculation*. Karger, Basel, 1971, pp. 156–159.

C14. Chien, S., Jan, K.-M. and Usami, S. Rheological considerations of colloid replacement, in *Les Solutes de Substitution (Reequilibration Metabolque) en Anesthesiologie et en Reanimation*. Societe Francaise D'Anesthesie, D; Analgesie et de Reanimation. Libraire Arnette, Paris, 1971, pp. 153–166.

C15. Chien, S. Present state of blood rheology, in *Hemodilution, Theoretical Basis and Clinical Application*, eds. Messmer, K. and Schmid-Schöenbein, H. Karger, Basel, 1972, pp. 1–40.

C16. Chien, S., Usami, S., Dellenback, R.J. and Magazinovic, V. Blood rheology in hemorrhage and endotoxin, in *Neurohumoral and Metabolic Aspects of Injury. Advances in Experimental Medicine and Biology, Vol. 33*, eds. Kovach, A.G.B., Stoner, H.G. and Spitzer, J.J. Plenum Publishing Corp., New York, 1973, pp. 75–93.

C17. Chien, S., Usami, S., Jan, K.-M. and Skalak, R. Macrorheological and microrheological correlation of blood flow in the macrocirculation and microcirculation, in *Rheology of Biological Systems*, eds. Gabelnick, H.L. and Litt, M. Charles C. Thomas, Springfield, Illinois, 1973, pp. 12–48.

C18. Chien, S., Usami, S. and Jan, K.-M. Fundamental determination of blood viscosity, in *First Symposium on Flow — Its Measurement and Control in Science and Industry, Instrument Society of America*, ed. Dowdell, R.B., 1974, pp. 1411–1417.

C19. Usami, S., Benis, A.M. and Chien, S. Measurement of blood viscosity in couette viscometer with different clearances, in *First Symposium on Flow — Its Measurement and Control in Science and Industry, Instrument Society of America*, ed. Dowdell, R.B., 1974, pp. 1419–1423.

C20. Chien, S. Biophysical behavior of red cells in suspensions, in *The Red Blood Cell, 2nd edn.*, ed. Surgenor, D. MacN. Academic Press, New York, 1975, Vol. 2, pp. 1031–1133.

C21. Chien, S., Usami, S., Jan, K.-M., Smith, J.A. and Bertles, J.F. Blood rheology in sickle cell disease, in *Proceedings of the Symposium on Molecular and Cellular Aspects of Sickle Cell Disease*, eds. Hercules, J.I., Cottam, G.L., Waterman, M.R. and Schechter, A.N. DHEW Pub. No. (NIH) 76-1007, National Institutes of Health, Bethesda, Maryland, 1977, pp. 277–300.

C22. Chien, S. Summing up I. Rheological methods. *Blood Cells* **3**: 427–430, 1977.

C23. Chien, S., Usami, S., Jan, K.-M., Smith, J.A. and Bertles, J.F. Blood rheology in sickle cell disease, in *Proceedings of the Symposium on Molecular and Cellular Aspects of Sickle Cell Disease*, eds. Hercules, J.I., Cottam, G.L., Waterman, M.R. and Schechter, A.N. DHEW Pub. No. (NIH) 76-1007, National Institutes of Health, Bethesda, Maryland, 1977, pp. 277–300.

C24. Chien, S. Red cell membrane and hemolysis, in *Cardiovascular Flow Dynamics and Measurements*, eds. Whang, N.D.C. and Norman, N.A. University Park Press, Baltimore, Maryland, 1977, pp. 757–798.

C25. Zarda, P.R., Chien, S. and Skalak, R. Interaction of viscous incompressible fluid with an elastic body, in *Computational Methods of Fluid-Structure Interaction Problems — AMD, Vol. 26*, eds. Belytschko, T. and Geers, T.L. American Society of Mechanical Engineers, New York, 1977, pp. 65–82.

C26. Skalak, R., Zarda, P.R., Jan, K.-M. and Chien, S. Theory of rouleau formation, in *Cardiovascular and Pulmonary Dynamics*, ed. Jaffrin, M.-Y. INSERM-Euromech 92, 1977, pp. 299–308.

C27. Chien, S. Homeostatic regulation of blood viscosity by red cell deformation, in *Studies and Essays in Commemoration of the Golden Jubilee of Academia Sinica*, ed. Chien, S.L. Academia Sinica, Taiwan, 1978, Vol. I, pp. 515–530.

C28. Chien, S. Transport across arterial endothelium, in *Progress In Hemostasis and Thrombosis*, ed. Spaet, T.H. Grune and Stratten, New York, 1978, Vol. IV, pp. 1–36.

C29. Chien, S., King, R.G., Schuessler, G.B., Skalak, R., Tözeren, A., Usami, S. and Copley, A.L. Roles of red cell deformability and aggregation in blood viscoelasticity, in *AIChE Symposium Series: Biorheology*, eds. Huang, C.-R. and Copley, A.L. No. 182, 1978, Vol. 74, pp. 56–60.

C30. Sung, K.-L.P. and Chien, S. Viscous and elastic properties of human red cell membrane, in *AIChE Symposium Series: Biorheology*, eds. Huang, C.-R. and Copley, A.L. No. 182, 1978, Vol. 74, pp. 81–84.

C31. Chien, S. Blood rheology, in *Quantitative Cardiovascular Studies*, eds. Hwang, N.H.C. and Gross, D.R. University Park Press, Baltimore, Maryland, 1979, pp. 241–287.

C32. Chien, S. Correlation of blood rheology in macro- and microcirculation, in *Proceedings of the 3rd Engineering Mechanics Division Specialty Conference*. American Society of Civil Engineers, New York, 1979, pp. 47–50.

C33. Chien, S., Sung, K.-L.P., Schmid-Schöenbein, G.W., Tözeren, A., Tözeren, H., Usami, S. and Skalak, R. Microrheology of erythrocytes and leukocytes, in *Symposium on Hemorheology and Diseases*, eds. Stoltz, J.F. and Drouin, P. Doin Editeurs, Paris, 1980, pp. 93–108.

C34. Weinbaum, S. and Chien, S. Vesicular transport of macromolecules across vascular endothelium, in *The Mathematics of Microcirculation Phenomena*, eds. Gross, J.F. and Popel, A.S. Raven Press, New York, 1980, pp. 109–131.

C35. Weinbaum, S., Arminski, L., Pfeffer, R. and Chien, S. Theoretical models for endothelial junction formation and vesicular transport, in *The Role of Fluid Mechanics in Atherogenesis*, ed. Nerem, R.M. University of Houston, 1980, pp. 31–33.

C36. Chien, S. Functional rheology of erythrocytes and leukocytes, in *The Rheology of Blood, Blood Vessels and Associated Tissues*, eds. Hwang, N.H.C. and Gross, D.R. Sijthoff and Noordhoof, Netherlands, 1981, pp. 118–136.

C37. Chien, S. Cerebral circulation, in *Principles of Neural Science*, eds. Kandel, E.R. and Schwartz, J. Elsevier, New York, 1981, pp. 660–666.

C38. Chien, S. Hemodynamics and rheology in the microcirculation, in *Symposium on Recent Developments in Microcirculation Research*, eds. Davis, E. and Marcel, G.A. Excerpta Medica, Amsterdam, 1981, pp. 54–69.

C39. Skalak, R., Schmid-Schöenbein, G.W. and Chien, S. Analysis of white blood cell deformation, in *White Blood Cells, Morphology and Rheology as Related to Function*, eds. Bagge, U., Born, G.V.R. and Gaehtgens, P. Martinus Nijhoff, The Hague, 1982, pp. 1–10.

C40. Chien, S., Sung, K.-L.P., Skalak, R. and Schmid-Schöenbein, G.W. Viscoelastic properties of leukocytes in passive deformation, in *White Blood Cells, Morphology and Rheology as Related to Function*, eds. Bagge, U., Born, G.V.R. and Gaehtgens, P. Martinus Nijhoff, The Hague, 1982, pp. 11–20.

C41. Schmid-Schöenbein, G.W., Skalak, R., Sung, K.-L.P. and Chien, S. Human leukocytes in the active state, in *White Blood Cells, Morphology and Rheology as Related to Function*, eds. Bagge, U., Born, G.V.R. and Gaehtgens, P. Martinus Nijhoff, The Hague, 1982, pp. 21–30.

C42. Chien, S. and Simchon, S. The sympathetic and central nervous systems in shock, in *The Handbook of Trauma*, eds. Altura, B.M., Lefer, A.M. and Schumer, W. Raven Press, New York, 1983, Vol. 1, pp. 149–166.

C43. Chien, S. Fluid mechanical factors in macromolecular transport by the arterial wall, in *Fluid Dynamics as a Localizing Factor for Atherosclerosis*, eds. Schettker, G. *et al.* Springer-Verlag, Berlin, Heidelberg, 1983, pp. 87–90.

C44. Chien, S. Introduction to the symposium on the correlation of microcirculation and macrocirculation. *Int. J. Microcirc., Clin. Exp.* 1: 347–350, 1983.

C45. Skalak, R., Chien, S. and Schmid-Schöenbein, G.W. Viscoelastic deformation of white cells: theory and analysis, in *White Cell Mechanics: Basic Science and Clinical Aspects*, eds. Meiselman, H.J., Lichtman, M.A. and La Celle, P.L. Alan R. Liss, New York, 1984, pp 3–18.

C46. Chien, S., Schmid-Schöenbein, G.W., Sung, K.-L.P., Schmalzer, E.A. and Skalak, R. Viscoelastic properties of leukocytes, in *White Cell Mechanics: Basic Science and Clinical Aspects*, eds. Meiselman, H.J., Lichtman, M.A. and La Celle, P.L. Alan R. Liss, New York, 1984, pp 19–51.

C47. Chien, S. Leukocyte motion in the microcirculation (commentary), in *White Cell Mechanics: Basic Science and Clinical Aspects*, eds. Meiselman, H.J., Lichtman, M.A. and La Celle, P.L. Alan R. Liss, New York, 1984, pp. 99–102.

C48. Chien, S. Blood rheology, in *Blood Vessels and Lymphatics in Organ Systems*, eds. Abramson, D.I. and Dobrin, P.B. Academic Press, New York, 1984, pp. 77–85.

C49. Chien, S., Usami, S. and Skalak, R. Blood flow in small tubes, in *Handbook of Physiology, Circulation. Section on Microcirculation*, eds. Renkin, E.M. and Michel, C. American Physiological Society, Bethesda, Maryland, 1984, pp. 217–249.

C50. King, R.G., Chien, S., Usami, S. and Copley, A.L. Biorheological methods employing the Weissenberg Rheogoniometer. *Biorheology* Suppl. I: 23–24, 1984.

C51. Usami, S., Reinhart, W., Danoff, S. and Chien, S. A new capillary viscometer for measurements on small blood samples over a wide shear rate range. *Clin. Hemorheol.* **4**: 295–297, 1984.

C52. Chien, S. Molecular biology in microcirculation, in *Microcirculation, Vol. 1*, eds. Tsuchiya, M., Asano, M., Mishima, Y. and Oda, M. Excerpta Medica, Amsterdam, 1987, pp. 35–50.

C53. Weinbaum, S., Pfeffer, R. and Chien, S. An hypothesis for the localization of atherogenesis and its relationship to endothelial regulation and transport, in *Proceedings of the International Symposium*, Hyogo, October 1987.

C54. Laboratory techniques: guidelines on selection of laboratory tests for monitoring the acute phase response. *J. Clin. Pathol.* 41: 1203–1212, 1988.

C55. Chien, S. and Simchon, S. Tissue hematocrit and hemorheology, in *Cerebral Hyperemia and Ischemia: From the Standpoint of Cerebral Blood Volume*, eds. Tomita, M., Sawada, T., Naritomi, H. and Heiss, W.D. Excerpta Medica, Amsterdam, 1988, pp. 44.

C56. Lai, M.-L., Lin, C.-Y., Liu, C.H.L., Lo, H.-J., Wang, M.E., Chen, L.B., Chen, W.-T., Chien, S., Ip, W., Lin, D.C., Lin, J.J.C, Sun, T.-T., Wang, E., Wang, J.L., Wu, C.-W., Wu, R. and Lin, S. Monoclonal antibodies against nuclear antigens associated with proliferation and differentiation of bovine esophageal epithelium, in *Biochemical and Structural Dynamics of the Cell Nucleus*, eds. Wang, E., Wang, J., Chien, S., Cheung, W.Y. and Wu, C.W. Academic Press, Orlando, Florida, 1990, pp. 1–16.

C57. Chabanel, A. and Chien, S. Blood viscosity as a factor in human hypertension, in *Hypertension: Pathophysiology, Diagnosis and Management, 2nd edn.*, eds. Laragh, J.H. and Brenner, B.M. Raven Press, New York, 1995, pp. 365–376.

C58. Chien, S. and Gallik, S. Rheology in normal individuals and polycythemia vera, in *Polycythemia and the Myeloproliferative Disorders*, eds. Wasserman, L.R. and Berk, P.D. W.B. Saunders and N.I. Berlin, Philadelphia, 1995, pp. 114–129.

C59. Chien, S. and Shyy, J.Y.J. Mechanisms of mechanochemical transduction in vascular cells, in *Ischemic Blood Flow in the Brain*, eds. Fujuuchi, Y., Tomita, M. and Koto, A. Springer-Verlag, Tokyo, 2001, pp. 13–19.

D. Publications on Science Policy and Other General Topics

D1. Chien, S. and Silverstein, S.C. Economic impact of application of monoclonal antibodies to medicine and biology. *FASEB J.* **7**: 1426–1431, 1993.

D2. Chien, S. Let our voices be heard and amplified. The time is now! [editorial]. *FASEB J.* **7**: 615–616, 1993.

D3. Nerem, R.M., Taylor, K.D., Arnold, F., Chien, S., Katona, P.G., Littell, C. and Young, W.D. *Support for Bioengineering Research: Prepared for the National Institutes of Health by The External Consultants Committee*, 1994.

D4. Chien, S., Cherrington, A., Cook, J.S., Metting, P., Raff, H., Valtin, H., Young, D.B. and Yool, A. The sun breaks through the clouds: a bright future for physiology. *The Physiologist* **39**: 375–388, 1996.

D5. Huntsman, L.L., Chien, S., Davis, R.W., Griffith, L.G., Hendee, W.R., Henry, S.A., Hubbell, J.A., Koonin, S., Phillips, W.M. and Engel, L.W. *Expanding Opportunities.* Report of The Working Group on Review of Bioengineering and Technology and Instrumentation Development Research, 1999.

D6. Chien, S. *The Interplays between Biology-Medicine and Humanities-Society.* Hao Ran Workshop Technology and Human Development, Hao Ran Foundation, Taipei, Taiwan, ROC, 2001, pp. 2–27.

D7. Chien, S. and Maynard, C.D. Newest member of the NIH family. *Science* **291**: 1701–1702, 2001.
D8. Hendee, W.R., Chien, S., Maynard, C.G. and Dean, D.J. The National Institute of Biomedical Imaging and Bioengineering: history, status and potential impact. *Ann. Biomed. Eng.* **30**: 2–10, 2002.

Section VII
Programs of Celebrations

A. La Jolla, 23 June 2001

Symposium in Honor of Professor Shu Chien's 70th Birthday (23 June 2001)

*Garren Auditorium, School of Medicine,
University of California, San Diego*

CONFERENCE PROGRAM

8:30–9:00 Registration and Continental Breakfast

Session I. Chair: Dr. Geert W. Schmid-Schönbein

9:00–9:10 **Introduction**
Dr. Geert W. Schmid-Schönbein

9:10–9:25 **Opening Remarks**
Dr. Robert Conn, Dean JSOE

9:25–9:40 **Shu Chien: The Founding Chair of UCSD Department of Bioengineering**
Dr. David Gough, Chair

9:40–10:00 **Some of Shu's Important Contributions to My Understanding of Blood Cells and Microcirculation**
Dr. Paul La Celle

10:00–10:20 Vascular Biology, Tissue Engineering, NIH and Shu Chien
Dr. Bob Nerem

10:20–10:50 Coffee Break

Session II. Chair: Dr. Paul Sung

10:50–11:10 Deforming Congo Snake Red Cells and Dancing Human Neutrophils
Dr. Harry Goldsmith

11:10–11:30 Mechanical Effects in Vascular Biology
Dr. Larry McIntire

11:30–11:50 A Tribute to Shu Chien
Dr. Van Mow

11:50–12:10 Potential Application of Robotic Technology to Improve Shu's Golf Game
Dr. Savio L.-Y. Woo

12:10–12:30 Chemotherapeutic Engineering: Polymeric Nanospheres for Clinical Administration of Anticancer Drugs
Dr. Si-shen Feng

12:30–1:30 **Lunch**

Session III. Chair: Dr. Kung-ming Jan

1:30–1:50 Skiing and Tip-toeing on the Endothelial Glycocalyx
Dr. Sheldon Weinbaum

1:50–2:10 *In Vitro* and *In Vivo* Views of Blood Rheology: Are They the Crouching Tigers and Hidden Dragons of Microvascular Function?
Dr. Herb Lipowsky

2:10–2:30 Electron-microscopy: The Way to Really See What's Going On
Dr. Ann Baldwin

2:30–2:50   **Coffee Break**

Session IV.   Chair: Dr. Julie Y.-S. Li

2:50–3:10   **Measuring 2D Kinetics Between Apposing Membranes**
Dr. Cheng Zhu

3:10–3:30   **Shu Is Not a Good Businessman: He Lends You $70 and You Are Only $30 in Debt**
Dr. John Y.-J. Shyy

3:30–3:50   **Constraining Shu: Is it Possible?**
Dr. Bernhard Ø. Palsson

3:50–4:00   **How Does Shu Shine Like a Rainbow in the Sky?**
Dr. Y.C. Fung

Cutting the birthday cake during the celebration banquet.

23 JUNE 2001 CELEBRATION ATTENDEES[*]

Akeson, Wayne
Bagge, Ulf
Baldwin, Ann
Bassingthwaighte, James
Baxter, Walt
Baxter, Amy
Bhatia, Sangeeta
Browning, Linda
Butler, Peter
Canteri, Laurence
Caro, Colin
Chen, Anne
Chen, Benjamin
Chen, Jenny C-G
Chen, Kuang-Tien
Chen, Peter
Chen, Richard
Cheng, Koun Ping
Cheng, Lilly
Chien, Anne
Chien, K.C.
Chien, May
Chien, Shu
Chuang, Agnes
Chuang, Alexander
Conn, Bob
Craig, Bill
Craig, Nancy
Dellenback, Geraldine
Dellenback, Robert
Eskin, Suzanne
Feng, Si-Shen
Field, Carmen
Fronek, Arnost

Fronek, Kitty
Fung, Luna
Fung, Yuan-Cheng
Gao, Jian
Goldsmith, Harry
Gong, Xing
Gough, Carol
Gough, Dave
Griffin, Jennifer
Haidekker, Mark
Hart, Mary
Hickman, Elizabeth
Ho, Chien
Hsu, Anna
Hsu, Chien Szu
Hsu, Chih Chao Chin
Hsu, Pin-Pin
Hsu, Sawyer
Hu, Amy
Hu, David
Hu, Huiping
Hu, Te Chiang
Hu, Winnie
Hu, Yan Jun
Hu, Yinghe
Hu, Yingli
Huang, Ernie
Huang, Ke
Huang, Su
Huang, Wei
Intaglietta, Marcos
Jan, Connie
Jan, Kung-ming
Jin, Gang

Johnson, Genevieve
Johnson, Paul
Kassab, Ghassan
Kaunas, Roland
Kim, Syngcuk
Knosler, Daniel
La Celle, Paul
La Celle, June
Lai, Linda
Lai, Michael W.
Lao, Jianmin
Lee-Karlon, Ann
Lemons, Daniel
Leonard, Edward
Leonard, Sheri
Li, Jane
Li, Julie
Li, Song
Lieber, Richard
Liew, Fah Seong
Liew, Polly
Lin, Kurt
Lin, Lily Yuli
Lin, Shao Chi
Lipowsky, Herbert
Liu, Shu Q.
Louie, Elise
Lu, Shaoying (Kathy)
Mackenna, Deidre
Mazzoni, Michelle
McCulloch, Andrew
McIntire, Larry
McKelvy, Jeffrey
Miao, Hui
Mow, Barbara
Mow, Van

Nerem, Bob
Nerem, Marilyn
Nguyen, Phu
Norwich, Gerard
Ou-Yang, Daniel
Portner, Peer
Price, Jeff
Ranney, Helen
Ratcliffe, Tony
Reinhart, Walter
Ritten, Kathryn
Ringle, David
Sah, Robert
Sangiorgio, Sophia
Schmid-Schöenbein, Geert
Schmid-Schöenbein, Renate
Scmalzer, Emily
Secomb, Clair
Secomb, Julian
Secomb, Timothy
Shaaban, Mona
Shah, Jangesh
Shiu, Yang-Ting Elizabeth
Shyy, John
Skalak, Anna
Smith, Loretta
Sobin, Sidney
Sobin, Venus
Stewart, MaryAnn
Subramaniam, Shankar
Sung, K-L. Paul
Sung, Lanping Amy
Tseng, Ellen
Usami, Shunichi
Usami, Tazuko
Wang, Alice

Celebration of Academician Shu Chien's
70th Birthday (8 August 2001)
Institute of Biomedical Sciences
Academia Sinica, Taipei, Taiwan

謹訂於九十年八月八日（星期三）下午六時整假生醫所前棟大廳舉行錢煦院士七十壽辰慶祝會，敬備菲酌。

恭　請

台光

中央研究院生醫所
國家衛生研究院　敬上
美洲華人生物學會

※ 非院內人員敬請務必攜帶邀請函，以憑入院。謝謝！
聯絡人：Tel: 2789-9001許職甄；Tel: 2789-9113陳華純

請柬

Invitation

Institute of Biomedical Sciences, Academia Sinica,
National Health Research Institutes, and
Society of Chinese Bioscientists in America (SCBA)
Request the pleasure of your company
For a birthday party of Dr. Shu Chien
At 18:00
On August 8, 2001
At Lobby of 1st Floor, Front Building, IBMS

錢煦院士七秩壽辰慶祝會
節目單

司儀：Dr. 潘文涵

時　間	地　點	節　目　內　容
6:00 – 6:40 pm	前棟中庭	Dinner 切蛋糕 音樂演奏：醫學對樂四重奏 （樂團資料附於後頁）
6:40 – 7:10 pm	B1C演講廳	精彩鏡頭回顧 (Dr. 李小緣)
7:10 – 8:10 pm		來賓致辭 (引言人：吳成文院長)
8:10 – 8:25 pm		節目表演：歌樂頌 (帶動唱歌詞附於表後)
8:25 – 8:30 pm		致贈禮物 (生醫所代表：陳垣崇所長) (國衛院代表：吳成文院長)
8:30 – 8:50 pm		錢煦院士致謝辭

來賓致詞順序

1. 李遠哲　院長
2. 陳長謙　副院長
3. 蔡作雍　院士
4. 何曼德　院士
5. 方懷時　院士
6. 鄔永齊　院士
7. 劉源俊　校長
8. 田玲玲　女士
9. 錢胡匡政　女士
10. 陳　笠　副院長
11. 王　弘　博士

夢勿忘——田園

一個人孤孤單單走到街角天邊　全世界　心相連　用愛舖一條大国圈
多希望有一天能再聽到你的談　夢該圓　再鑑遠　心裡音樂就會團圓
全世界　走一邊　不任來人間兒一回　韓階慢從前回味成長的酸和甜
夢可以　做返返　丘島樂爾部不改變　再多的苦吉　抱著歲月　心甘情願

Program for the Celebration of Academician Shu Chien's 70th Birthday

Master of Ceremony: Dr. Wen-Han Pan

6:00–6:40 pm   Dinner
Birthday Cake
String Quartet Performance

6:40–7:10 pm   Review of Precious Moments
(Dr. Eminy H. Y. Lee)

7:10–8:10 pm   Speeches by Invited Speakers
(Speakers Introduction: Director Cheng-Wen Wu of NHRI)
1. President Yuan T. Lee (Academia Sinica)
2. Vice President Sunney I. Chen (Academia Sinica)
3. Academician C.Y. Cai
4. Academician Monto Ho
5. Academician H.S. Fang
6. Academician Tommy Yung-Chi Cheng
7. President Yuan-Tsun Liu (Soochow University)
8. Ms. Ling Ling Tien Chien (Mrs. Fredrick F. Chien)
9. Ms. Kuang-Chung Hu Chien (Mrs. Shu Chien)
10. Vice President Zhu Chen (Chinese Academy of Sciences)
11. Dr. Danny Ning Wang (IBMS)

8:10–8:25 pm   Entertainment Performance: Song of Joy

8:25–8:30 pm   Gift Presentations
(Representative from IBMS: Director Yuan-Tsung Chen)
(Representative from NHRI: Director Cheng-Wen Wu)

8:30–8:50 pm   Academician Shu Chien's Closing Remarks

Sing Along — "Reunited"

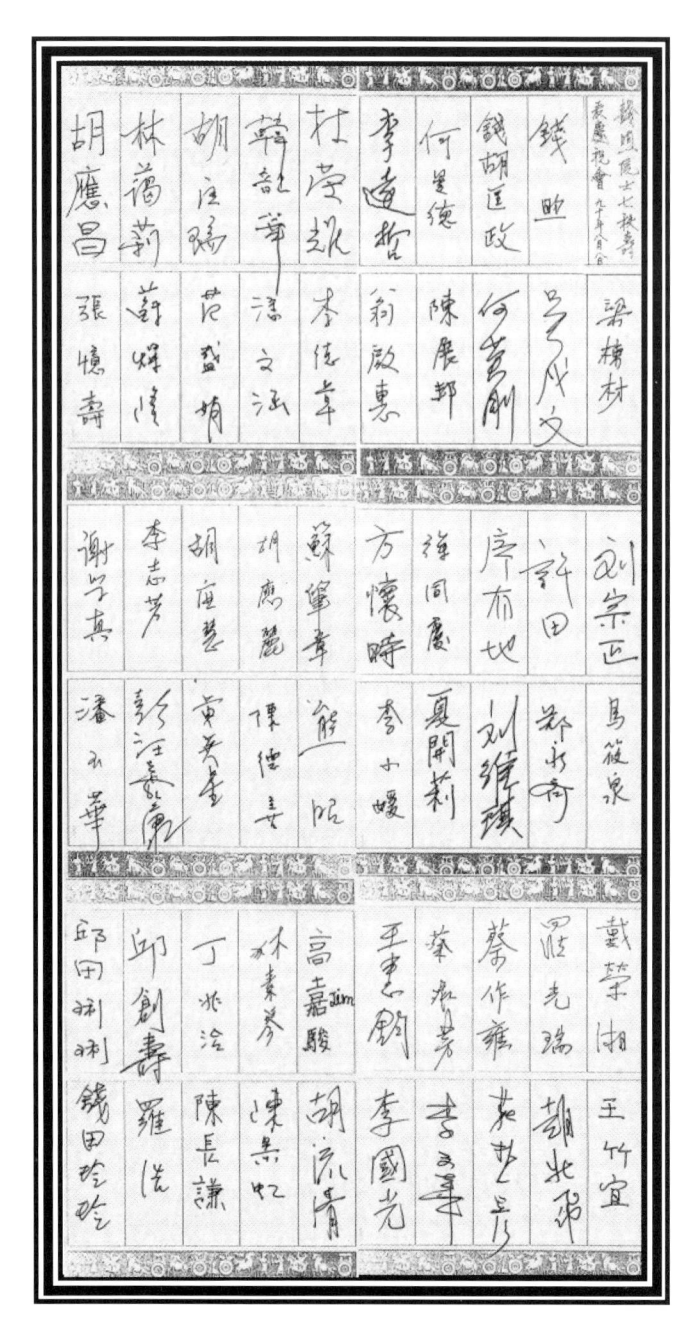

Signature of guests at 8 August 2001 celebration in Taipei.

Signature of guests at 8 August 2001 celebration in Taipei.

Postscript by Mrs Kuang-Chung Hu Chien

In going over the first proofs of the Tribute Book for Shu, I have the great pleasure of reading the precious articles in the book one more time. The tremendous joy, excitement and gratitude I feel deeply in my heart cannot be described in words. The book depicts 70 years of Shu's life filled with love, energy and vitality. I am deeply touched by the wonderful articles written by Shu's seniors, relatives, friends, colleagues and students which span nearly a century in time and the whole world in space. The warm blessings and words of wisdom from seniors and youth, from near and far, have generated an extraordinary resonance. Having been together with Shu for half a century, I particularly appreciate and treasure these articles about him. These articles vividly portray the wonderful person he is and enables me to relive all the happy years we spent together. Shu has always been himself; with all his successes he is still the same Shu Chien, loving and beloved, giving and sharing, and enjoys and appreciates life.

I wish to take this opportunity to thank everyone who has contributed to the book. Shu and I are forever grateful to you. I would like to particularly thank Dr. Amy Sung. Amy spent many long hours collecting, arranging, editing, proof-reading, amongst many other things, bringing the book together. Without her, the book would not have been completed the way it is. I would like to express my sincere gratitude to Amy.

In advising his students on how to improve their presentations and writing, Shu often says "When you have become sick and tired of practicing the presentation or revising your draft, this is when you should do it at least one more time." I have enjoyed reading the wonderful articles in this Tribute Book every time and never gotten tired of doing it. I appreciate immensely the opportunity of reading it again in the proofs format. I am sure that I will read the published book again and again with even greater enjoyment every time.

Kuang-Chung Hu Chien
錢胡匡政
(3 September 2003)

Artwork by
Mrs Kuang-Chung Hu Chien

Author Index

Artmann, Gerhard M., 266

Baldwin, Ann L., 299
Baldwin, Wendy, 199
Berns, Michael W., 174
Bhatia, Sangeeta, 244
Brånemark, P.I., 382
Busch, May Chien, 23, 302
Butler, Peter J., 216

Caro, Colin G., 118
Chabanel, Anne, 94
Chai, C.Y., 314
Chan, Sunney I., 312
Chang, Chao-ching, 490
Chang, Nicki Shu-Chuan, 334
Chau, Lee Young, 322
Chen, Hua Chin, 52
Chen, Jin-Jer, 343
Chen, Peter C.Y., 164
Chen, Richard Y.C., 81
Chen, Yu-Lian, 356
Chen, Yuan-Tsung, 361
Chen, Zhu, 372
Cheng, David H., 116
Cheng, Koung-Ping, 226

Cheng, Lilly, 226
Cheng, Peter Kuang-Hun, 41
Cheng, Tommy Yung-Chi, 364
Chi, Lang-Ming, 351
Chien, Fredrick Foo, 13
Chien, Kuang-Chung Hu, 16, 569
Chien, Robert Chun, 10
Chien, Shih-liang, 492
Chien, Shu, 389
Chin, Chih Chao, 50
Chu, Amy Hsiao Hsuan, 337

Dellenback, Robert J., 294

Fang, H.S., 268, 369
Feng, Si-Shen, 287
Frank, Martin, 189
Fronek, Arnost, 242
Fronek, Kitty, 242
Fu, Pauline, 366
Fung, Luna, 305
Fung, Yuan-Cheng, 137, 305

Garrison, Howard H., 194
Giddens, Don P., 183
Goldsmith, Harry, 281

Gough, David, 149
Guidera, Ann Chien, 27, 302

Han, Pei Pin, 37
Ho, Chien, 297
Ho, Monto, 133
House, Steven D., 96
Hsieh, Jerry Hsyue-Jen, 345
Hu, Chi, 3
Hu, David Kuang-Jeou, 30
Hu, Margaret, 6
Hu, Shirley Tsae-Shene, 336
Hu, Tsu Won, 6
Hu, Yingli, 213
Huang, Ernest Chun-Ming, 228
Huang, Huei-Jen Su, 228
Huang, Natalie, 228
Huber, Gary, 159

Jackson, Michael J., 192
Jain, Rakesh K., 259
Jan, Kung-ming, 72
Jin, Gang, 208
Johnson, Paul C., 162

Katona, Peter G., 184
Kaunas, Roland, 221
Kirschstein, Ruth L., 245
Kuo, Cheng-Den, 353

La Celle, Paul, 275
Lee, Eminy H.Y., 318
Lee, Shaotsu, 44
Lee, Sho-Tone, 363
Lee, Yuan T., 129
Leonard, Edward F., 66

Li, Daming, 292
Li, Julie Y.-S., 203
Li, Song, 201
Lieber, Richard, 172
Lin, Annie Su-Chin, 331
Lin, Shing-Jong, 348
Lipowsky, Herbert H., 83
Liu, Pung Show, 48
Liu, Yuan-Tsun, 366
Loh, Kung Chen, 268

MacKenna, Deidre, 160
McCulloch, Andrew, 160
McIntire, Larry V., 181
Mow, Van C., 122

Nerem, Robert M., 278
Noble, Denis, 253
Norwich, Gerard, 205

O'Connor, Kevin W., 197
Ou-Yang, H. Daniel, 54

Palsson, Bernhard Ø., 152
Pierson, Richard N. Jr, 62
Price, Jeffery H., 170

Rudee, M. Lea, 135

Sah, Robert L., 155
Schmalzer, Emily, 262
Schmid-Schönbein, Geert W., 90

Secomb, Timothy W., 299
Shih, Chia Hui, 235
Shiu, Yan-Ting Elizabeth, 217

Shyy, John Y.-J., 290
Silverstein, Samuel C., 121
Simchon, Shlomoh, 238
Skalak, Anna L., 108
Skalak, Thomas C., 166
Sobin, Sidney S., 240
Subramaniam, Shankar, 158
Sung, Lanping Amy, 76
Sung, K.-L. Paul, 98

Tang, Tang, 325
Tang, Phoebe Yueh-Bih, 358
Taylor, Aubrey E., 186
Tseng, Ellen, 206
Ts'o, Muriel, 131
Ts'o, Paul O.P., 131
Tu, Ivy Su-Ching, 335

Usami, Shunichi, 70

Wang, Danny Ling, 328
Wang, Hsueh Hwa, 59

Wang, Mamie Kwoh, 56
Wang, Mark Yong-Qiang, 33
Wang, Tom, 33
Wang, Xiao-Chuan, 33
Wang, Xiong, 223
Wang, Yingxiao, 218
Wang, Yun Peng, 377
Watson, John, 177
Weibel, Ewald R., 247
Weinbaum, Sheldon, 111
Woo, Savio L.-Y., 284
Wu, Cheng-Wen, 126
Wu, Jackson Chieh-Hsi, 340
Wu, Yun Peng, 377

Yang, R.F., 377
Yu, Winston C.Y., 338
Yuan, Suli, 236

Zhu, Cheng, 105
Zhuang, Fengyuan, 379
Zweifach, Beatrice, 146